P. Brimblecombe C. Pfister (Eds.)

The Silent COUNTDOWN

Essays in European Environmental History

With 37 Figures

Springer-Verlag
Berlin Heidelberg New York
London Paris Tokyo
Hong Kong Barcelona

Dr. PETER BRIMBLECOMBE
School of Environmental Sciences
University of East Anglia
Norwich NR4 7TJ, United Kingdom

Prof. Dr. CHRISTIAN PFISTER
Historisches Institut
der Universität Bern
Engehaldenstrasse 4
3012 Bern, Switzerland

ISBN 3-540-51790-1 Springer-Verlag Berlin Heidelberg New York
ISBN 0-387-51790-1 Springer-Verlag New York Berlin Heidelberg

Library of Congress Cataloging-in-Publication Data. The Silent countdown: essays in European environmental history / P. Brimblecombe, C. Pfister, eds. p. cm. ISBN 0-387-51790-1 (U.S.: alk. paper) 1. Human ecology – Europe – History. 2. Pollution – Europe – History. 3. Environmental policy – Europe – History. I. Brimblecombe, Peter, 1949– . II. Pfister, Christian. GF540.S55 1990 304.2′8′094 – dc20 90-9633 CIP

Printed in Germany

Typesetting: International Typesetters Inc., Makati, Philippines
2132/3145(3011)-543210 – Printed on acid-free paper

This book is dedicated to the WERNER REIMERS FOUNDATION and to the people who support it. In 1988 they helped to bring into being the European Association for Environmental History.

Preface

There is a growing need for cooperation between disciplines, not only to deal with the burning problems of the present, but to study the interaction of societies and their ecosystems in the past. In the 1970s studies in Environmental History were largely confined to North America. Recent years have brought about a vast increase in the amount, the quality and the scope of scholarship on historical interactions between human (social and economic) development and the biosphere in Europe, both East and West. This broad interest in environmental history may have been heightened and sharpened by the dangers of unbridled technology and unlimited growth, which are becoming more and more manifest. However, for several reasons it is still difficult to become familiar with the different approaches to this new and interdisciplinary field of study. Many fields of thought – biology, anthropology, geography, sociology and history – are involved; the relevant books and articles are hard to find and a coherent theoretical framework is still lacking, because the key issues have yet to be submitted to a thorough scholarly debate. It is hoped that the present volume will make a contribution towards overcoming those shortcomings.

The book includes papers which were presented at the first International Workshop on European Environmental History in early March 1988. The meeting was sponsored by the Werner Reimers Foundation in Bad Homburg (Federal Republic of Germany). This institution is named after the forward-looking industrialist, Werner Reimers, who gave so much to promoting dialogue among academic disciplines and the publication of results for a broad readership. We are indebted to Konrad von Krosigkh, who is the trustee of the foundation, and his team for creating the proper ambience for discussion and study.

In order to maintain a proper forum for the continuing debate on the interaction of man with the natural environment in the past, an European Association for Environmental History was formed in Bad Homburg.

Dr. Wolfgang Engel has agreed to include this book in the program of Springer, Heidelberg. We are grateful to him and to his assistant Gustava Hess for giving useful hints and providing the organizational framework for this publication.

CHRISTIAN PFISTER
PETER BRIMBLECOMBE

Contents

Urban and Industrial Impacts

Environmental Consciousness

Contributors

You will find the addresses at the beginning of the respective contribution

Introduction

C. Pfister[1] and P. Brimblecombe[2]

Today, the public increasingly perceives humanity as a part of the biosphere where continued rapid transformation and modification of the global environment could endanger the human race. A new type of threshold has been reached in many sectors of the economy: neither capital, market, nor labor, but the stress placed upon the environment has become the limiting factor (Thober et al. 1985). The current public debate focuses on the formulation and justification of a policy which is likely to be appropriate and sufficient for containing and avoiding this threat. How can a society be brought about to change its abuse of the environment? Do humans respond to rational arguments about environmental issues? How far can environmental education and appeals to common values modify human behaviour? To what extent do we need incentives and penalties? The answers which are given to those questions will draw from a variety of images of humankind and make implicit assumptions about the kind of interactions which have taken place between societies and their natural environment in the past.

On the one hand, there is the belief that the present environmental crisis is a rather recent transitional feature which will be overcome as soon as adequate technologies will have been put into operation. That science and technology can define and solve almost all human problems has become an axiom of modern industrial societies. On the other hand, pessimists refer to the known cycles of deforestation in European History and the degree of pollution in medieval and early modern towns in order to demonstrate that anthropogenic impacts and devastations of natural environments have a long tradition (Brüggemeier and Rommelspacher 1987). Often given less prominence is the observation that the effect of human interference need not always be deleterious. It is also possible to restore and improve land, and to create new, productive ecosystems. Human management of the land without the result of degradation is not only possible in a great majority of environments, but has been frequently accomplished in human history (Blaikie and Brookfield 1987).

In this situation the historian must participate in the debate and try to give an answer as to (1) the roots of the present crisis when fundamental thresholds have been crossed, and (2) under which conditions past societies have managed to maintain a dynamic stability in their environments. Unlike their colleagues in

[1]Historisches Institut, Universität Bern, Engehaldenstr. 4, 3012 Bern, Switzerland
[2]School of Environmental Sciences, University of East Anglia, Norwich NR 4 7TJ, Great Britain

Brimblecombe, Pfister (Eds.)
The Silent Countdown

North America, historians in Europe have long been reluctant to explore the dimensions of past relations between societies and their natural habitat. However, the recent years have brought about a vast increase in the amount, the quality, and the scope of scholarship on historical interactions between human (social and economic) development and the biosphere (Pfister 1988). In Europe, where a network for exchanging information and ideas was still lacking, the time was ripe for bringing the relevant scholars and fields together.

Twenty-one scholars from 11 European countries, east and west, gathered from 29 Feb–3 March, 1988 in Bad Homburg (Federal Republic of Germany) at a first international workshop on European Environmental History which was sponsored by the Werner Reimers Foundation. Three scientists – Frank Oldfield (physicist), Peter Brimblecombe (chemist), and Yrjö Vasari (biologist) – joined the group of historians.

One of the greatest problems for the understanding of ecological problems is the lack of communication between the social and natural sciences. Environmental history may serve as an area for dialogue between the two disciplines, which traditionally have their own methods of analysis and theories about how the world works (Worster 1984; Leibhardt 1988). The Bad Homburg meeting was a small step towards improving this situation, in the sense that common fields of interest could be identified. However, in order to initiate a true interdisciplinary exchange, more effort and more time will be required. A common focus needs to be identified, an appropriate body of "disciplinary" knowledge has to be previously collected, and both disciplines must be sufficiently familiar with each other to couch the results of the "other" side in the known set of theories and methodologies (Kaufmann 1987). Regarding environmental history, this means that the researcher should be familiar with both human and natural history.

The essays contained in the present volume were written in the context of the Bad Homburg symposium. They provide a good illustration of the different approaches to understanding and writing environmental history. Two conceptual frameworks that may be distinguished from the essays could be labelled "holistic" and "complementary".

The concept of *human ecology* or *population ecology* provides the conceptual framework for a *holistic* analysis of the complex links between people and nature within ecosystems. Holism expresses the idea that the whole is greater than the sum of its parts. The ecosystem is seen as a spatial concept, which implies the notion of boundaries perceived as zones of transition between adjacent systems. It is a heuristic unit that helps us carry out holistic research on given problems. It does not, however, represent a real physical unit (Moran 1984). Although the concept of human ecology was set up in the context of recent anthropological studies, it has already been applied to the study of the past (e.g., Netting 1981). An ecological approach to history emphasizes the two-way character of exchanges between productive and reproductive processes and the natural environment, although the relative influence in reciprocally causal relationships is never equal and may be very unequal (Kaplan and Manners 1972, in Ellen 1982). Within this ecological framework, the emphasis in historical research is shifted from the societies and individuals to the study of ecosystems (Sieferle, this Vol., 1988).

However, as history is an anthropocentric discipline, the historian will only consider those elements in the non-human natural world, which are in one way or another related to human activities. Studying plants and animals for their own sake is the object of "natural history", though human impacts must be taken increasingly into account, allowing opportunities for interdisciplinary debates. Ecological history or historical population ecology may perceive basic patterns in the history of civilizations in the longer term which "are consistent with the general direction and the dominant dynamics of the evolution of life in the biosphere – and likewise with the evolution of matter in the cosmos" (Laszlo 1987).

One of the most fundamental elements in ecosystemic analysis is the *flow of energy*. A web of energy and material relations allows the ecosystems approach to draw together the natural and human processes. This focus on common components in an interconnected system provides an important means of specifying how nature and society are interrelated (Sack 1990). Ellen (1982), for example, considers human society, along with other biological systems, in terms of energy production, utilization, and exchange, as these provide "the material basis of human existence". The special value of energy flow models is in aiding us to compare ecosystems in space and over time, and to identify their functions despite ecosystemic differences. By translating the biomass and the rate of energy flow into common units (usually joules), a general model can be built that quantifies dynamic relationships, elucidates functional relationships, and provides a basis for the comparison of ecosystems (Shugart et al. 1975, in Moran 1984). The major difficulties with an energy flow approach to ecological study seem to be the quantification of all the relevant data – energy production, energy consumption, and energy expenditure – and the decision as to which are the major functions and which are less significant (Moran 1979). Quantification is therefore an important ingredient of the energy flow studies. In this volume the "holistic approach" is represented in studies by, Rolf Peter Sieferle (Neckargemünd), Jean-Paul Deléage (Paris), Christian Pfister (Bern), Pier Paolo Viazzo (Cambridge), and Karlheinz Blaschke (Friedewald GDR). Sieferle focuses upon the transitional period from the solar energy system to the fossil energy system in order to show to what amount the characteristics of energy systems define the scope of ecologically relevant behaviour of society. Deléage provides a global view of ecological history in the very long term, which includes the social dimension and ends up with a review of present environmental problems. Pfister demonstrates, using the example of the Canton of Bern, that the success of the measures which were aimed at improving the carrying capacity of the agrosystems in the short term was the cause of the instability of the ecosystem in the long term. Viazzo asks for conditions under which local resource stability has been obtained in the alpine area. Blaschke raises some questions of general concern to environmental historians.

The "complementary" conception perceives environmental issues rather as an additional facet of history than a contribution to a new theory of societal evolution. Within this context, two different approaches may be distinguished: one approach describes the impact of processes of production and consumption upon the social environment and on the natural habitat. Perhaps the most important factor that has altered the face of the earth in many parts of the world is the *clearing of the forests* that once covered large parts of it. The forest has been subject to a sustained and steady attack by mankind throughout the centuries, not only through clearing, but also through pollution emissions. Clearing has been regarded as an essential, but incidental by-product to agricultural expansion and timber exploitation, i.e., for obtaining energy in form of food and fuelwood, and materials for building and fencing. It has been regarded as part of the "natural" process of development. This process is described in a general medieval context by Teresa Dunin-Wasowicz (Warsaw). While deriving inspiration from the theory of Schove she reminds us of the need to test its validity. Istvan Kiss (Budapest) has focussed on the forest policy of Hungary from the Middle Ages to the present. Yrjö Vasari (Helsinki) makes an ecological analysis of the forests of Kuusamo. Agricultural activity within the traditional three-field system has promoted the process of *erosion*, as Jean Vogt (Strasbourg) demonstrates from examples in Central Europe.

Emissions of toxic substances into the air, the water, or the soil affect people's health, comfort, and property and induce undesired changes in the natural environment. Oldfield is able to derive a quantitative picture of acidification over the last 200 years by examining the fossil diatom record in lake sediments. In addition to describing the source and the nature of the emission, this issue involves discussing ways in which "nature" or the affected population responded to it. A survey of the older and the more recent works on "environmental history" has allowed distinguishing several qualities (and quantities) of pollution (Mieck 1986): Le Roy Ladurie (1973) has coined the term *pollution microbienne*. It designates breeding grounds for microbes which were created in the densely populated environment of towns in the absence of any efficient sanitation. Nicole Lange shows that in Hamburg the link between health and the cleanliness of the water and the air was recognized in the 16th century, leading to a large number of conservational decrees. In addition to the effects of high population density, certain trades such as paper-mills, tanneries, and dye-works were sources of *pollution artisanale* in the form of smoke, stench, and water pollution. Legislation in nineteenth century Leyden refers to this localized type of pollution (Diederiks, Jeurgens). The term *pollution industrielle* is reserved for large-scale impacts of industrial plants such as iron works and chemical industries, which could only be supported by the use of fossil fuel. The transition between *pollution artisanale* and *pollution industrielle* is hard to define: initially it was mainly a matter of scale, although more recently qualitative differences have also become apparent (e.g., the production of synthetic chemicals such as dioxin).

Battles between the industries, the local councils, the experts, and the suffering people are described in the papers by Brimblecombe, Brüggemeier, and Schramm. It should be stressed that authorities in the 19th century were not concerned about pollution of the environment. They acted only when the health of people was threatened (Diederiks). There is a large amount of 19th century legislation, and though it lead to improved urban health, appropriate technologies for reducing air pollution were not developed. In England, for instance, even where local governments were anxious to abate smoke, they found it impossible to give local industry satisfactory guidance (Brimblecombe). A similar difficulty in laying down precise regulations to prevent pollution existed in the Ruhr. In Germany, the industries were only forced to compensate for its damages, rather than to abate the nuisance, and only to the extent that their economic existence was not jeopardized (Brüggemeier). Expert advice about environmental damage from smelter smoke had already drawn attention to complex models of causality which, however, were not suitable to gain legal support (Schramm, see also MacLeod 1965). The Ruhr emerges as a key area for demonstrating the essence of *pollution industrielle*, where there were so many chimneys and sewage pipes that the effect of any individual source was unrecognizable in the overall "sea of pollution". In this region, industry was protected, not the environment, though the situation has been improved over the last 15–20 years (Brüggemeier).

As a common denominator, all types of pollution are connected to concentrations of energy which exceed the natural productive capacity of the environment on which they are deposed. Therefore, when food and fuelwood had to be brought into a medieval town, *pollution microbienne* and *pollution artisanale* emerged as a consequence. *Pollution industrielle* results whether foodstuffs have to be bought for industrial husbandry, or whether fossil or nuclear fuels are used as sources of energy. Industrial pollution, related to intense utilization of fossil fuel, has characterized the last 150 years. In the last decades, the possibility of importing large amounts of cheap fodder to raise livestock at high densities has increased. Liquid manure leads to a new type of pollution: *pollution agricole* (Pfister). Further novel forms of pollution are apparent: *pollution accidentelle* (when toxic or radioactive substances are set free through an accident) and *pollution globale* (when the breakdowns of local and regional ecological equilibria have broadened to include the planet).

Another approach to environmental history focusses upon the changing perceptions and the *role of the natural environment in individual and group consciousness*. Consciousness is the totality of one's thoughts, feelings, and impressions, the awareness of one's acts, symbols, and images of nature. In different historical epochs, particular characteristics dominate a society's forms of consciousness. The world is perceived, understood, and interpreted through these forms of consciousness. They are socially constructed and subject to change (Merchant 1987). The critical unit of human ecology is not "man", but different types of human culture. In his contribution, François Walter (Geneva) has identified five components in the relationship of western societies with their natural environment. Using the case of Switzerland, he demonstrates how the image of nature became an important element of cultural politics. Similarly, Brimblecombe and Schramm draw attention to a surprisingly early public concern of a loss of natural environment in the face of industrialization. Thelander (Lund) explores the difficulties of even defining environmental problems and the scientification and inaction that may arise as a result of this process.

The study of environmental history is still in its infancy and many topics and relationships have yet to be explored. It is our hope that readers will be challenged by these gaps and will be moved to make contributions toward bridging them. Whether environmental history will go the way of demographic history which, after a couple of decades, was recognized as a subdiscipline of its own, will depend on the stability of institutions that support it, the creation of consistent theories, and its value to other fields of history.

References

Bennett J (1990) Behaviour, institutions, and organizations in the human response to the physical environment. In: Kates R, Turner W (eds) The world as transformed by human action. Univ Press, Cambridge (in press)
Blaikie P, Brookfield H (1987) Land degradation and society. Methuen, London
Brüggemeier FJ, Rommelspacher T (1987) Besiegte Natur. Geschichte der Umwelt im 19. und 20. Jahrhundert. Beck, München
Ellen R (1982) Environment, subsistence and system. The ecology of small-scale social formations. Univ Press, Cambridge
Kaufmann FX (1987) Interdisziplinäre Wissenschaftspraxis. Erfahrungen und Kriterien. In: Kocka J (ed) Interdisziplinarität. Praxis – Herausforderung – Ideologie. Suhrkamp, Frankfurt
Laszlo E (1987) Evolution. The grand synthesis. New Science Library, Shambhala, Boston London
Le Roy Ladurie E (1973) Un concept: l'unification microbienne du monde (XVIe–XVIIe siècles). In: Schweiz Z Gesch 23:627–696
Leibhardt B (1988) Interpretation and causal analysis: theories in environmental history. Environ Rev 12:23–36
MacLeod RM (1965) The alkali acts administration 1863–1884: the emergence of the civil scientist. Victorian Stud 8:86–112
Merchant C (1987) The theoretical structure of ecological revolutions. Environ Rev 11:265–274
Mieck I (1986) Industrialisierung und Entwicklung des Umweltschutzes im 19. Jahrhundert. In: Leidinger P (ed) Historische Ökologie und ökologisches Lernen im historisch-politischen Unterricht Schöningh, Paderborn, pp 21–43
Moran EF (1979) Human adaptability. An introduction to ecological anthropology. Westview, Boulder
Moran EF (1984) Limitations and advances in ecosystems research. In: Moran EF (ed) The ecosystem concept in anthropology. Westview, Boulder, Cal, p 3
Netting R (1981) Balancing on an Alp. Ecological change and continuity in a Swiss mountain community. Univ. Press Cambridge, New York
Pfister C (1988) Ansätze und Ergebnisse historischer Umweltforschung: ökologische und soziale Dimensionen der Tragfähigkeit am Beispiel von Solarenergie – Gesselschaften des schweizerischen 18. und 19. Jahrhunderts. In: Jahreskongr Schweiz Ges Wirtsch Sozialgesch, 27. Nov 1987, Lausanne, pp 25–41
Sack R (1990) Social theory and environmental research. In: Kates R, Turner W (eds) The world as transformed by human action. Univ Press, Cambridge (in press)
Sieferle RP (1988) Perspektiven einer historischen Umweltforschung. In: Sieferle RP (ed) Fortschritte der Naturzerstörung. Suhrkamp, Frankfurt
Thober H, Lieth H, Fabrewitz S (1985) Modelling of the socio-economical and ecological consequences of high animal waste application (MOSEC). In: Hall DO, Myers N, Margaris NS (eds) Economics of ecosystem management. Junk, Dordrecht, pp 71–85
White L (1967) The historical roots of our ecological crisis. Science 55:1203–1207
Worster D (1984) History as natural history: an essay on theory and method. Pac Hist Rev 53:3–25

Holistic Conceptions

The Energy System – A Basic Concept of Environmental History

R.P. Sieferle[1]

The imminent ecological crisis, perceived during the last 20 years in Europe and in America, offers a challenge not only to explain the present problems, but also to understand their historical evolution. There are great advantages in the historical reconstruction of ecological conditions and ecological crises in the past: we can observe them over longer periods of time, we are not obliged to offer "solutions", and we can look upon social systems and social values as functionally integrated parts of a human-ecological whole, not being obliged to share opinions, values or interests of the past. On the other hand, it will be necessary to integrate the results of different natural scientific disciplines into historical explanation, from whence severe problems of interdisciplinary cooperation are to be expected.

For history as an academic discipline, this means that a new set of viewpoints and methods has to be incorporated into conventional wisdom and procedures. In my opinion, the main challenge of environmental history lies in a change of point of view: from anthropocentrism to the ecosystem concept. The language of "ecosystem" can be used as a heuristic concept, which allows the use of models of explanation, developed within the framework of general system theory, for the explanation of complex life processes. Concepts such as "self-regulation", "balance", "resilience", or "feedback" must thus be understood as mere heuristic devices, not as essential qualities of "nature".

As we can learn from the science of ecology, the basic characteristics of ecosystems are flow of energy and circulation of materials. The biosphere can be seen as an energetically open system, operating through a flux of energy, which allows the synthesise of biomass, converting it into motion and maintaining the metabolism of organisms. Complexes with high information content can thus be organized and transmitted. Matter and materials are circulating within the system, while energy flows through it, building its motor.

"Balance" and "stability" of ecosystems are mere matters of perspective and time horizon. Ecosystems usually maintain certain levels of organization and regularity, showing a measure of resilience when confronted with disturbances. In the long run, however, the systems are not stable or balanced, but evolving. From the usual historical and anthropomorphistic point of view, nature appears as a balanced and unchanging entity, because the evolution of ecosystems takes place in periods of time not accessible to elementary human experience.

[1]Adalbert-Seifriz-Strasse 13, 6903 Neckargemünd, FRG

Brimblecombe, Pfister (Eds.)
The Silent Countdown

The special role of humankind results from its specific "strategy" of survival: cultural evolution. Changes of behaviour and ecologically relevant characteristics do not any longer exclusively result from the process of variation and selection of genotypes, as in all other species, but can be produced by neural ways of information processing, transmitted by symbolic communication. The "Darwinian" method of evolution is replaced by a "Lamarckian" method, allowing an enormous acceleration of human cultural evolution, when compared with the normal inert pace of molecularly fixed biological evolution. It is this difference from whence man's potential as a highly disturbing factor within his natural environment results. Cultural evolution allows a velocity of change, orders of magnitude faster than that which happens within the time frame of biological evolution. Both are not synchronized. In the biological arms race, other species (apart from microorganisms and certain generalists) have no chance against man.

The critical unit of human ecology is not the human race, but different types of human culture. In the process of cultural evolution, certain patterns evolve, which contain an intrinsic social logic of development. "Man" as a biological species is only a formal requirement of human cultural systems, which evolve with a momentum of their own and win characteristics which can not simply be deduced from their "human" elements. The subject of ecological history, therefore, is the relationship between cultural formations and their specific ecological niche, which is moulded by them, but which also sets certain limits for their progression.

This interaction of cultural systems and ecosystems can be demonstrated by the example of socially modified energy systems. In ecological anthropology, it has become usual to analyze the flow of energy within certain types of societies, in analogy to the flow of energy in natural ecosystems (cf. White 1943, 1959; Cottrell 1955; Kemp 1971; Rappaport 1971; Adams 1975, 1978; Hardesty 1977). As any other natural species, man has to participate in the natural flow of energy. This is necessary, firstly because he can only maintain his physical structure and can only win the potential to move and to work, when he successfully converts energy for his purposes. Furthermore, the social/cultural/institutional framework of his existence can only be built up and maintained, when large amounts of energy can be mobilized. This is far more than what is needed for his physical metabolism in a narrow sense.

For this reason, it makes sense to analyze certain types of societies from the point of how they organize their energy flows, and which amounts of energy they use. Usually, three stages of social evolution and three types of energy use can be distinguished:

1. Hunter/gatherer societies and an unmoulded solar energy system;
2. Agricultural societies and a progressively moulded solar energy system;
3. Industrial society and a fossil (or nuclear) energy system.

"Primitive" hunter/gatherer societies participate in the natural flow of energy analogous to other predator species, without severely modifying it. They use solar energy, which has been fixed by the photosynthetic process of green plants or has been converted into the biomass of herbivores. They use energy primarily through their metabolic process, which allows them to convert the chemical energy of their food into mechanical energy, enabling them to work. In this continual conversion process, temperatures and levels of energy density are low, with the exemption of the use of fire being the same as an accelerated conversion of energy chemically fixed in biomass into heat. The use of fire allows them to change the physical and chemical structure of food, making things digestable which were not so while uncooked. Moreover, fire helps to create a microclimate, allowing existence in habitats to which man is not physiologically adapted. Thus primitive man can live in quite different regions of the world, spreading from the tropics to the Arctic regions of the high north.

Agricultural societies are also limited to the use of solar energy, but they succeed in changing the flow of energy considerably. This is the basic strategy of agriculture itself, i.e., clearing the land, tilling the soil, fighting noxious weeds and vermin, breeding new varieties of animals and plants, and changing the geomorphological structure of the earth such as building terraces, irrigating, draining etc. Agriculture makes it possible for more people to live on a certain space than under the conditions of hunting and gathering. The resulting higher density of human populations is the direct result of the modified energy flow: since the amount of solar energy, which reaches a certain area, is given, and since the energy density of this solar energy is rather low, it has to be collected and concentrated by plants over a period of time. A human society which succeeds in monopolizing the greatest amount of energy flowing into its biosphere for its own purposes can afford to be more populous than a society which has to share this energy with other life forms. Thus the removal or extermination of useless and noxious animals and plants has to be one of the basic characteristics of agriculture.

Agricultural societies have a high potential for innovation. The higher density of population allows them to develop social differentiation and division of labour, producing a surplus. Specialized trades can emerge. Certain groups can be set free from physical labour, enabling them to concentrate on fields of knowledge. Social and political power are concentrated, mobilizing resources for certain cultural purposes, such as the building of cities, pyramids, and cathedrals. Civilized agricultural societies develop a peculiar "anthroposhere", growing into the biosphere, with the tendency to out-grow it (cf. Boulding 1985). The social-institutional organizations steer the energy flow through the whole system in such a way, that a new world of artefacts arises, stabilizing the culture and winning a momentum of its own. In this process of innovation and development the whole population can grow, and so can the per capita consumption of energy.

Most important for developed agricultural civilizations is, that they succeed in using new methods of converting solar energy beyond the traditional conversion by human or animal labour. The first thing is the use of wind, mainly for sailing, so that transportation over long distances becomes possible. The next thing is the use of water power, allowing the mechanization of simple work processes such as milling or hammering. But all these methods of energy use remain within the narrow framework of a traditional solar energy system: man participates in the natural flow of energy, which he can skillfully modify. But he is not able to use more energy than is *continually* flowing trough the ecosystem. He can use the stored biomass of forests, but in the long run he can not use greater amounts of energy than are flowing into the system during the same period of time from its only source: the sun.

This is the reason why agricultural societies are faced with a fundamental dilemma, stemming from their basic ecological condition: on the one hand, they can develop rather complex social and cultural systems with sophisticated mechanical devices, resulting in a peculiar dynamic of economic and demographic growth. On the other hand, they are fundamentally restricted within the framework of a continual energy flow. Over a certain period of time, they can not, on the whole, use more energy than is delivered by the sun. Moreover, since this energy is flowing in over an area, they must always spend large amounts of materials and labour to collect and concentrate this energy. This is the basic reason why energy is always scarce in agricultural economies, and why there are ever-recurring complaints about imminent energy (mainly fuel) shortages.

There has been a long debate in the history of forestry, whether or not the European societies in the wake of industrialization have been facing a severe shortage of wood (cf. Radkau 1986). Wood, of course, has been a central resource in these economies, not only as a fuel, but as a material for a whole variety of purposes. Structurally, the problem of preindustrial fuel shortage has the following scope: when an agricultural society expands, when its population and its industrial output grow, more agricultural area is needed, while at the same time more forest products, mainly wood, are required. As the amount of available land is given, a problem of alternative land use arises. Shall the surface of a country be used for forests or coppice woods for timber or fuel, or for fields for grain, or for grassland, to win meat or wool?

Of course, there was always the opportunity of technological change. When new methods of husbandry are developed, it becomes possible to raise more food from a given area. The same is the case with forestry. But the basic problem remains unchanged: a solar energy system depends on the surface of land on which energy can be collected and stored. From this basic characteristic, the following tendencies of agricultural civilizations can be deducted:

1. A consumer who wants to use fuelwood must collect this fuel from an area, before he can burn it. The relation between his expenditures and the returns remains in all cases linear. Economies of scale are not possible, on the contrary: the farther the distance between the woodland and the consumer, the more difficult it becomes to transport this bulky commodity. This means that the size of a productive unit is energetically restricted. As for the smelting of iron, an ironwork in the 18th century could not produce more than about 2000 t a year. Similar restrictions are to be found in water mills, whose maximum amount of available mechanical energy are limited by the capacity of the river and by the problems of transporting mechanical energy (cf. Hills 1970; von Tunzelmann 1978). A preindustrial solar energy system thus had to remain decentralized.

2. As the energy income of the biosphere is the only energy source of a solar energy system, a certain country cannot in the long run use more energy than is flowing onto its surface. Of course, there are potentials for technological innovation, so that the ultimate limit may be rather flexible. Finally, however, there must be limits to growth in such an economy. Moreover, economic growth can not accelerate very much in a system which is always on the brink of energy shortage. An agricultural economy, therefore, is always confronted with the dismal perspective of a stationary state of production, consumption and population. If there is a tendency toward social change and economic/demographic growth, arising from the dynamics of the social and economic system itself, as was the case in 18th-century Europe, this growth would have been throttled, had there not been a fundamental change of the energy system.

These general considerations make clear that something of the kind of an "industrial revolution" could not have been possible on the basis of a solar energy system. The process of industrialization seen by Europe during the last 200 years was necessarily linked with the change of the energy system. Or, to put it the other way round, an energy system had to be developed which bore physical properties, giving room for the unfolding industrial dynamics. The historical switch from the moulded solar energy system to the fossil energy system was thus functionally connected with industrialization. Different uses of energy form different system requirements of social and economic formations. Energetically, the transition from agricultural to industrial economy is *identical* with the transition from solar to fossil energy system. The relative abundance of energy generated by the use of fossilized biomass was the driving force in the new industrial economy developing since the late 18th century. The basis of the new processes of chemistry, metallurgy, transportation, agriculture etc. is the breaking of the bounds set by the traditional solar energy system.

This is not the place to retell the story of the historical change from the solar to the fossil energy system (cf. Sieferle 1982). In short, it originated out of certain local energy shortages, which were always inherent in the agricultural economy, but which could be overcome in England, where exceptional environmental conditions allowed the use of coal. This opened a self-increasing and self-intensifying process, beginning with local and sporadic uses of fossil fuel, in the course of which a new energy system was formed, showing unpreceded proprieties and qualities which fitted into the requirements of the industrial economy. In the course of this process, many technological problems had to be solved; some solutions were extremely difficult to find. At the end of the 18th century, however, the key innovations in mining, metallurgy, and transport had been achieved, so that the fundamental characteristics of the new energy system became totally visible and, for the non-English world, the development of a fossil energy system became a matter of necessity. The fossil energy system formed a new evolutionary stage of human ecology; it was no longer possible to remain within the bounds of the traditional solar energy system, unless a country was content to be marginalized or, in the long run, to be taken over by a more advanced world economy based on fossil fuel.

In the case of England, it can be easily demonstrated what the change of energy systems meant ecologically. When wood was substituted by coal as a fuel, the whole pattern of land use could be changed. In an agrarian economy there must always exist a certain proportion between fields, pasture, and forests/coppice woods. When this economy succeeded in substituting coal for wood, it could use the woodland area for other purposes, e.g., to raise sheep or grain. Land for energy plantations – and coppices are such from an energy point of view – is not needed any longer, when there is an alternative energy resource. Thus is becomes possible to compare a certain amount of coal with an amount of fuel wood, which was not needed any more, this fuel wood being the yield of a certain woodland area.

I would like to illustrate this with some rough data. A coppice of 1 ha yields annually about 5 m^3 wood. As 1 t of coal has an energy content equivalent to about 5 m^3 wood, by the use of 1 t coal an area of 1 ha woodland is set free for other purposes. On the basis of this data it is possible to assess the amount of space England won by using coal as a fuel. For this purpose, I would like to compare the historical development of British coal production with substituted wood and space equivalents (Table 1).

The whole area of England and Wales being about 150 000 km^2, we get an impression what the change of the energy system meant in terms of land use. The 210 000 t coal which were annually produced in the years between 1551 and 1560 replace a coppice area of about 2000 km^2; that is just 1.4% of the whole area of England and Wales. These data illustrate that the early modern coal use had nothing to do with a severe wood shortage, but that it was only a sporadic and local matter of substitution. Coal was still used within the general framework of the solar energy system, having some slight advantages in some professions. By 1681–90 about 30 000 km^2, about one-fifth of the whole area of England and Wales, should have been used as coppices had there been no coal. Here we are at the point of transition to the new energy system. In the early 19th century, however, the whole area of England should have been planted with wood for

Table 1. British coal production and its conjectural wood and area equivalents (cf. Sieferle 1982 for further details)

Year average	Coal in Mio t	Wood in Mio m^3	Forest area in 1000 km^2
1551/60	0.2	1.0	2.0
1681/90	2.9	14.5	29.0
1751/60	4.3	21.5	43.0
1781/90	8.0	40.0	80.0
1801/10	13.9	69.5	139.0
1821/25	22.6	113.0	226.0
1841/45	46.3	231.0	463.0
1913	287.4	1437.0	2874.0

energy purposes, had there been no coal. In other words, England of the early industrial revolution could use a land equivalent twice as large as its own surface, thanks to fossil fuel.

The same can be shown for single industrial branches, e.g., iron smelting, since this was a key industry for modern development. Traditionally, iron was smelted and refined in charcoal furnaces. To produce 1 t pig iron, 800 ft^3 of charcoal were needed (Hammersley 1973), for refining another 1000 ft^3 of charcoal were necessary. The whole amount of charcoal to win 1 t wrought iron had to be produced from 50 m^3 wood. This was the perpetual annual yield of a coppice of 10 ha extension. Thus we get a rough equation of 1 t wrought iron = 10 ha land. From these data we can expolate the area equivalents of British steel production (Table 2).

From these data we can see that the development of British steel production is closely connected with the development of coal production. It has been a long-discussed matter that British steel production stagnated from the 17th century to the middle of the 18th because of charcoal shortage (Hammersley 1973; Hyde 1977; Flinn 1978). I cannot go into the details of this debate here, but one point seems to be clear: about 1820, already one-third of the whole area of England and Wales would have to have been turned into coppice woods, had the British steel production of this time been obliged to use charcoal as its only fuel. In the middle of the 19th century, the area of woods necessary for fuel would have been more than equal to the whole surface of the country. This makes clear, again, what the change to the fossil fuel system meant in terms of land use: it was as if Britain had won another British island.

The use of coal has thus been a fundamental requisite for industrialization; on the other hand, the formal condition for this use did not lie in the hands of human societies. The mere existence of coal-beds arises out of an geological accident. There are parts of the earth where there is no coal at all; there are other

Table 2. Conjectural area equivalents of British steel production (calculated after Hammersley 1973; Mitchell 1973)

Year average[a]	Steel production (1000 t)	Area equivalent (1000 km^2)
1620	19	1.9
1690	23	2.3
1720	25	2.5
1781/90	69	6.9
1800/14	127	12.7
1820/24	669	66.9
1850/54	2716	271.6
1900/04	8778	877.8

[a] The early and later data are not exactly compatible, since they may not contain the difference between pig iron, wrought iron, and steel, and since there may be severe regional differences. These data only give rough orientation marks, showing the tendency and the order of magnitude.

parts, where the first mining of coal, or the first extraction of other fossil fuels such as petroleum are so difficult, that even great energy shortages cannot induce a change of energy systems. There are two basic geographical conditions for a coal economy: (1) the existence and accessibility of coal and (2) suitable transport facilities for this bulky commodity. From this it becomes clear that only in England could the transition to a fossil energy system begin, apart from its indigenous social and economic dynamics of capitalist production.

In the case of Italy and Greece, both countries are easily accessible by water, so that there was never a problem of transport. Unfortunately, there was no coal, so that it was impossible to overcome the permanent fuel shortage. This may have been a severe obstacle for an industrial development in this region, although they did not lack a commercial or technological spirit. On the other hand, in the Ruhr valley or in Silesia there were huge coal deposits, reaching partly to the surface, already accessible in the Middle Ages, or even back in Roman times. But since they could not be transported over land, the great centres of population and industry could not easily be supplied with them.

Only in Britain did both conditions come together: coal, lying at the surface of the Tyne valley, and cities, which could easily be reached by water transport. If anywhere in the world, or at least in Europe, it was in Britain where the new energy system could come into existence. Once the fundamental threshold was overcome, this development could be copied in other countries, at least when overland transport by rail and steam became available.

One could speculate what would have happened if there had been no coal in England. Obviously, the industrial revolution would have had severe shortages not only of fuel, but also of iron and steel. It has been estimated that in the middle of the 18th century about 30–50% of the iron used in England was absorbed by agriculture (Bairoch 1978, p. 491). The great breakthrough in agrarian technology in the 18th century meant that increasing amounts of iron went into agriculture. Supposing that the European iron production would have become stationary at the turn of the 19th century, most of the iron would have gone into food production to rear a growing population. There would not have been much left for steam engines or railroads.

Of course, one could argue that wood could have been imported to England, since there were still huge forest areas in the Baltic or in Scandinavia. As a matter of fact, there were high imports of forest products to England, mainly of materials for ship building such as timber, masts, planks, tar, and pitch, not to mention iron or potash. In the 18th century, annual British timber imports lay between 0.04 M m^3 and 0.2 M m^3 (Aström 1970, p. 20). This was a huge amount, but it was nothing compared with the wood equivalents of coal used at the same period. Since the shipping of a bulky commodity like timber was rather expensive (cf. Albion 1926; Davis 1962), it was limited for building materials, especially for ship-building. Imported wood was far too expensive to be used merely as a fuelstuff.

What can be said for wood goes for energy-intensive commodities in a similar way. To illustrate this point, we can compare the weight of certain products with the weight of fuelwood necessary to produce them (Table 3).

Table 3. Weight proportions between products and fuelwood used (Gleitsmann 1981)

Salt	1:7
Pig iron	1:15
Wrought iron	1:30
Silver	1:200
Glass	1:2400

Similar proportions are found in other materials or chemical products. In this respect it seems quite improbable that fuelwood would have been transported to England just to burn it with the intention to smelt or process a stuff which could itself more easily be transported. There is one striking example: wrought iron could be economically produced with coke only since the last third of the 18th century, because there were severe technological problems. The British iron industry almost stagnated after the late 17th century, as it was cheaper to import iron from Scandinavia. A similar thing would have happened with other branches of industry where large amounts of fuel were needed. So it may be no exaggeration to hold the view that the European industrial evolution would have been much more decentralized (and slower), had there been no coal in England. The country would not have become the workshop of the world, except, perhaps, for the textile manufacture, which before the comparatively late introduction of steam power was based completely on solar energy: biomass, water, wind (sails).

The data presented above show the sheer amount of energy which became available through fossil fuel. Of course, coal, too, contains chemically fixed solar energy, but there are critical differences between a solar energy system and a fossil energy system. In the latter, energy is used which has been stored over periods of time which are orders of magnitude larger than the time of consumption. We may compare this with the first use of a primeval forest. It may contain an amount of energy fixed by photosynthesis over 300 years. When such a virgin forest is first used, we get a large amount of stored energy within a short time. From a forest managed by forestry, however, we can only harvest the annual yield, being the same amount as is fixed within the same period of time. To use fossil energy means in this perspective that we can use something like a very large, albeit not inexhaustible forest. A society changing to a fossil energy system thus moves into a state of energy abundance; this is historically a pioneer situation. Large, almost infinite energy resources allow the transformation of a whole set of ecologically relevant parameters. This pioneer situation, however, is not perpetual. One day the "subterranean forest" must be exhausted, and contrary to a real forest, it is not possible to switch to a "husbanding", i.e., continual harvest of perpetual yields.

From the basic characteristics of the fossil energy system, some of the physical features of industrial economy can be explained. First, the number of human beings could multiply almost tenfold. Then, the flow of materials running through the social and economic system could multiply in even larger proportions. The natural cycle of materials could be broken up. Material resources began an accelerated move, from concentrated ores through higher concentrations in industrial processes to finite dispersion over the whole surface of the earth and the oceans. In this perspective, the transition of the energy system is the formal precondition of industrial pollution, since mobilizing such gigantic amounts of materials, effectively disturbing large natural systems such as the global atmosphere, requires quantities of energy, so large that they could never be available within the bounds of a traditional solar energy system.

In this perspective it is interesting to see what happened in agriculture itself. Within the framework of the solar energy system, agriculture was a central part of the energy system itself. It could be defined as a means to collect energy for human consumption. This meant that preindustrial agriculture had to have a positive energy yield. It was impossible that the energy input for food production was (on the average and overall) greater than its energy output. But this is what has happened in modern agriculture: the aggregated energy inputs in tilling, fertilizing, plant protection, harvesting, processing, and conservation are far greater than the energy content of food. Agriculture thus changed from an energy-yielding to a material-transforming system, with positive inputs of fossil energy (cf. Leach 1976).

Some additional features of industrial economy can be explained by basic characteristics of the fossil energy system. While solar energy radiates over large areas and has first to be collected and concentrated before it can be used, coal and petrol are found in high concentrations and locally fixed. Thus the labour needed to make energy available is comparatively low. When coal is mined, it appears in great quantities on the pit mouth, so that it is sensible to build canals or railroads to transport it (which would have been prohibitively inefficient for a woodland). These new transport systems allow the concentration of industries as well as a swelling of sheer size. Economies of scale get into operation, especially in rolling mills, in furnaces, and in chemical manufactures. Concentration and centralization of industrial activities become possible and, soon after become almost essential for works who want to survive industrial competition.

Another important effect of the fossil energy system is the technological acceleration of many productive processes. When pig iron was fined into wrought iron in a preindustrial charcoal furnace, the whole process took about 2 weeks. The puddle process of the late 18th century, based on coked coal, was finished after 2.5 days. The Bessemer or Thomas process of the 19th century was only a matter of 20 min. This shows that even if it had been possible to considerably increase the number of charcoal furnaces, the industrial development would have slowed down when such a key process could not be speeded up.

Taking all these points together, it can hardly be imagined that industrialization with its characteristic acceleration of cultural, technological, and economic evolution could have been possible without the use of fossil energy resources. The huge reservoir of energy, becoming available through the utilization of coal, petrol, and natural gas, catalyzed the emergence of totally new patterns of behaviour shown by socio-economic systems. This means, too, that these patterns can only be conserved as long as this period of energy abundance lasts.

It is the intention of this argumentation to show to what extent the characteristics of energy systems define the scope of ecologically relevant behaviour in human societies. This remains, of course, a rather global and abstract argument. What ecological anthropologists have tried to demonstrate for "primitive" societies should vice versa be possible for more differentiated civilized societies in the past. The patterns of energy use have here been shown on a very aggregated level, e.g., energy potentials in Britain. Historical reality, however, is far more complex, and what may be true for the whole, may be quite different in some local parts. So I see in these systematic remarks more a matter of question than a set of answers.

In conclusion, I would like to put the energy system analysis into the framework of the more general problems of environmental history in an age of ecological crisis. Environmental history can try to give an answer to the question, as to the fundamental features of this crisis. It can do so, when it takes the perspective of universal history seriously, asking what are the peculiarities of man-environmental relations, taking into account that it is the social system with its own logic of differentiation and evolution that brings "the human race" into specific contact with "nature".

In this perspective, the present ecological crisis results from a radicalization of the primitive human potential for ecological disturbance. Environmental problems have a long tradition, but they have won a new quality through the novel dimensions of industrial economy and modern society. The following factors can be mentioned: growth of human populations; synthesis of new materials; development of huge areas; omnipresence of human artefacts; emissions of materials in unpreceded concentration or amount (pollution problem); destruction of various ecosystems and species; change of elementary parameters of the biosphere, such as the gas composition of the atmosphere, the acidity of precipitations, of the radiant milieu and even – prospectively – of the climate.

The core of the present and future environmental problems can thus be defined in the following way: it is improbable that the multiple and rapidly expanding ecologically relevant human activities are neutralizing each other in such a way, that exactly those environmental conditions are preserved which existed previously and whose further existence is desirable or necessary for human survival. It is more likely that unwanted, unforeseen, uncontrolled, and uncontrollable effects are produced. On the other hand, modern technology is far from being sophisticated enough to repair damages and to restore mechanisms of natural self-regulation which produce favoured conditions. Technology may be competent enough to transcend the self-regulating capacities of natural systems, but may fail to restrain itself on a level below self-destruction.

Therefore, the critical question of environmental history is, how and when has the threshold been passed, beyond which this dynamic of self-destruction has been set free. When exactly has the point-of-no-return been reached, beyond which the present tendency toward an ecological crisis has become possible or even probable? It has been suggested that the transformation of energy systems played a major role in this process; this of course will not be the whole story.

Below this level of universal ecological history there is, I think, a whole set of further questions. Is it possible to distinguish different phases of ecologically relevant activities, correlated to different types and evolutionary stages of society? Which are the structural determinants other than energy systems which define evolutionary conditions of the man-nature-relationship? Are there key activities, key industries, key technological tendencies? What is the role of social values, preferences, cognitions, types of social organization and stratifications, or political grievances? An environmental history which will not just add some interesting remarks within the traditional framework of social and economic history needs to address these questions.

References

Adams RN (1975) Energy and structure. University of Texas Press, Austin

Adams RN (1978) Man, energy, and anthropology. Am Anthropol 80:297–309

Albion RG (1926) Forests and sea power. Univ Press, Cambridge

Aström SE (1970) English timber imports from northern Europe in the 18th Century. Scand Econ Hist Rev 18:12–32

Bairoch P (1978) Agriculture and the industrial revolution, 1700–1914. Fontana economic history of Europe 3. Fontana, Glasgow, pp 452–506

Boulding KE (1985) The world as a total system. Beverly Hills

Cottrell F (1955) Energy and society. McGraw-Hill, New York

Davis R (1962) The rise of the English shipping industry. Macmillan, London

Flinn MW (1978) Technical change as an escape from resource scarcity. England in the 17th and 18th centuries. In: Natural resources in European history. Resources for the Future, Inc. Washington, pp 139ff

Gleitsmann RJ (1981) Aspekte der Ressourcenproblematik in historischer Sicht. Scr Mercat 15:33–89

Hammersley G (1973) The charcoal iron industry and its fuel, 1540–1750. EHR 26:593–613

Hardesty DL (1977) Ecological anthropology. Wiley, New York

Hills RL (1970) Power in the industrial revolution. Kelley, New York

Hyde CK (1977) Technological change and the British iron industry, 1700–1870. Univ Press, Princeton

Kemp WB (1971) The flow of energy in a hunting society. Sci Am 224:55–65

Leach G (1976) Energy and food production. JPC Business Press, Guildford

Mitchell BR (1973) Statistical appendix, 1700–1914. Fontana economic history of Europe 4.2. Fontana, Glasgow, pp 738–820

Radkau J (1986) Zur angeblichen Energiekrise des 18. Jahrhunderts VSWG 73:1–37

Rappaport RA (1971) The flow of energy in an agricultural society. Sci Am 224:116–33

Sieferle RP (1982) Der unterirdische Wald. Energiekrise und Industrielle Revolution. Beck, München

Tunzelmann GN von (1978) Steam power and British industrialization to 1860. Clarendon Press, Oxford

White LA (1943) Energy and the evolution of culture. Am Anthropol 45:335–56

White LA (1959) The evolution of culture. McGraw-Hill, New York

From Ecological History to World Ecology

J.P. DELEAGE and D. HEMERY[1]

"From space, we see a small and fragile ball dominated not by human activity or edifice, but by a pattern of clouds, oceans, greenery and soils. Humanity's inability to fit its doings into that pattern is changing planetary systems, fundamentally. Many such changes are accompanied by life-threatening hazards. This new reality, from which there is no escape must be recognized – and managed" (World Commission of Environment and Development).

It is no longer possible to close one's eyes to the obvious: man does not use the planet with impunity; he does not dominate it; he is a part of it. It is no longer possible to grasp and understand the ecological determinations of society's evolution without an in-depth examination of the influence of natural determinants on very-long-term history. This sort of historical thinking, however, is barely emerging from its childhood, and ecological thought often ignores the scant, but available, information on the environmental past of great civilizations. Yet the value of ecological thinking based on a systematic body of duly analyzed historical data need no longer be demonstrated; it is today one of the key conditions for mastering our own history (Debeir et al. 1986; Ruffolo 1988).

1 The Temporalities of History and the Temporalities of Nature

Societies and their ecosystems, all the biotopes and physical environments in which they are inserted and from which they draw their resources, form living, interactive wholes in motion. An ecological time therefore exists in history, side by side with the economic, cultural, political, and other times. Any attempt at historical ecology must therefore interpret the relations between human populations and their environment in an evolutionist perspective, and apply different time scales in considering social ecosystems, and the mechanisms which guarantee the latter's stability as well as the processes which, to the contrary, undermine their properly ecological foundations (Febvre 1922; Crosby 1986; Ruffolo 1988).

Duration is therefore a decisive modality in the various ecological regulations of human demography. Alter their duration and you interrupt these processes of limitation or regulation. The global stability of human ecosystems is only ap-

[1]UFR de Didactique – Université Paris 7, 2, place Jussieu, 75271 Paris, Cedex 05, France

Brimblecombe, Pfister (Eds.)
The Silent Countdown

parent; the value of historical ecology, a discipline to be created, would precisely lie in locating the major disturbances which destroy these ecosystems' ability to stabilize. In the present state of our knowledge, the history of relations between societies and their ecosystems is characterized by a permanent race between homeostatic situations which generate relative stability by reproducing these relations, and break-off situations which jeopardize or, to the contrary, heighten the capacity of these societies to adapt to definitive changes of their environment. In fact, there exist only dynamic socio-ecological equilibria of longer or shorter periods.

Three major data characterize the historical temporalities of the environment.

1. There is hardly any comparison between their duration and that of the life of an individual, of generations, or even of great civilizations. The time of biophysical processes is beyond the concrete experience of humans, who, for a long time, only knew and used its phenomenal manifestations. Thus, until the development of nuclear reactors, social production of energy consisted in tapping a very small part of the energetic effects of very complex natural cycles (vegetation cycles, the water cycle, the cycle of winds, geological cycles) and in making extremely fragmented use of certain links or moments in the operation of very long energy chains. It required the mastery of phenomenal, derived (animal, vegetal, mineral, etc.) forms of solar energy. What characterizes ecological periods is that they include both the infinitely short and the infinitely long, hence the impression of great stability in which all generations to this day viewed their relations with their surroundings: "A slow-flowing history, quite often made of insistent returns, of endless cycles repeated" (Braudel 1966). The notion of resources, which suggests ecosystems' availability to social needs, or social productivity, acquires a scientific significance only if it is considered in the very long run, according to a time scale far greater than that of human life. It is estimated that 400 years are needed to regenerate completely the tropical primary rain forest which is now receding at a yearly rate of 1% of its surface! The ozone layer, which protects the planet from ultraviolet rays and is now threatened, particularly by the emission of halogenated organic compounds (chlorofluorocarbons) into the atmosphere, took about 2 billion years to form through the photosynthetic activity of plants.

2. The great physicochemical cycles unfold according to extremely rigid temporalities and temporal constraints, which weigh heavily on the future of societies. "Man is a prisoner for centuries of climates, vegetations, animal populations, crops, an equilibrium slowly built up, from which he can deviate only at the cost of jeopardizing everything" (Braudel 1958). Climatic constraints have been extensively studied by geographers and their historical operation is now relatively well known. They were the object of much research published since the 1950s (Leroy-Ladurie 1967; Wigley et al. 1981; Alexandre 1987; Pfister 1988). This research shows that there exists a very high correlation between meteorology and the economic conjuncture of agricultural societies based on the succession of

agricultural revolutions which began in the neolithic age. In the case of France, Leroy-Ladurie showed that the six uninterruptedly rainy years from 1646–1651 were one of the causes of the profound social and economic malaise expressed in the Fronde. In subsequent years, there is a contrast between the warm vegetative years of 1652–1687, and the cool springs and summers of 1687–1717 which made subsistence goods scarcer and dearer in the last two decades of the Grand Siècle. More recently, research by C. Pfister (1988) has shown the importance of climatic variation on the price of cereals and, more generally, on economic cycles in preindustrial societies, and beyond them, in continental Europe down to the construction of rail networks.

A series of major climatic fluctuations, related to the lower atmosphere's zonal eastward flow of air masses, have now been identified. Europe's time has been marked by natural cycles with a constant alternation of optimum and pessimum phases: the great optimum warmth of late prehistory (from 5000 B.C.–2300 B.C.) which helped the first land clearances; the long "sub-Atlantic deterioration" of Antiquity and the first millennium of our era, apparently coupled with gains of natural vegetation and forests; the brief lukewarm optimum of the year 1000 (9th–12th centuries) which coincides with the establishment of Western Europe's rural space; the "short glacial age" of 1590–1750; and the renewed warmth of the 18th–20th centuries, followed by the humid sequence of the 20th century.

Likewise, in another geo-historical area, the decline of Indian irrigated agriculture in the Southwest of the present United States, beginning in the 13th century, has been correlated with the desertification of immense regions in the Colorado, Gila and Rio Grande basins and the long drought of the second half of the 13th century; the latter caused a prolonged demographic decline from which the Indian peasantry never recovered.

3. There exists no absolute reversibility on the scale of nature's time. As mentioned earlier, the dynamic of human productions can be gravely shaken by this, sometimes immediately, sometimes after a delay. But this production has a return effect on ecosystems, which, added to endogenous physical degradation and entropic factors, so weighs down these ecosystems as to make their effects irreversible. When the temporalities of human history get the upper hand over the temporalities of ecological history, thresholds may be crossed in the nonreproduction of ecosystems or their march towards growing entropy. The humanization of nature deprives the wilderness of its second chance.

But long and slow processes are not always required for humans to cause the crossing of the thresholds of irreversibility. The history of ecosystems displays sharp discontinuities between periods in which they accumulate and balance elements, and moments in which their constitutive structure suddenly and punctually breaks down. The former prepare the latter, but the latter are easier to identify because they are reflected in devastating ecological catastrophes: cataclysmic floods, breakdown of river dikes, prolonged drought in desertified regions, etc. Today, the Earth's middle atmosphere is the site of a phenomenon of accumulation of this type. It is the product of the photosynthetic activity of living creatures and took several billion years to form. But now, the increasing use of fossil fuels for slightly over two centuries – the equivalent of 10 billion metric tons of coal by 1987 – has progressively accumulated in its lower and middle layers a growing quantity of atmospheric pollutants, particularly CO_2, whose concentration in the air increases by 1.3 ppm/year, and SO_2, of which 110 million tons are dumped every year (Trabalka and Reichle 1986; Ramade 1987).

Specificity of ecological time, historical weight of the constraints to which it subjects societies, irreversibility of the accelerations or ruptures caused by the latter: we see that the longest times of nature, its "sheets of slow history" (Brandel 1958) mingle in fact with (all things considered) the quite brief duration of the history of human structures. Real time, that of "societies in nature," relocated in their ecosystems, remains to be constructed. Eco-history is not only an urgent need, but now, thanks to the victory of long duration in contemporary historical thinking, a definite possibility. It alone can articulate social temporalities and ecological temporalities.

2 A Social, Universal and Multi-Millennial Conflict with Nature

Ecological determinations influence the entire field of society; they are not limited to particular areas of this field. The forms of social exploitation, the processes of unequal appropriation of the means of production, land, livestock, water, and mineral resources, emerge only through the development of new systems of exploration of nature, new forms of agricultural and industrial production. "Everywhere, there emerges an intimate link between the way of using nature and the way of using Man" (Godelier 1984). In every society, it is through appropriating nature than humans cooperate or exploit each other, that they organize and transform their relations of production and social relations. There is no crisis of the use of nature that is not also a crisis of Man's way of life.

From prehistorical times, human predatory and productive activities have generally reduced and continually transformed natural and seminatural ecosystems in certain multi-secular, or even multi-millennial, patterns. Current ecological tensions therefore have a very distant origin; they are the result of cumulated environmental crises. No civilization has been ecologically innocent. Long before modern European industrialization, human activity proved highly destructive to the original ecological fabric and inflicted irreparable alterations upon it, the most ancient and general of which is deforestation. The latter was the underside and condition for the development of hunting-gathering, agriculture, herding, the crafts, and proto-industrial activities. In fact, everywhere in the world, disorganized natural ecosystems were replaced with agro-systems, incorporating many natural species, of course, which became the keystone of all the present ecological complexes. Their creation and subsequent generalization resulted in the irreversible destruction of the primal natural equilibria and their replacement by unstable secondary equilibria.

Forests had to bear the brunt of preindustrial societies' destructive levies. The process can be grasped very early in China, where the lack of new land was several times accompanied by growing ecological imbalances and physical shortages. Many regions were cleared of forests under the direct or indirect impact of the continuous expansion of the cereal system. J. Needham notes this occurred with dire effects in the Shaanxi and Gansu highlands as early as the 16th century (Needham 1965); the process was temporarily slowed down by the political energy with which reforestation was implemented in these provinces around 1580, but resumed in the late 18th century and progressed rapidly in the regions north of the Yangzi. In fact, the destruction of the forest of the upper valleys is ancient. In the Yellow River basin, it probably dates back to the first centuries of the Empire. It may in fact be one of the causes of the slow shift of Chinese civilization from the northwest to the southeast between the Tang and Sung periods. The Sung Empire then based its brilliant urban and naval achievements on the development of the southern mountain ranges' wood reserves and on imports of Japanese wood. With respect to the modern period, S. Ashead has formulated the hypothesis that a prolonged energy crisis prevailed in China from 1400–1800 (Ashead 1974).

In the Moslem world, a wood shortage has been brought to light, which occurred in the 7th century following increased demand caused by the splendid urban expansion in Islamic lands. Fuel had to be sought further and further afield, on the Christian side of the Mediterranean, at increasing costs. As early as the 11th century, "while the West, still smothered under its forests, was beginning to use them for its ships, construction and industries, the Moslem world was declining and giving way" (Lombard 1972).

This crisis also affected Europe, but to a far lesser extent. In Atlantic and Middle Europe, clearance advanced from the beginning of the High Middle Ages. According to Roman authors, the Hercynian forest of Germany spanned a 60-days' march. It began to shrink rapidly in the 6th century. Temperate primal forests were degraded into humanized, controlled forests as a result of the space given over to crops and the demand for fuel and construction wood. Large-scale clearance reached its apex between the mid-11th century and the late 13th century. In the 14th century, European ecosystems reached their saturation point; western Europe had become a truly full world. The hour of the great ecological crises had struck, heralding the so-called Industrial Revolution – a revolution which spanned several centuries.

Beginning in the 16th century, there began in England, and probably in the Netherlands, a grave and prolonged crisis of wood, which would later spread to the continent (Nef 1932, 1978; Dion 1938; Trenard 1966; Woronoff 1984; Debeir 1986). England was the first place, where fossil fuels were massively substituted for vegetal fuels, and the energy revolution of modern times occurred, because it was the first place to experience wood shortages. Wood shortages were dramatically evident in the Elizabethan and Stuart ages, from 1550–1700, although anthropic deforestation had destroyed the forests of the north of the island much earlier, particularly the immense forest of Caledonian pines which covered the

Scottish Highlands in the beginning of our era. They were caused by demographic growth – the population of England and Wales nearly doubled between 1530 and 1690 (from 3 to close to 6 million people) –, and amplified by the increased demand from cities then in full expansion. Under Elizabeth I (1558–1603) and James I (1603–1625), the rise in wood prices was dizzying and overtook that of other goods. In 1776, Adam Smith wrote that in Edinburgh, his native city, there probably did not exist a single piece of Scottish wood (Smith 1776).

The wood crisis did not only strike in England. Somewhat like China at the same epoch, all Western civilization between the 16th and 18th centuries entered a situation of environmental instability, a process of growing distortion between the increased demand for means of subsistence generated by its growth during the three centuries at hand and the possibilities of the environment. It ran up against an ecological limit of which the constraints of deforestation were the central element (Debeir et al. 1986).

3 Ecological Crises and Social Crises

History, in its present state, is not yet able to establish precise correlations between ecological constraints and the fate of past civilizations, except perhaps with respect to the consequences of great natural cataclysms and climatic ruptures, which cannot be considered negligible (Delumeau and Leguin 1987).

Thus, climatic fluctuations have weighed heavily on the socio-political crises of agrarian societies. In France, for instance, the four very hot summers of 1778–1781, optimal for cereal and wine production, were one of the main factors of the famous crisis of agricultural prices which led to the revolution of 1789, studied by Ernest Labrousse. Conversely, the heat and drought of summer 1788 burnt the corn ears; things were made worse yet by the well-known giant hail storm of July 13 of that year, in the northern Parisian basin. The ensuing catastrophic harvests made grain very expensive and fostered peasants' anxiety in the Great Fear of the early summer 1789; their anxiety was further aggravated by soil degradation in many regions of Europe (Blaikie and Brookfield 1987).

Have the great forest and ecosphere crises played a role in the collapse or decline of ancient civilizations? This hypothesis was put forward to explain the sudden fall of Maya civilization, in the early 10th century, in the present Peten department of Guatemala and in Honduras: the degradation of the forest ecosystem and water cycle ruined Maya agriculture, which had been based on the "milpa" system (slash-and-burn corn cultivation with four- to eight-year fallows). However, this hypothesis has not yet been proved decisively (Thomson 1973).

By contrast, it is certain that, the disorganization of hydrology played an important role in weakening Mesopotamian civilizations, in particular the soil erosion caused by deforestation and overgrazing on the hillsides feeding the Tigris and Euphrates. Accelerated deforestation to meet needs for fire and construction wood and the expansion of arboriculture on the slopes provoked a growing alluvial overload of both rivers. The Babylonian and Assyrian Empires were unable to stop the clogging of the giant irrigation network of Lower

Mesopotamia, and civilization in the Fertile Crescent entered a slow process of decline which culminated with the destruction of the irrigation canals by Mongol invaders in the 11th century (Ramade 1987). Over the last three millenia, the two rivers' common delta has gained 250 km on the sea through alluvial deposits. A similar explanation pointing to the silting of the "baray" (artificial reservoir) system, and the saturation of cultivable land, has been proposed to account for the collapse of the Angkorian Empire (Groslier 1979).

In the Mediterranean zone, primal natural equilibria were also upset from the earliest phases of antiquity, and deforestation affected vast regions from the Roman era. In the Islamic world, the threat of wood shortages loomed from the 8th century. It did not become dramatic as long as the Islamic states of the Mashrek and Maghreb still received a regular supply of Sudanese gold which enabled them to pay their wood purchases advantageously. But with the arrival of Hillalian Arabs, the Transaharan gold routes, whose mastery had brought prosperity to Fatimid Egypt, were cut and the entire Moslem East weakened (Lombard 1972).

All things considered, the conclusion reached by the admittedly too few historical studies now available is that, the degradation of the environment, quite ancient in many geographical areas, only came into play in conjunction with other social (economic, technical, cultural, and other) factors and rather as an overall boundary than as an immediate and direct cause. Considering the evidence it seems that every grave ecological crisis leads after a delay to the temporary or lasting decline of the civilization, a decline which leads to its collapse or to a mutation of its innermost structures. But this decline is generally completed over long periods of time, and only if the society cannot devise mechanisms to compensate the ecological crisis, such as the development of long-distance trade or the invention of the technical and economic means for a different sort of growth (Pomian 1976; Debeir et al. 1986).

In most identified historical cases, the representations elaborated by societies to account for their relation to the environment seem to have played a significant role, as these were the starting point from which they could act upon the latter. These representations either encouraged or held back the major devastating depletions (Haudricourt 1962). In general, until the massive industrialization of Europe between the 16th and 19th century, the relationship to the ecosphere was conceived as an "exchange" with natural forces, often regarded as sacred in a series of myths and religious cosmologies, rather than as a "transformation of Nature." Not to mention a "transformation" of human "nature," for humans were considered as a component of cosmic nature. This type of outlook, of course, did not perceive nature through its tangible aspects alone, but as a set of invisible forces and powers commanding the fate of the human group. "All forms of concrete activity invented by man to appropriate natural realities, contain and combine at once and necessarily gestures and 'material' behaviors to act on their visible and tangible aspects, and gestures and behaviors that we now call 'symbolic' to act on their invisible background" (Godelier 1984).

In such cultures, where man is the main productive force, adapting to the ecosystem is the fundamental principle of society's functioning, but a principle which operates in a wide range of concrete modalities. In this regard, the development of agricultural systems in the Neolithic certainly represented an important historical threshold, because it opened several different possible ways of thinking about the Man/Nature relationship. André Haudricourt has shown that "with respect to the vegetal and animal world, beginning in the Neolithic, Man is no longer simply a predator and consumer, but henceforth assists (author's emphasis) and protects (author's emphasis), and coexists for long periods with species which he has 'domesticated'. New relations have been established on the 'friendly' mode, not unlike those of men to each other inside the group" (Haudricourt 1962). From then on, several models for dealing with nature crystallize. As shown by A. Haudricourt, one can counterpose the more direct and domineering action of Western farmers to methods based on indirect action on plants developed in the framework of rice paddies or root crops of the Melanesians. "There is no possible friendship between inanimate things, Aristotle teaches us, any more than there is between men and horses or cattle or between the master and the slave as a slave" (Haudricourt 1962).

In fact, all ancient societies which practised agriculture accumulated immense empirical and encyclopaedic knowledge about nature, on which they built the first great intellectual systems. One can now perceive the causal relations between the major ecological options and these great systems of representation of nature. But the real environmental situations remained hidden. Their understanding was confined to the levels of cosmology, metaphysics or theology. It is therefore hardly appropriate to attribute overall environmental strategies to these cultures.

4 From the Emergence of Industrial Capitalism to the Advent of "World-Ecology"

The advent and expansion of industrial capitalism since the 16th century has brought about a genuine revolution in the collective representation of Nature. The idea of a solidarity between Man and the Physical Universe was replaced by that of the domination of Nature by Man (White 1967).

The Promethean view of Nature's submission to the human became hegemonic in Western culture beginning in the 16th century. Between 1500 and 1800, the precept of Genesis (I, 28) "Replenish the earth and subdue it" became, under the cumulative influence of the Reformation, Counter-Reformation, and scientism of the 18th and 19th centuries, one of the founding axioms of this culture: an absolute antropocentrism established itself. "Man, if we look to final causes, may be regarded as the center of the world," Francis Bacon proclaimed in the 16th century. We must force nature to restitute its wealth, Isaac Newton wrote two centuries later. Henceforward, Man held absolute sway over the ecosphere, and when Nature was celebrated, it was only as a space for leisure, reveries, nostalgia, a humanized space (Merchant 1983; Thomas 1984). Referring to Capital's "great civilizing mission," Marx expressed in positive terms the cultural and ideological degradation of nature which, for three centuries, had everywhere legitimated industrialization processes: "It [Capital] rises to such a social level that all previous societies appear as purely local developments of humanity and as idolatry of nature. . . Nature becomes an object for Man, a useful thing" (Marx 1858).

This period of upheaval in people's outlook was also the time when Europe undertook to explore and conquer the world. Together with technological innovation, this was Europe's answer to its pressing shortage of energy, land and resources. Breaking out of the old geographical boundaries of the medieval West, thanks to a formidable upsurge in science and technology (Brockway 1979; Haedriek 1981, 1988), it had, by the dawn of the 20th century, taken over almost all the lands of the planet (1800: 35%, 1878: 67%; 1914: 84%). By resorting to substitutes for raw materials and energy on a massive scale, Europe opened the historical transition of human societies from the solar age to the fossil fuel age. This was the way that, the first industrialized societies grappling with grave environmental problems broadened their geopolitical room for maneuver on a scale never seen before.

The story is well known but, some of its episodes deserve recounting (Debeir et al. 1986). England's wood crisis was alleviated at first through reforestation and massive imports of Scandinavian and North American wood, and later resolved by tapping wholesale the immense fossil fuel reserves of the British Isles. From the early 17th century, the use of coal became widespread thanks to easy sea-borne transport from Newcastle to London, and in the 18th century, thanks to the construction of a canal network which drastically cut coal prices in the hinterland. At the time of the Civil War, around 1640, despite the stinking smoke it released, coal shipped via this route ("sea coal") had become indispensable for heating the houses of London. The volume of local seaborne shipments of coal was multiplied by 20 from 1550–1700, and London's imports by 30, so that by 1700, the United Kingdom produced at least 3 million tons of coal yearly. The era of fossil fuels, which was to wreak havoc on the world, had begun and with it, that of large-scale industrial pollution (Nef 1932; Debeir et al. 1986).

In France, the problem of transportation costs was resolved in the 19th century with the expansion of railroads throughout the country. One can therefore state that the mobilization of scientific and technical know-how was responsible to a large extent for the substitution phenomenon which made it possible to overcome the forest crisis. In England, this substitution generated as early as the 17th century extraordinary technical advances in most fields of industrial production: use of coal in glass manufacturing around 1610, use of reflector ovens for smelting nonferrous metal ores around 1660, coal-fueled brick ovens, invention of the Darby process for coke smelting in 1709, of puddling in 1784, and finally, development of a revolutionary energy converter with Newcomen's (1712) and Watt's (1784) steam engines. In a sense, large-scale industry was born of the ecological tensions. But the sort of answer given to these tensions brought new forms of environmental degradation and a formidable displacement of the latter in time and space (Debeir et al. 1986; Brimblecombe 1987). It is no exaggeration to state that, a historical threshold was crossed in the degradation of ecosystems with the advent of industrial capitalism. With massive industrialization, pollution and environmental degradation tended to become general; they became a genuine feature of civilization and acquired planetary dimensions in the 20th century, particularly after World War Two.

As we approach the end of the 20th century, the creation of a world productive space has laid the basis for the ecological unification of the world. In the first place, the artificialness of vegetal ecosystems has won the day. Forests and grass are now cultivated, plant species selected and uniformized, tall trees grown as garden plants. Artificiality has spread to freshwater fauna and will soon encompass sea fauna. Artificial agrosystems are now assigned the role of "nature". All these processes feature considerable energy requirements (Pimentel 1980), great fragility and delayed effects on land and sea environments (Blaikie and Brookfield 1987; Rosenberg 1988). One of the potentially most dangerous factors of the modification of ecosystems stems from the improvement of plants which makes them dependent on constant inputs of genetic variation. Until World War Two, the bulk of these inputs came from centers of variability located in Third World countries. But it is precisely in the latter countries that improvement centers have now been established and new high-yield varieties substituted for the old diversified local populations.

Many cultivated plants, particularly cereals, have reached the point of no return. Indeed, demographic growth has been such that the massive use of new high-yield crops is an irreversible fact if chronic food shortages are to be avoided. This situation is fraught with enormous ecological risks: (1) the diversified populations being replaced may be lost forever if they are not collected and conserved in due time. (2) The substitution of a uniform genetic material to the adapted mixed local populations is an invitation to disastrous epidemics of various diseases and parasites (Harlan 1975).

In the "new countries," forests were destroyed at an unprecedented rate beginning in the 19th century for clearance or to provide fuel. In the United States, of the 170 million ha covered by the vast primal deciduous-tree forest that stretched from the Atlantic to the Mississippi, only about 10 million survive today. In Canada, the conifer forest only covers one-third of the area it did before Europeans' arrival. Clearance was not suited to the soils of the American Midwest and triggered the cycles of wind erosion which created the conditions for the formation of "dust bowls." As we approach the end of the 20th century, deforestation has reached such a frenzied pace that it has overtaken the whole planet's forest renewal rate by far. In the Siberian forest, the rate of wood cutting is 150% higher than the average yearly growth of so-called production forests. Since the early 1980s, the net destruction of the tropical forest has progressed, according to the FAO, at a rate of 15.7 million ha yearly; that is the equivalent of France's forests, the largest in Europe (Boulbet 1984; Guppy 1984; World Commission of Environment and Development 1987; Dumont 1988).

Beginning in the mid-19th century, one can observe a continuous growth of the frequency of ecological catastrophes of anthropic origin (Chambers 1973; Timberlake and Wijkman 1984; Ramade 1987), related to the forward flight into uncontrolled technical innovation: Minamata disease in the 1950s, long-term effects of the dissemination of DDT-type insecticides and plastifying agents in the biosphere, multiple oil slicks, proliferation of destructive algae (Rosenberg 1988).

Most notably, the pollution of the biosphere and atmosphere has reached disastrous proportions. Until the mid-20th century, it remained confined to the industrial cities where rickets, a disease caused by an ultraviolet ray deficiency and apparently unknown among prehistoric peoples, developed among the working-class population, beginning in the 18th century. It is mentioned for the first time in England in 1650 and a map of the scourge drawn in 1889 shows that its main centers were Aberdeen, Dundee, Glasgow, Edinburgh, Newcastle, Manchester, Liverpool, Birmingham, Cardiff, and London (Chambers 1973; Brimblecombe 1987).

Sulfur dioxide pollution is ancient in industrialized countries, but until the middle of this century, it had been confined to the immediate environment of the dischargers. One of the best known cases is that of the Sudbury, Ontario, smelting plant; it began operating in 1888 and has now destroyed the 200 000 hectares of pine forest which surrounded it. In the 1950s, the industrial system began to function truly as an increasingly active range of artificial volcanoes. Two types of pollutants were to have a particularly destructive impact on the entire biosphere:

1. The first is the colossal volume of gases discharged by the combustion of various forms of fossil carbon (coal, brown coal, oil, gas). They are gas byproducts of carbon: carbon dioxide (CO_2), is the top pollutant by volume, of which these combustions have dumped 25 billion tons yearly into the atmosphere in the 1980s, causing a ceaseless rise of its concentration in the air, which in turn, could provoke an artificial warming of the Earth's climate in the 21st century (Trabalka and Reichle 1986). Our planet has long been about 35 degrees Celsius warmer than it would be without water and carbon dioxide (CO_2) in its atmosphere. This "greenhouse effect" is intensifying because these gases are being released in even greater amounts by deforestation (Baumgartner 1984) and industrial and domestic activities. Other byproducts of carbon are carbon monoxide, discharged at an annual rate of 350 million tons; unburned hydrocarbons; nitrogen oxide, sulfur dioxide (SO_2) and sulfur trioxide, discharged at an annual rate of 100 million tons, and particularly involved in acid rains (Baumgartner 1984; World Commission of Environment and Development 1987).
2. The effect of the atmospheric micropollutants produced by countless reactions is no less threatening, i.e., peroxi-acyl-nitrates and particles. Atmospheric pollution has progressively spread to the entire globe as shown by the rise of solid particles in the recent layers of Antarctic inland ice, despite their remoteness from the polluting sources. To this long list should be added the planet-wide diffusion of radioactivity following military nuclear tests and accidents in civilian nuclear plants, of which Chernobyl has become the disquieting symbol (Bojcun and Haynes 1988).

One of the most spectacular consequences has been the phenomenon of acid rain. Until the Industrial Revolution, rainwater in continental environments was nearly neutral. The pH of melted snow water obtained by coring Greenland pack ice stood between 6 and 7.6 before 1800. It is precisely between 1950 and 1979 that acidity levels suddenly shot up in Swedish lakes, after remaining remarkably stable for the previous 12 500 years since the Würm glaciation. Studies conducted since 1960 (Ramade 1987) have shown that the average pH of rain underwent a spectacular drop in all central and northern Europe (today's pH is below 5), a concrete measure of its rise in acidity.

In the United States, pH fell to 5.9 in 1939 and 4.2 today during the summer (Ramade 1987). In fact, the acidification of rain affects the entire northern hemisphere and the iso-pH 5 line has included Singapore since 1980. It is an unprecedented environmental disaster, implying the ecological death of hundreds of lakes and rivers in Europe and North America as aquatic biocenoses disappear, and the degeneration and sometimes the death of immense forest areas, now including certain tropical rain forests in China and Brazil. A new German word is entering most European languages "Waldsterben," i.e. "death of the forest".

But the ecological crisis of the 20th century has been most alarming in the Third World, because the latter concentrates the environmental rupture phenomena of both the preindustrial and industrial ages, in relation with the growth of the "overpopulation/underdevelopment twin" and with the sort of industrialization now underway there, by no means free of imperialist spoliation mechanisms. The Aswan dam bears telling witness to this: by withholding the Nile's alluvia, it ended the river's famous floods, but ruined maritime fishing in the delta.

As for environmental crises of industrial origin, they are perhaps worse today in Third World countries than they were some time ago in the early industrializers. This reality was symbolized long before 1983 by the ultra-modern Bhopal factory, surrounded by an immense slum inhabited by poor peasants. Moreover, while the frequency of natural disasters in the Third World has not increased, their human effects have steadily worsened as the population explosion drives landless peasantries to migrate towards high-risk zones, particularly fertile volcanic soils. The impressive growth of environmental degradation is in fact more and more of an active factor in natural catastrophes. As the upper valleys of rivers are stripped of wooded areas covering their slopes, floods become more frequent. This correlation is observable throughout the Third World. Thus, in the Indian subcontinent, the deforestation of the lower slopes of the Himalaya led to a multiplication of the floods in the plains downstream: between 1955 and 1975, the annual cost of flood damage was multiplied 14-fold (Timberlake and Wijkman 1984; Ramade 1987; George 1988). In Manaus (Brazil), flooding has become a very great problem in recent years. It is believed that this is caused by the loss of tree cover in nearby forest areas.

More generally, the increase of under-development has had the effect of impoverishing ecosystems and opening in certain regions a lasting "scissors crisis" between the decreasing natural resources and increasing population. The numerical growth of ethnic groups practising slash-and-burn cultivation in the tropical rain forest, still involving one-third of their surface in 1975, has led to a general imbalance of their agriculture: over a half-century, the average fallow period has been shortened to such an extent that it is now below 15 years. Additional factors are alteration of the soils, clearance of hillside woods, irreversible erosion of the soil, etc. In particular, the implementation of deliberate policies combining large-scale export of wood by forestry companies, the development of extensive ranching whose profits go to opulent corporations of the North, and the "transmigration" of landless peasantries has brought the rainforests of the intertropical zone into a hallucinating cycle of continuous clearance throughout the 20th century, and particularly since the 1950s (Becker 1982; Guppy 1984; Dumont 1988; George 1988).

In the Third World, it is also in the 20th century, that old-style environmental tensions have reached a paroxysm. Deforestation, overexploitation of cultivated lands, and overgrazing continue to cause desertification as they have for millennia. In 1984, it was estimated that one-third of the surface of the continents and over 850 million people, that is over 20% of the world population, were affected (Blaikie and Brookfield 1987; Ramade 1987). Since 1930, the Thar desert has expanded northeastward into the Indian peninsula an average 13 000 ha per year. Over the last half-century, in the Sahel – the area between the 100 mm and 500 mm isohyetal lines – desert has gained an average 2 km per year. Brazil has lost half the forest surface which it had before European settlement and the pace of the devastation encouraged by the authorities in the Amazonian forest is such that it might be entirely wiped out by the early 21st century (Becker 1982; Baumgartner 1984; Guppy 1984).

As for China, it retains only an estimated 6–13% of its surface in forests and lost between one-fourth and one-fifth of its already inadequate wooded area between 1949 and 1980 (Smil 1983; Henry 1985). Over the same period, Sichuan, now China's most populous province with 100 million inhabitants, lost one-third of its forest; the authorities now believe this is one of the main factors causing the swelling of the Yangzi's floods, which, it is feared, may become a "Second Yellow River" as it now carries about 2.5 billion tons of alluvia per year with a rate of increase higher than the Huangho's.

Finally, in all epochs, the expansion of poverty – according to the World Bank, over 1 billion men now live in a state of absolute poverty – has been an essential factor of vulnerability to ecological risks. Today, "these absolute poor live more or less all the time in the context of catastrophe. . . They can only settle on land of little value, often unhealthy; the 'favelas' of South America built on sharp inclines and therefore subject to landslides, the 'slums' of Calcutta sprawling on riverbeds and land subject to flooding, the miserable peasants of Bengal over ten million of whom now live in delta zones, whose altitude is below three meters above sea level – making them vulnerable to any hurricane or tidal wave" (Ramade 1987).

With the end of the 20th century, we are indeed facing a cumulative ecological crisis, a major and present danger for humanity. Industrialization, by triggering a fantastic demographic explosion through the Pasteurian revolution, the therapeutic discoveries of the last century, the spectacular increase of world agricultural output, have created a growing distortion between the productive capacity of the biosphere and its equilibria on the one hand, and its human load on the other. While it took 2 million years for humanity to reach 4 billion people in 1975, only 40 will be needed at the present rate of demographic growth to double this figure and for world population to reach 8 billion beings. This presupposes that, world agricultural output will increase during this period by the same amount as it did since the Neolithic age! The carrying capacity of ecosystems has already been reached in many areas (Dumont 1988).

Industrialization in its present form not only induces irreversible processes of deterioration of existing ecosystems and has proven incapable of stopping the major classical processes of environmental degradation, but jeopardizes after a time the very existence of the biosphere. In this respect, a threshold seems to have been crossed around 1960. Until then, with world population standing below the 3 billion mark, the production of the main biological resources increased in parallel fashion to the number of humans. After that date, the correlation was reversed and in several geographic areas, human numbers grew faster than the supply in resources. While ecological crises generated by imbalances between the population and productive capacities used to be regional or local, with the worldwide nature of industrial capitalism, the prospect of a critical ecological situation of planetary dimensions can no longer be discarded (World Commission of Environment and Development 1987; Dumont 1988). The situation is a source of particular concern inasmuch, as none of the socialist attempts to break with the dominant mode of production and consumption has succeeded (Komarov 1981; Blaikie and Brookfield 1987).

Considered from the standpoint of historical ecology, the growth of industrial civilization was fostered by a succession of local and regional breakdowns of ancient ecological equilibria. Now, the range of these breakdowns has broadened to include the planet. By driving them to the threshold of a "world economy", industrial capitalism, in its "classical" historical versions as well as in its various "socialist" reincarnations, has thrown human societies into a new relationship to nature: that of "world-ecology". As the century comes to a close, the natural sciences have enabled human society to acquire a power comparable to that of the geological and biological forces which have shaped the planet. The same forces which have peered into the secrets of the natural world have disclosed dizzying perspectives to our societies, leading history to raise some fundamental questions about itself.

Acknowledgement. With thanks to John Barzman for this English version.

References

Alexandre P (1987) Le climat en Europe au Moyen Age. EHSS, Paris
Ashead SA (1974) An energy crisis in early modern China. Ch'ing-shih Wen-t'i 3 (Dec 1974)
Baumgartner A (1984) Effects of deforestation and afforestation on climate. Geojournal 8, 3:283–288
Becker B (1982) A geopolitica da amazonia. Paz e Terra, Rio de Janeiro
Blaikie P, Brookfield H (1987) Land degradation and society. Methuen, London New York

Bojcun M, Haynes V (1988) The Chernobyl disaster. Hogarth, London
Braudel F (1958) La longue durée. Ann ESC 13:5–6
Braudel F (1966) La Méditerranée et le monde méditerranéen à l'époque de Philippe II, 2nd edn. Colin, Paris
Boulbet J (1984) Forêts et pays. EFEO, Paris
Brimblecombe P (1987) The big smoke. Methuen, London
Brockway LH (1979) Science and colonial expansion. Academic Press, New York London
Chambers LA (1973) Air pollution in historical perspective. In: Detweiler R, Sutherland JN, Werttsmann MS (eds) Environmental decay in its historical context. Glenview, Ill
Crosby AW (1986) Ecological imperialism, the biological expansion of Europe, 900–1900, Univ Press, Cambridge
Debeir JC, Deléage JP, Hémery D (1986) Les servitudes de la puissance, une histoire de l'énergie. Flammarion, Paris
Delumeau J, Lequin Y (eds) (1987) Les malheurs du temps, histoire des fléaux et des calamités en France. Larousse, Paris
Dion R (1938) Usines et forêts. Conséquences de l'ancien emploi du bois comme combustible industriel. Rev Eaux For Oct 1938:771–782
Dumont R (1988) Un monde intolérable, le libéralisme en question. Seuil, Paris
Febvre L (1922) La Terre et l'évolution humaine. Introduction géographique à l'histoire. Albin Michel, Paris
George S (1988) A fate worse than debt. Penguin, Harmondsworth, UK
Godelier M (1984) L'idéel et le matériel. Fayard, Paris
Groslier BP (1979) La cité hydraulique angkorienne: exploitation ou surexploitation du sol. BEFEO 65, Paris
Guppy N (1984) Tropical deforestation: a global view. Foreign Affairs, Washington 62, Spring 1984:928–965
Haedrick DR (1981) The tools of empire. Oxford Univ Press, New York Oxford
Haedrick DR (1988) The tentacles of progress. Oxford Univ Press, New York Oxford
Harlan J (1975) Crops and man. Am Soc Agron, and Crop Soc Am 677 South Segoe Road, Madison WI 53711
Haudricourt AG (1962) Domestication des animaux, culture des plantes et traitement d'autrui. L'Homme II, 1:40–50
Henry C (1985) La spécificité de la politique chinoise de l'environnement. Rev Fr Admin Publique, juillet-septembre 1985
Komarov B (1981) Le rouge et le vert. Le Seuil, Paris
Leroy-Ladurie E (1967) Histoire du climat depuis l'an mil. Flammarion, Paris
Lombard M (1972) Espaces et réseaux du haut moyen age. Mouton, Paris La Haye
Marx K (1858) Fondements de la critique de l'économie politique. Anthropos, Paris (French edn, repr 1967)
Merchant C (1983) The death of nature – women, ecology and the scientific revolution. Harper & Row, New York
Needham J (1965) Science and civilisation in China, IV, Physics and physical technology 3. Univ Press, Cambridge
Nef JU (1932) The rise of the british coal industry, 2 vols. Routledge London
Nef JU (1978) Les conséquences d'une crise historique de l'énergie." Pour Sci Feb 26, pp 92–101
Pfister C (1988) Fluctuations climatiques et prix céréaliers en Europe du XVIe au XXe siècles. Ann ESC 43, 1:25–53
Pimentel D (1980) Handbook of energy utilisation in agriculture. CRC, Boca Raton, Fla
Pomian K (1976) Les limites écologiques des civilisations. Soc Sci Inf 15:1
Ramade F (1987) Les catastrophes écologiques. McGraw-Hill, New York Paris
Rosenberg R (1988) Silent spring in the sea. Ambio 17, 4:289–290
Ruffolo G (1988) Potenza e potere, la fluttuazione gigante dell'Occidente. Laterza, Bari
Smil V (1983) Deforestation in China. Ambio 12:226–231
Smith A (1776) The wealth of nations, new edn 1970. Penguin, Harmondsworth, UK
Thomas K (1984) Man and the natural world. Penguin, Harmondsworth, UK
Thomson JES (1973) Grandeur et décadence de la civilisation Maya. Payot, Paris

Timberlake L, Wijkman A (1984) Natural disasters, acts of god or acts of man? Earthscan, London Washington

Trabalka JR, Reichle DE (1986) The changing carbon cycle, a global analysis. Springer, Berlin Heidelberg New York Tokyo

Trenard L (1966) Le charbon avant l'ère industrielle. In: Charbon et sciences humaines. Mouton, Paris La Haye

White L Jr (1967) The historical roots of our ecological crisis. Reprint in: Mitcham C, Mackey R (eds) (1983) Philosophy and technology. Free Press, New York

Wigley TML, Ingram MJ, Farmer G (eds) (1981) Climate and history: studies in past climates and their impact on man. Univ Press, Cambridge

World Commission of Environment and Development (ed) (1987) Our common future. Oxford Univ Press, Oxford New York

Woronoff D (1984) Forges prédatrices, forges protectrices. Rev Geogr Pyrénées Sud-Ouest 55, 2:213–218

The Early Loss of Ecological Stability in an Agrarian Region

C. PFISTER[1]

1 "Solar" Agriculture: A Source of Economic Growth

The environment of a society is both natural and social; exchange of energy, materials, and information takes place within the sociosphere which is made up of other societies and their infrastructure, and also through processes of production and reproduction within the biosphere and its various ecologies which are all submitted to natural laws of evolution. In its broadest sense, human ecology provides the conceptual framework for the most realistic description of the complex links between people and nature. Ecology and ecosystems stress connectivity and mutual causality among the natural and human components. This approach emphasizes the two-way character of causality, although the relative influence in reciprocally causal relationships is never equal and may be very unequal (Kaplan and Manners 1972, quoted in Ellen 1982). One of the most important and elemental is the flow of energy. A web of energy and material relations allows the ecosystems approach to draw together the natural and human processes. This focus on common components in an interconnected system provides an important means of specifying how nature and society are in fact interrelated (Sack 1990).

Anthropologists have associated cultural development and economic growth with the control of progressively greater amounts of energy (White 1949; Cottrell 1970; Moran 1979). According to Ervin Laszlo (1987), a natural scientist, society has moved along an evolutionary axis, sparked by technology, toward the exploitation of increasingly dense and abundant energies. The economic dimension of ecological energetics has been discussed from the early 19th century (Martinez-Alier 1987) but until the work of Georgescu-Roegen (1971) it had not been couched into a coherent theory.

Debeir et al. (1987) have coined the term of *energy system*. An energy system includes not only the sources and the converters of energy and their efficiency, but also the social structures in which the converters are created and maintained and in which the energy is appropriated. Thus it is more than just an ensemble of producers, transformers, conductors, and reservoirs for calories or joules. In order to understand how it works, its social, technological, political, economic, and cognitive dimensions have to be taken into account.

[1]Historisches Institut der Universität Bern, Engehaldenstrasse 4, 3012 Bern, Switzerland

Brimblecombe, Pfister (Eds.)
The Silent Countdown

Schnaiberg (1980) who discusses human and other biological systems in terms of energy, notices that one important area wherein human systems diverge from other biological systems is in the "creation and disposition of surplus energy". Whereas the ecosystem reaches a steady-state by permitting the growth of just enough species and populations to offset the surplus, societies, particularly industrial societies, tend to use the surplus to accumulate still more economic surplus in the future. The question then to be answered by environmental history is: when, where and why has this quest for accumulation begun?

In human history there have been three basic types of energy systems: (1) the unmodified solar system of hunter-gatherers; (2) the modified solar system of agricultural societies and (3) the fossil energy system of industrial societies (Sieferle, this Vol.). Up to the present, ecological analysis has focussed upon the transitional period between those systems and the period of precipitous environmental transformation in the last three decades, where major shifts in human relations with nonhuman nature are known to have occurred. (Sieferle, Deléage, this Vol.). Less attention has been paid to the energetic dimension of the modifications which are known to have occurred *within the solar energy system*, e.g., the invention of the harness and the water- mill, and the introduction of new crops and rotations in order to raise agricultural output (Sieferle 1982; Debeir et al. 1987). This latter innovation was at the origin of a fundamental shift in carrying capacity, the magnitude of which may be compared to the Neolithic Revolution.

Following a period of prolonged stagnation. the population of Europe grew by 90% between 1750 and 1850 (Köllmann 1965), i.e., over a period in which grain or guano could still not be imported in substantial quantities from overseas. The challenge of increasing agricultural production to feed 120 million additional people and of releasing much income and labour for secondary and tertiary industries was successfully met by an agriculture still lacking significant industrial inputs (Grigg 1980). The "new" agriculture came into practice in 17th-century Flanders and was then transferred to England. By the mid-18th century, it had already raised agricultural productivity quite considerably in this country (Slicher van Bath 1963; Jones and Woolf 1969). Continental Europe was somewhat slower to follow. In Denmark (Osterud 1978), in Hannover (Ulbricht 1980) and in some other regions of Germany (Abel 1978), and in Switzerland (Pfister 1984) the take-off dates from the late 18th century, in France from the early 19th century (Désert and Specklin 1976). Jones (1974, p. 128) suggests that "developments on the production side of agriculture were instrumental in bringing the earliest advanced countries to the brink of industrialisation". Bairoch (1976) mentions the growth of population and purchasing power, the demand for steel for agricultural implements, and the investment of "agricultural" capital into industry, which all were connected to the "agricultural revolution" as being the decisive factors for the success of the industrial revolution.

2 Sinks and Holes in the Capturing Network

Flow of energy and cycles of matter are aggregate measures appropriate to macroecosystem description. While energy flows continually in and out of systems, matter essentially cycles from one state to another. Cycles of matter can be illustrated by noting how plants make use of elements and how human beings enter into these processes. As with energy, the plant producers are the key element in the material exchange cycles. The energy converted and stored in plant biomass powers a series of other organisms of the ecosystem in what has been described as a food chain (Moran 1979).

The solar energy system of a preindustrial Europe comprised two kinds of conversion: (1) the chemical conversion of solar energy by plants into biomass; and (2) the conversion of rainfall and wind into mechanical wind or water energy (Fig. 1). This figure uses the energy flow symbols suggested by Odum (1971) which have become standard in ecosystem analysis.

While the energy flow from the sun is quasi-constant at the outer space, it varies from year to year on the ground according to changing meteorological conditions, in particular in middle and high latitudes. To take an example: in Central Europe (Zürich) the duration of sunshine over the vegetative period from March through October may fluctuate between 1142 and 1813 h. (Schüepp 1962). The amount and the distribution of rainfall are also highly variable. The short-term variability of the energy flows originating from the climate system which, in Fig. 1, is illustrated with the switch symbol, underlies the year-to-year fluctuations of yields, of the growth-rate of trees, and of the mechanical energy available from wind and water. In the long term, the management of land consists of applying known or discovered skills to land use in such a way as to minimize or repair degradation, and ensuring that the capability of the land is continued beyond the present crop and other activities, so as to be available for the next. Natural processes fall into two main groups, (1) the mainly biological/biochemical and (2) the mainly physical. The basic problem is to cope with the fact that purposive plant growth and removal for use tends to extract mineral and organic elements from the soil faster than they can be reproduced (Blaikie and Brookfield 1987, p. 8). The level of biomass production varies according to a set of variables which reflect the natural characteristics of the system and the technological level of the society. Its basic elements are the surface under cultivation, the composition of the crop mix and the forms of rotation, the input of soil nutrients, the input and the quality of animal power and human labour, and the specific photosynthetic properties of the crops grown. Almost no input of labour or matter was provided for maintaining the forest ecosystem.

The net in which agriculture collects solar energy has sinks and holes. Part of the energy is lost in harvesting. The amount of losses depends upon the weather during the harvest. When the mature grain crop is exposed to the rain for a long time, the ears will sprout and the flour content of the crop will be deficient. In addition, moisture will increase losses in storage from insects, mould, and fungi. In Third World countries, today these losses are estimated to be 10% (Blanckenburg 1986), in Ancient Régime Europe they could go up to one-third of

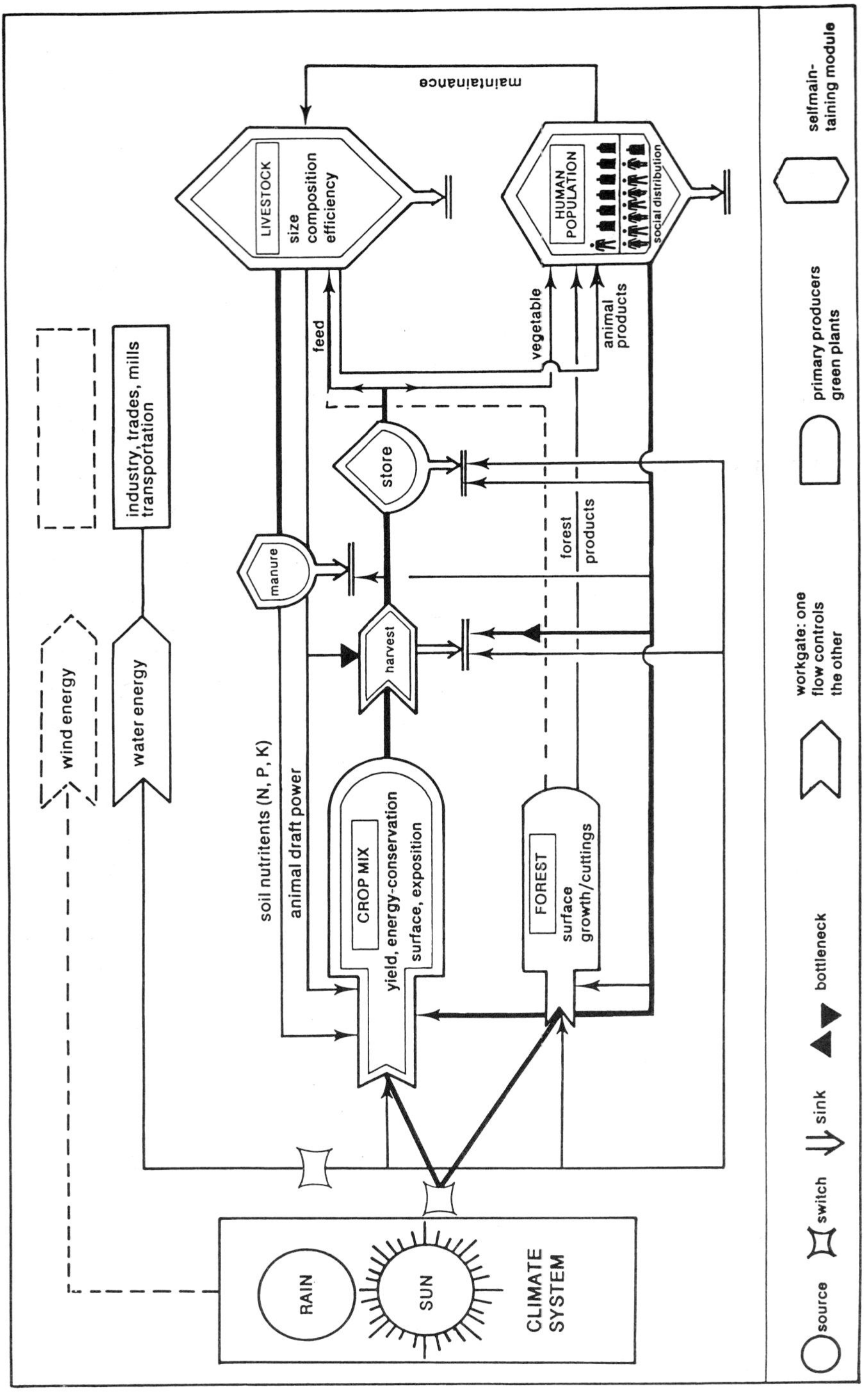

CLIMATE SYSTEM
RAIN
SUN
wind energy
water energy
industry, trades, mills transportation
soil nutritents (N, P, K)
animal draft power
CROP MIX
yield, energy-conservation
surface, exposition
FOREST
surface
growth/cuttings
harvest
manure
store
feed
vegetable
animal products
forest products
LIVESTOCK
size
composition
efficiency
HUMAN POPULATION
social distribution
maintainance
source
switch
sink
bottleneck
workgate: one flow controls the other
primary producers green plants
selfmain-taining module

the harvest, when the harvest was not properly dried in the field. The amount of energy which is contained in the hay crop also depends primarily on the meteorological conditions. Hay which is repeatedly washed out by the rain loses up to 70% of its energy content. Prior to the early 19th century, when the productivity of the cows was still very poor, variations of energy in the feed had a profound impact upon milk yields (see Fig. 4; Pfister 1984). In order to minimize energy losses, agrarian societies had to use their working capacity and animal draft power to the fullest during the harvest peak. Often the number of hands decided whether the crop was sheltered in dry condition, or whether it got drenched.

Manpower and draft animal power, which were abundant during the slack season, were short during the harvest peak. In Fig. 1 this shortage is represented with a "bottleneck" symbol (created for this paper). Unequal access to basic resources is a fundamental property of human societies. Because it is often overlooked in the traditional ecological "per capita" approach to carrying capacity, adequate symbols have been inserted in the module which represents human population in Fig. 1. As for livestock, its composition, its size, and its yield are the important parameters which all mirror the quality and quantity of feed. In modeling ecosystemic relationships, the biomass which is put in for animal feed must match the known properties and yield of the livestock. The manure is a very valuable part of the produce, because it contains the essential soil nutrients (nitrogen, phosphorous, and potash) which could not otherwise be obtained in an agriculture lacking manufactured inputs. The amount of nutrients actually recycled depends on how much of the manure is actually collected, how long it is stored, and how it is managed. Poor recycling technologies were a basic cause for the inherent low productivity of "traditional agriculture".

Sieferle and Deléage (this Vol.) have argued that a "solar" economy is bound to a stationary state of production, consumption, and population, because the energy input of the sun is limited. This may be true for an agrosystem such as that of the Chinese, in which an appropriate crop mix, a high input of labour, and adequate recycling strategies have pushed the use of land, material, and manpower near the theoretical limit of carrying capacity. Most of the agrosystems in

Fig. 1. A model of the flows of energy and materials in solar energy agrosystems of Central Europe. The *sun* is the unique source of energy which powers the system. In temperate latitudes the energy input (radiation, water power) at the ground is variable according to weather conditions (*switch*). The energy content of the biomass collected and the soil nutrients are leaking out the agrosystem to some extent (*sink*). Losses in biomass are primarily caused by wetness patterns during the harvest period. Their magnitude depends partly from the animal draft power and the working capacity available (*bottleneck*). The product itself is unevenly distributed within the society. This is essential in considering carrying capacity.

During the 'agricultural revolution' the input of soil nutrients was increased by growing leguminous fodder crops (which obtain nitrogen from the air) and by improved recycling technologies (stall-feeding of cattle in summer and collecting liquid manure). As a consequence more biomass was harvested. Population and livestock expanded in a self-reinforcing process (*double lines*), whereas forest output remained constant

Ancient Regime Europe, however, were far from this limit. They had a wide margin within which the efficiency of the systems could be improved. The duplicate symbols in Fig. 1 express the fact that over the period under investigation, biomass production was manipulated through deliberate interventions in the material cycles and energy flows. The application and expansion of new technologies is contained within the institutional framework of a society. Institutions may be entitled to block, regulate, or unleash the introduction of technological innovations. These actions or non-actions are justified with reference to the set of values and norms which underlie the coherence of a society. In Ancient Regime Europe, institutions were more attuned at preserving the static flow of resources in favor of a closed group than at promoting economic growth. In most cases, new technologies involved modifications in the control of land and resources which interfered with the interest of the ruling classes.

Because the output of the systems was stagnating in the long run and variable in the short run, every agrarian society had to cope with three basic problems:

1. It had to find ways for adapting its numbers to the carrying capacity of the energy system, i.e., to prevent population from increasing to an extent that would have endangered the cohesion and the survival of society.
2. It had to preserve the energy converting property of the system from degrading, providing for certain inputs of human and animal work in order to get and harvest the biomass needed for food, heating, and cooking, and to maintain the installations which were transforming water-power into mechanical energy.
3. It had to protect its members from the effects of short-term climatic fluctuations by accumulating stocks of grain or large herds to provide food which could be consumed in periods of climatic hardship.

3 A Testing Ground for Ecological History

The quantitative analysis of energy flows within an agrosystem requires dependable data on agricultural energetics. The various flows of energy and cycles of matter must be identified and quantified, in order to get reliable estimates for the chemical transformation of solar energy into biomass and its gradual diffusion and loss through a food web (Moran 1979; Fluck and Baird 1980). This involves measuring a number of key variables such as land use, agricultural and forest production, water-power, losses through harvesting and in storage, livestock size and composition, and population size (Fig. 1).

Most of the variables needed to investigate agricultural energetics have been set up in the administrative context of the Swiss Canton of Bern. Before it became a member of the new Swiss Confederation in 1848, Bern was an independent republic of 6000 km^2 with a population of 400 000 (1850). Its territory roughly makes up one-seventh of the population and the surface of Switzerland. The canton is not an ecotope in the sense of a holistic land unit (Naveh and Lieberman 1984). It comprises "ecozones" at altitudes between 350 m and the line of permanent snow, such as a vine belt and a grain belt in the lowlands, a zone of mixed farming in the hills, and a dairy belt in the Alps (Pfister 1989). This diversity makes the canton an ideal ground for testing the complex interactions between population and its natural habitat. Moreover, it was basically self-sufficient, although items such as cloth, wine, coffee, and sugar were imported in exchange for cattle, timber, and cheese. In years of below-average harvests, grain was also imported. But it has been shown from the example of 1847 that shortage and hunger must be attributed to unequal distribution of the resources which were still sufficient to cover the demand of the population on a per capita basis (Pfister 1990).

This evidence for analyzing energy flows is being collected, verified, interpreted, and stored in a large data-bank called BERNHIST (Pfister and Schüle 1989). At present, only labour inputs have not yet been quantified. The ecosystem concept serves mainly as a didactic device to emphasize the interaction between living and nonliving components of a system, such as is has originally be used by Odum (1953). Boundaries are drawn according to the data available for the analysis.

For a long time, ecosystem research and modelling efforts have ignored historical change, with the exception of the chemist Martin Frissel (1977) who has compared the nutrient balance for dutch farms in 1800 and 1972. The present paper tries to shed more light on the transformations which have been brought upon by purposeful human interventions in the cycles of material from the period around 1760 –1890, i.e., during the final phase of the solar economy and the beginning of the transitional period from the solar to the fossil energy system.

For the first period from 1760–1885 evidence is available for the county of Büren (Stampfli and Frey 1987). This county of 88 km^2 is part of the so-called Kornland, where grain was the dominant crop within the three-field system. For the 1880's an energy balance is set up for the entire canton.

4 Dynamic Stability: The County of Büren Before 1760

Land use and productivity around 1760 are represented in the left part of Fig. 2. Net grain yields (seeds deduced) were only 850 kg/ha, estimated milk yields were some 1000 liters per cow and 250 liters per goat. A comprehensive starting point for analyzing the flows of energy and the circulation of matter is the question of limiting factors. Land was in excess – 35% of the area was only marginally used. A part of it, the commons, belonged to the community; approximately one-third of the arable land was laid fallow within the three-year rotation. Insufficient manuring is the reason which is most frequently given in the sources for the low productivity. This suggests a shortage of biologically available nitrogen which determines the production of nearly all food and the growing of plants used for other products in the vast majority of agricultural systems (Bolin and Arrhenius 1977; Postgate 1980). Manure was in short supply because the number of cattle was too low compared to the surface cultivated in grain. This was explained by a shortage of hay, which was a consequence of the small acreage of meadows in

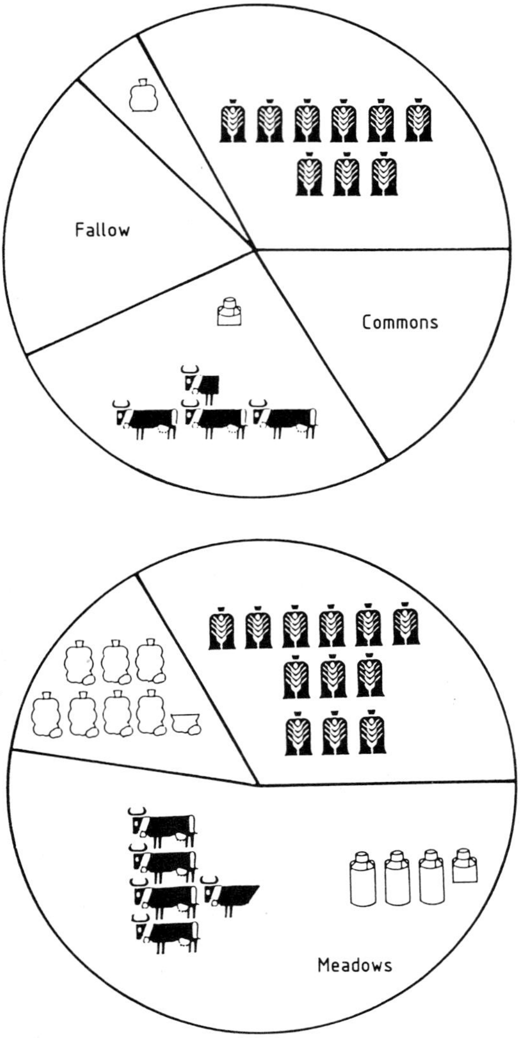

Fig. 2. The slices of the pie diagram are proportionate to the amount of land used for different cultivations or fallow. Land use in 1760 was estimated from tithe maps and records. For 1847 the amount of arable land is known from statistics, the surface of meadows is estimated. Each entire *production symbol* stands for a net production of 5.8 million J/day: *ears* for grain production, *sacks* for potato production, *milk cans* for milk production. *Partial symbols* denote proportion of food energy produced

proportion to the acreage of arable fields. This blockage has to be interpreted in the context of a political system which was tuned to maintain the status quo. Arable land could not be converted into meadows without the triple consent of the central authorities, the owners of the tithe, and the village community, who affected a loss in temporary pasture.

Moreover it has been estimated that only one-fifth of the dung and urine produced by the livestock was actually recycled to the fields (Table 1). In making this estimate, animals of different size had to be aggregated to cattle-units and the lower body weight of animals in the 18th century had to be taken into account in

Table 1. Land use, nitrogen inputs and food production in Büren country (88 km²)[a]

	1764		1847		1885	
	Land use	N-input	Land use	N-input	Land use	N-input
Cultivated surface	2750 ha	–	4000 ha	49 kg/ha	4552 ha	108 kg/ha
Manured surface	750 ha	25 kg/ha	4000 ha	49 kg/ha	4552 ha	108 kg/ha
Population	3640		7544		8162	
Cattle Units	3130[b]		4080		4985	
N from dung	48 t[c]	14 t(28%)	75 t[d]	60 t(80%)	116 t	93 t(80%)
N from urine	48 t[c]	5 t(10%)	75 t[d]	60 t(80%)	116 t	93 t(80%)
N from legumes[f]	–		77 t[g]	77 t	308 t	308 t
N total	96 t	19 t	227 t	197 t	540 t	494 t
Total net food produced yr^{-1}	17×10^{12} J		48×10^{12} J		54×10^{12} J	
Per capita food produced d^{-1}	13×10^{6} J		17×10^{6} J		18×10^{6} J	

[a] Without the parish of Pieterlen.
[b] Back projection from the proportion of 86 cattle-units per 100 people computed from the cattle statistic of 1790 (3602 cattle units).
Estimated nitrogen production per cattle unit:
[c] 1760 31 kg = 66.6% of the 46.5 kg obtained from the statistic of 1876 for cattle units of 370 kg each (Müller and Grete 1885).
[d] 1847 37 kg (80% of the values for 1876).
[e] Losses according to Krämer (1902).
[f] N production: 300 kg/ha (Postgate 1980).
[g] Assumed surface 226 ha (25% of the 1025 ha known for 1885).
[h] Seeds deduced.
Data: Stampfli 1987; Pfister and Schüle 1989.

estimating manuring capacity. The estimates obtained from the figures in Müller and Grete (1885) are in agreement with those obtained by Huggel (1979). Though it is known that the content of the dung and the urine in nitrogen (N), phosphorus (P) and potash (K) depends upon the quality of the feed, it was assumed that the quality did not change prior to the late 19th century. This has the advantage of not underestimating the manuring potential of a cattle unit in the 18th century.

Nitrogen losses are difficult to evaluate. It is known that the urine, which contained 50% of the nitrogen, trickled through to the loamy floor of the stable into the groundwater. In summer it was sprinkled on the common pastures which, however, were very poorly managed. The major part of the biomass was trampled down by the grazing animals. The dung was only properly collected during the 5 months when the animals were fed in the stable. For the remaining time it was wasted across the common pastures and on the streets. To make things worse: the manure which had been collected during the winter months was not brought onto the fields before the following autumn. During the whole summer, it was exposed to the sun and to the rain. This involved losses of nitrogen which may be estimated from experiments carried out at the station of Hohenheim in the early 1880's. The nitrogen content of manure which was exposed to the weather had dropped by

57% within a year, while losses in a sheltered box were only 32% (Müller and Grete 1885). For the 18th century the estimated loss of 35% for the period from early March to late September may rather be below than above the real figure.

All the available manure was applied to the winter crop within the three-field system (Table 1). It was estimated from the model that the nitrogen input in the case of Büren county was only 28 kg per hectare. The summer crop, grown in a second year of the rotation, was not manured. Accordingly, yields were considerably lower. In the third year of the rotation, fields were left fallow. This may be interpreted as a strategy for avoiding the risk of loosing scarce seed grain, because net yields might often have been negative without any input of nitrogen for 2 consecutive years. This practice is not at all related to the restoration of soil fertility as it is still described in many textbooks. On the contrary, soil fertility was further degraded, because the fallow land was ploughed repeatedly in order to suppress the weeds. The exposure of fallow soils to wind and rain resulted in a depletion of soil nitrogen by erosion (Bolin and Arrhenius 1977; Vogt, this Vol.).

To summarize the argument: the recycling of organic matter was inadequate, as a consequence, crop density and biomass production were low. From the ecological point of view, we have to evaluate the long-term stability and viability of the system, its *carrying capacity*. Ellen (1982, p. 41) has defined this term as being the "maximum population that a particular environment can support indefinitely without leading to degradation". If the term "support" is understood to mean providing enough nutrition for production and reproduction as well as enough energy for cooking and heating, this yields a workable definition. Stampfli and Frey (1987) have shown that in Büren county the per capita daily output per hectare was adequate according to the standards of the 20th century, despite the low productivity. To which extent the county could still cover the demand for timber and fuelwood may not be drawn from the sources. But it might have been partly imported from the zone of the hills, where acute shortages had not yet developed. The population which had been more or less stable for many decades did not press upon the resources in any case. Given the low intensity of land-use, many local ecosystems were still quasi-natural, particularly the marshes and along river margins. A broad genetic diversity of fauna and flora is described in contemporary reports. Man and nature may still have been in that kind of dynamic long-term equilibrium for which Waddington (1975) has coined the term "homeorhesis" (Naveh and Lieberman 1984, p. 61).

5 A Successful Management of Cycles and Flows

Economic history explains the take-off of European agriculture in the early 19th century in terms of market incentives, which became effective through the additional demand for food of a growing population. It is connected to the triumph of liberalism and individualism which removed the institutional and intellectual barriers to economic growth.

Ecological history raises the question how this take-off is associated to changes in the material cycles and energy flows, what the consequences were for the environment, and how they affected the long-term stability of the ecosystem. The transformations will be demonstrated from the example of Büren County (lower part of Fig. 2 and Table 1) which is typical for Central Europe. They involved expanding the cultivated area at the expense of the fallow and the commons, changing the crop mix, and increasing the input of soil nutrients and yields. In Büren County, most of the additional land had been converted to meadows. Net food energy had almost tripled. This was a consequence of a triple output of milk, of a fourfold expansion of the area cultivated with potatoes, and an 42% increase in grain yields. The doubling of population density and the 40% increase of per capita food output in 1847, compared to 1760, was the direct result of the modified energy flow. The transformation must be couched in the context of major human interventions in the nitrogen cycle. The strategic innovations involved a more efficient recycling of soil nutrients, a biological way for tapping the large reservoir of nitrogen contained in the air, a more productive form of converting solar energy into biomass, and the combination of those sources in cascades, chains, and couplings.

Attempts were made to fully exploit the dung potential of the livestock. Covered basins were built below the stables in order to collect the liquid manure. It was carried to the meadows, at first by men in a wooden pail on the back, later on special horsedrawn containers. The method became known in the twenties and thirties of the 18th century and was at first applied only in the German-speaking part of Switzerland. Propagated by farmers, and subsequently by the first agricultural scientists, it spread to other countries (Hauser 1974). Progress in stable-manuring farming was accomplished by keeping the livestock indoors all year round. As a consequence, more manure could be recycled to the fields and it was carried out more frequently, at least twice a year in spring and in autumn, which may have reduced losses to one-fourth (Table 1).

Leguminous fodder plants such as clover obtain nitrogen from the air by biological fixation. They were sown into the fields and fallows from the mid-18th century. A rapid diffusion from the late 1830's is described in the annual reports which the chief administrative officers in the districts had to submit to the cantonal government. For 1847 the acreage is not known. It has been set to 25% of the acreage known from 1885, not to overestimate the importance of this crop. In 1885 the energy yield of legumes was almost 50% higher than that of ordinary meadows and 25% higher than that of grain and vegetables (Fig. 3). This is attributed to a higher rate of photosynthesis (Kleiber 1967, p. 278). In 1885, legumes had become the dominant source of nitrogen (Table 1).

Through higher input rates of nitrogen, processes of positive self-amplifying feedback were induced. The shortage of nitrogen is an important reason why potatoes had not been grown in larger quantities before the early 19th century, although they had been known for almost a century by that time. The potato has a high demand for nitrogen – it takes 35 kg to produce 100 kg of tubers (Finck 1979, p. 308). It was only when the manuring gap was closed that the high energetic productivity of this crop (Fig. 3) was fully used. Part of the potatoes were fed to the livestock. As more animal feed was produced, the size of the herds was expanded, the animals were better fed, they were bigger in size, and produced

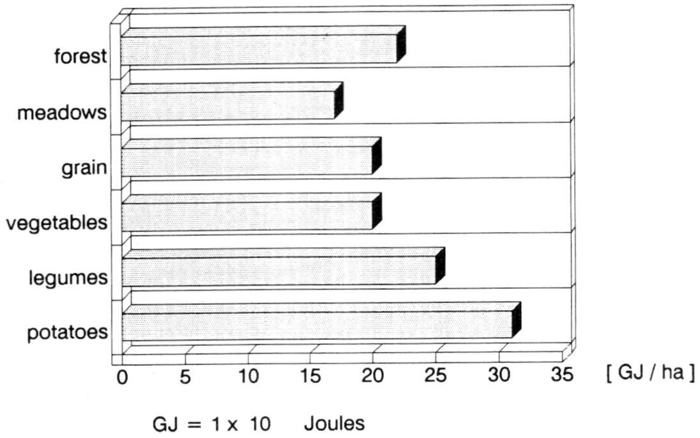

Fig. 3. Net yields of energy per hectare for different crops in the Canton of Bern in the 1880s. Land use and yield data have been obtained from agricultural statistics. For yield data a 5-year average was used. For grain and potatoes the amount required for seeds are deducted

more manure with a higher content in soil nutrients. Thus the production of animal feed and manure was mutually self-reinforcing. Compared to 1760, nitrogen inputs in the system may have been multiplied by ten in 1847 (Table 1). Given the weakness of some of the basic assumptions, this is a mere guestimate which nevertheless provides some rough idea on the magnitude of the transformations in the materials cycles which were involved in the growth process.

Milk production per animal increased out of proportion at the beginning of the take-off period, because the initial level of production had been very low. This is connected to the energetic properties of milk production. The maintainance of the metabolic rate of an animal requires a certain amount of nutrients which must be invested, before the first liter of milk may be obtained from it. The better the feed the higher the proportion of the nutrients which are converted into milk, therefore, the lower the energy input per liter. Within 100 years from the late 18th century the energy cost of milk was cut in half (Fig. 4). The relative abundance of milk had two consequences: (1) cheese manufacturing was introduced in most lowland communities in order to market excess production; and (2) nutrition became more rich in proteins, which may have improved resistance to disease and therefore promoted population growth and the capacity to work.

Population growth and the intensity of cropping were self-reinforcing in the first half of the 19th century. The cultivation of the fallow required a higher *input of work*. Labour-intensive crops such as potatoes were inserted in the rotation, greater amounts of manure were brought out, and a larger amount of biomass was harvested. Though population more than doubled from 1760–1847, most of the labour force was still engaged in agriculture or in trades connected to agricultural production in mid-century (Stampfli and Frey 1987).

When appropriate drainage technologies were available after 1850, the cultivated area was expanded at the expense of wilderness. Marginal land such as fallow, marshes, and hedges, which had been refuges for a broad variety of species,

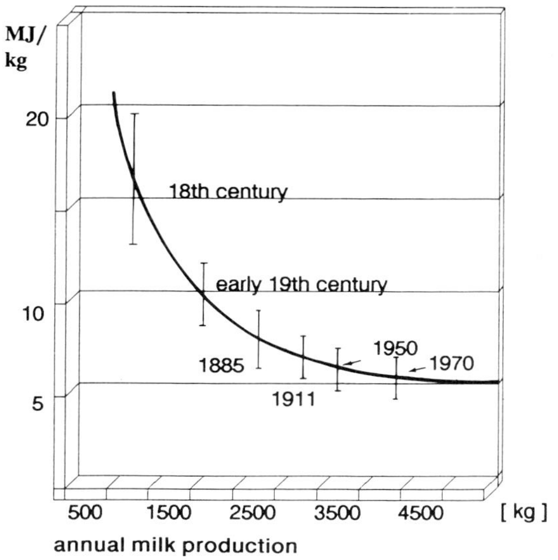

Fig. 4. The energetic efficiency of milk production. Input of net energy in 10^6 J as a function of the yearly average milk yield per cow. (Krämer 1888; Brugger 1979, 1985; Huggel 1979; Pfister 1984)

was diminished. The shift towards high-energy crops such as legumes and potatoes was intensified from 1847 to 1885. Given the imports of cheap grain by rail it paid to bear the high energetic losses which are involved in the production of meat and milk. Through the multiplying effect of herd size and legume production the rapid increase of nitrogen inputs into the agrosystem was maintained (Table 1).

Contemporary observers noted that the frequent application of liquid manure promoted the growth of grasses at the expense of Compositae and varieties of clover (Stebler and Schröter 1887). Though agriculture was still in its 'solar' phase, human interventions in the nitrogen cycle had already become important enough to change the floral composition of meadows. On the other hand, they were not yet sufficiently large to jeopardize human health, because the number of hogs or chicken, which a single farmer would support, were still related to the productive capacity of its farm. *Pollution agricole* only became an environmental hazard after the 1960's, when cheap imports of fodder and new technologies paved the way for large-scale industrial husbandry. In some areas in the Swiss Cantons of Lucerne and Bern, and in the South Oldenburg region, hogs and battery chicken have been so highly concentrated that the groundwater has become unfit for drinking from the seeping of nitrates (Thober et al. 1985). Nitrogen and phosphorus, which once had been limiting factors for agriculture, became a source of pollution.

6 Out of Balance: The Energy-System in the Canton of Bern in the 1880s

For the years 1885–89 the evidence contained in the BERNHIST data-bank allows setting up an energy balance for the entire canton. It includes the production of solar and fossil energy in different forms (biomass, water-power,

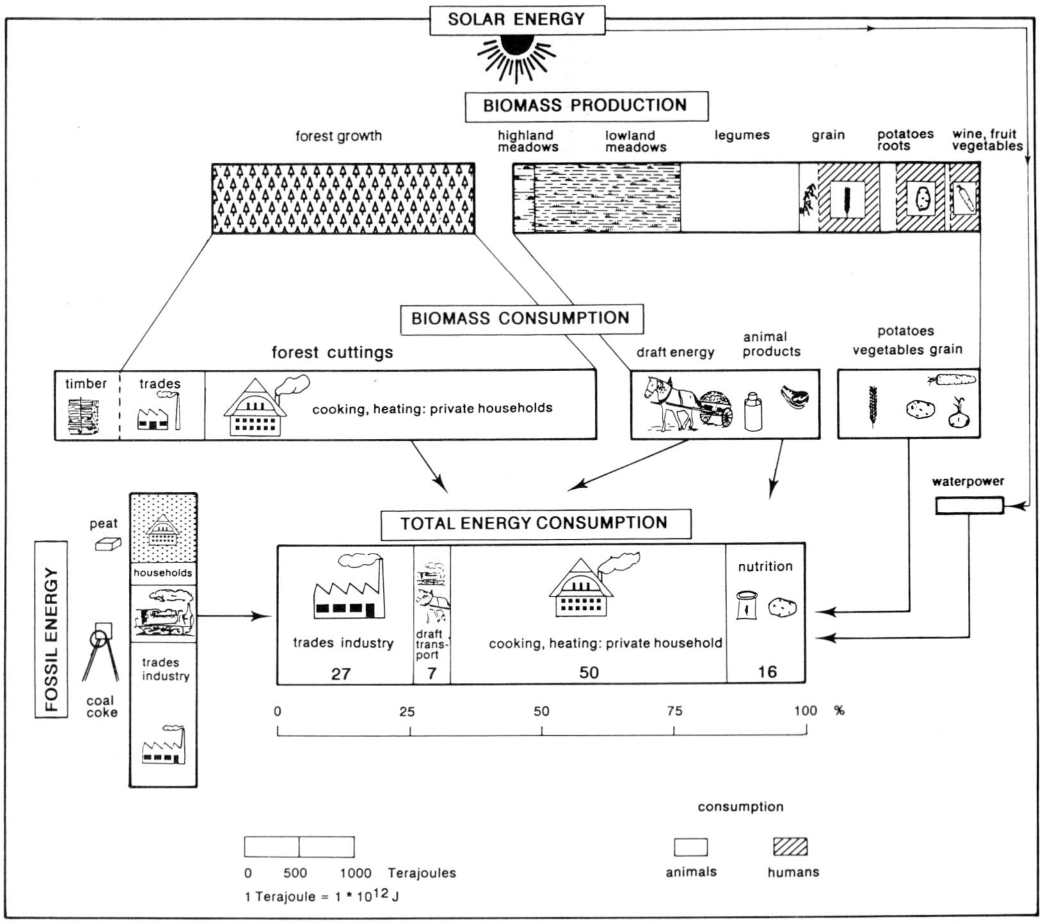

Fig. 5. The energy balance of the Canton of Bern around 1885. The data have been obtained from a variety of statistics taken in the 1880s. They were combined in a model (Pfister and Schüle, in preparation)

draft-energy, peat, coke, and coal) and the consumption of individuals (nutrition), of households (cooking, heating), of livestock, of transportation, and of trade and industry. The results will be discussed in greater detail in a forthcoming book.

Figure 5 displays the situation of a "dual ecology" in the transitional phase between the solar and the fossil energy system. The basic needs of the population (nutrition, cooking, heating) as well as the traditional occupations (agriculture, trades) were still met within the "solar sector" which comprised 72% of energy production. The new kinds of transportation and occupation, railroads and industries, belonged to the "fossil sector" of the economy. An exception is the domestic fossil fuel, peat, which had been mined from the 18th century as a result of the growing shortage of fuelwood. The most substantial part of the biomass was eaten up by the livestock in return to draft energy and animal products. Two-thirds of feed energy were required to maintain the metabolic rate of animals.

Most remarkable fact from an ecological point of view: there was an acute shortage of heat energy. The annual growth of trees hardly made up for half of the wood which was extracted from the forest. In the Bernese Oberland, where the per capita demand for heating was highest for climatic reasons, and where many trees were crippled by the teeth of goats, the annual growth of trees covered only 38% of the yearly consumption. It has been demonstrated from the example of Büren county how this energy crisis came into being. The agrosystems in the lowlands had been successfully attuned to feed the growing population, which in return provided the work energy for the more intensive cultivation. In contrast to earlier periods of demographic expansion, the take-off of the 19th century did not involve the clearing of forests. When enough feed became available from fields and meadows, forests in the lowlands were no longer used for pasture, which may even have improved their condition, although it took a long time before a sharp boundary between fields and forests came into being (Radkau and Schäfer 1987). However, the size and the yield of forests did not change significantly over the period. On the other hand, it may be inferred from the figures for 1889 that the demand for heat energy since the 18th century may have risen twice as fast than the demand for food as a result of population growth. In 1889, the annual consumption of fuelwood for cooking, heating, trades, and industry was 1.5 m^3 (10 GJ) per person, while the overall consumption of food energy was 4 GJ. Of course it must be considered that consumption per person may have somewhat declined in 1885, because in this year average household size was 0.5 persons larger than in 1760 and, according to Radkau and Schäfer (1987), new technologies for a more economical use of cooking and energy had been developed. But even assuming a consumption of 12 GJ per person in 1760, and taking the forest area and the yield of 1889, the annual growth of trees may have been adequate to meet the demand for firewood around 1760, when population size was 200 000, though local shortages had already developed. *The success of the measures which aimed at improving the carrying capacity of the agrosystems in the short term must therefore be considered to be the very reason why the ecosystem became unstable in the long term.* In order to maintain ecological stability, the areas under cultivation should have been reduced in favor of the area under forest when population began to grow. The threshold beyond which population density exceeded the carrying capacity of the forest may have been crossed around the end of the 18th century. The ongoing over-exploitation of the forests made the ecosystems more vulnerable for the impact of climatic extremes: soil erosion, landslides, and floods became more frequent during the 19th century (Der Hohe Schweizerische Bundesrath 1862). A similar dichotomy between food resources and fuel resources has developed in many parts of the Third World over the last few decades as a consequence of population explosion. China, for example, has become able to feed its population. However, the productive forest which was reduced by a third between 1949 and 1979 no longer meets the demand of the rural population for fuelwood (Aubert 1985). Today, nearly 60% of the wood removed from the world's forests is used to generate energy, overwhelmingly for cooking and keeping warm in the countries of the developing world. On the average, 0.45 m^3 of wood (2.9 GJ) is consumed per person, and up to 80% of their total energy requirements are supplied by wood (Williams 1990).

7 Concluding Remarks

Most of the "solar" agrosystems were far from optimal returns which could be expected from a careful management of the given resources. Through the application of soft technologies, the organization of feedback loops and energy cascade outputs could triple without any manufactured inputs. This inherent potential for growth has not been appropriately evaluated in most analyses, because the relevant hard data are not easily obtained. Moreover, yield figures, which are widely used as a yardstick for productivity, are often interpreted without considering the areas to which they refer and computing net returns (seeds deduced). Most fundamentally, the energetic content rather than market prices should be applied as a common denominator to measure and compare different kinds of agricultural produce in an ecological analysis, because the use-values of different agricultural products are not always related to their energy content (Martinez-Alier 1987).

Among those studies which attempt at measuring carrying capacity, the focus has been rather exclusively upon nutrition (e.g., Grigg 1980; Viazzo, this Vol.). Carrying capacity refers to long-term overall stability of human ecosystems and must therefore include the production and consumption of forestry resources. In most studies, it has been overlooked that per capita energy demand for cooking and heating rises much faster than that for nutrition as a result of population growth. It may be argued that the demand of private households for heat energy was one of the major constraints which have promoted the transition from the solar to the fossil energy system. It was met in the context of the railway network, which was itself operated by fossil fuel and allowed the import of coal and coke. For Sieferle (this Vol.) the development of a fossil energy system was a matter of *economic necessity*, unless a country would run the risk of economic marginalization. This is only half of the story. From the example of the canton of Bern, which may be typical in this respect, it must be inferred that the substitution of fuelwood by coal and coke was also an *ecological necessity*. If we disregard the use of nonrenewable energies, there were no alternatives left in the late 19th century for covering the energy needs of a population which had continuously grown for over 100 years, except by further plundering the forests. The price which had to be paid for economic growth was ecological instability. Of course, timber and fuelwood might have been imported from Scandinavia and Russia or from overseas, but it may be hypothesized that large segments of the population would not have been able to bear the price of transporting this bulky commodity over long distances. The rate at which the Bernese forests were exploited at the end of the 19th century suggests that, without the coming of fossil fuel, the Alps and the hilly regions of Central and Western Europe might have gone the way of the Mediterranean mountains, which became denuded and sterile as a consequence of deforestation.

Acknowledgement. This research is supported by the Swiss National Science Foundation.

References

Abel W (1978) Geschichte der deutschen Landwirtschaft vom frühen Mittelalter bis zum 19. Jahrhundert. Ulmer, Stuttgart

Aubert C (1985) La Chine est-elle surpeuplée? In: Gourou P, Etienne G (eds) Des labours de Cluny à la révolution verte. Techniques agricoles et population. Univ Press, France, Paris, pp 99–136

Bairoch P (1976) Die Landwirtschaft und die industrielle Revolution 1700–1914. In: Cipolla C (ed) Europäische Wirtschaftsgeschichte, vol 4. Die industrielle Revolution. Fischer, Stuttgart, pp 297–332

Blaikie P, Brookfield H (1987) Land degradation and society. Methuen, London

Blanckenburg P (1986) Welternährung. Gegenwartsprobleme und Strategien für die Zukunft. Beck, München

Bolin B, Arrhenius E (1977) Nitrogen. An essential life factor and a growing environmental hazard. Rep Nobel Symp 38. Ambio 6/2–3:96–105

Brugger H (1979) Die schweizerische Landwirtschaft 1850 bis 1914. Huber, Frauenfeld

Brugger H (1985) Die schweizerische Landwirtschaft 1914 bis 1980. Agrarverfassung, Pflanzenbau, Tierhaltung, Außenhandel. Huber, Frauenfeld

Cohen MN (1977) The food crisis in prehistory: overpopulation and the origins of agriculture. Yale Univ Press, New Haven

Cottrell F (1970) Energy and society. The relation between energy, social change and economic development. Greenwood, New York

Debeir J, Deléage J, Hémery D (1986) Les servitudes de la puissance. Une histoire de l'énergie. Flammarion, Paris

Der Hohe Schweizerische Bundesrath (ed) (1862) Bericht an den hohen schweizerischen Bundesrath über die Untersuchungen der schweizerischen Hochgebirgswaldungen, vorgenommen in den Jahren 1858, 1859 und 1860. Bern

Désert G, Specklin R (1976) Victoire sur la disette. In: Juillard E (ed) Apogée et crise de la civilisation paysanne. Histoire de la France Rurale, vol 3. Seuil, Paris

Ellen R (1982) Environment, subsistence and system. The ecology of small-scale social formations. Univ Press, Cambridge

Finck A (1979) Dünger und Düngung. Grundlagen, Anleitung zum Düngen von Kulturpflanzen. Chemie, Weinheim

Flohn H, Fantechi R (eds) (1984) The climate of Europe: past, present and future. Natural and man-induced climatic changes: an European perspective. Reidel, Dordrecht

Fluck RC, Baird RD (1980) Agricultural energetics. Avi, Westport
Frissel M (1977) Application of nitrogen fertilizers. Present trends and projections. Ambio 6:152–156
Georgescu-Roegen N (1971) the entropy law and the economic process. Harvard Univ Press, Cambridge
Georgescu-Roegen N (1971) The entropy law and the economic process. Harvard Univ Press, Cambridge
Grigg D (1980) Population growth and agrarian change. An historical perspective. Univ Press, Cambridge
Hauser A (1974) Güllewirtschaft und Stallmist – zwei große Erfindungen der Landwirtschaft. Schweiz Landwirtsch Forsch 13:15–26
Huggel S (1979) Die Einschlagsbewegung in der Basler Landschaft. Gründe und Folgen der wichtigsten agrarischen Neuerung im Ancien Regime, 2 vols. Katonaler Lehrmittelverlag, Liestal
Jones EL (1974) Agriculture and the industrial revolution. Blackwell, Oxford
Jones EL, Woolf SJ (1969) The historical role of agrarian change in economic development. In: Jones EL, Woolf SJ (eds) Agrarian change and economic development. The historical problems. Methuen, London
Kleiber M (1967) Der Energiehaushalt von Mensch und Haustier. Ein Lehrbuch der Tierenergetik. Parey, Hamburg
Köllmann W (1965) Bevölkerung und Raum in neuerer und neuester Zeit. Bevölkerungs-Ploetz. Raum und Bevölkerung in der Weltgeschichte 4. Ploetz, Würzburg
Krämer A (1888) Landwirtschaft. In: Furrer A (ed) Volkswirtschafts – Lexikon der Schweiz 3. Halbband Schmid, Francke, Bern, pp 228–320
Krämer A (1902) Die Landwirtschaft im neunzehnten Jahrhundert. Mit besonderer Berücksichtigung schweizerischer Verhältnisse. Huber, Frauenfeld
Laszlo E (1987) Evolution. The grand synthesis. New Science Library, Shambhala Boston London
Martinez-Alier J (1987) Ecological economics. Energy, environment and society. Blackwell, London
Merchant C (1987) The theoretical structure of ecological revolutions. Environ Rev 11:265–274
Moran EF (1979) Human adaptability. An introduction to ecological anthropology. Westview, Boulder
Moran EF (1984) Limitations and advances in ecosystems research. In: Moran EF (ed) The ecosystem concept in anthropology, Westview, Boulder, p 3
Müller NN, Grete NN (1885) Düngerwesen. In: Furrer A (ed) Volkswirtschafts-Lexikon der Schweiz. 2. Halbband. Schmid, Francke, Bern, pp 454–465
Naveh Z, Lieberman AS (1984) Landscape ecology. Theory and application. Springer, Berlin Heidelberg New York
Odum E (1953) Fundamentals of ecology. Saunders, Philadelphia
Odum EP (1971) Fundamentals of ecology. 3rd ed. Saunders, London
Osterud O (1978) Agrarian structure and peasant politics in Scandinavia. A comparative study of rural response to economic change. Univ Press, Oslo
Pfister C (1984) Das Klima der Schweiz von 1525 bis 1860 und seine Bedeutung in der Geschichte von Bevölkerung und Landwirtschaft. Haupt, Bern
Pfister C (1989) Metaquellen als Grundlagen zur Abgrenzung und Typisierung historischer Agrarzonen. Itinera 10, pp 28–57
Pfister C (1990) Food supply in the canton of Berne, 1847. In: Newman L (ed) Perspectives on the history of hunger. Blackwell, Oxford, pp 282–303
Pfister C, Schüle H (1989) BERNHIST – a laboratory for regional history. Building up a dynamic data-basis for the spatial analysis of population, economy and the environment in the Canton of Berne (Switzerland) 1700–1980. In: Denley P, Fogolvik S, Harvey O (eds) Proc 2nd Conf Computing and history. Univ Press, Manchester, pp 280–285
Postgate JR (1980) The nitrogen economy of marine and land environments. In: Blaxter K (ed) Food chains and human nutrition. Applied Science Publishers, London, pp 161–186
Radkau J, Schäfer I (1987) Holz. Ein Naturstoff in der Technikgeschichte. Rowohlt, Reinbeck bei Hamburg

Sack R (1990) Social theory and environmental research. In: Kates R, Turner W (eds) The world as transformed by human action. Univ Press, Cambridge (in press)

Sandgruber R (1982) Die Anfänge der Konsumgesellschaft. Konsumgüterverbrauch, Lebensstandard und Alltagskultur in Österreich im 18. und 19. Jahrhundert. Verlag für Geschichte und Politic, Wien

Schnaiberg A (1980) The environment: from surplus to scarcity. Oxford Univ Press, New York

Schüepp M (1962) Sonnenscheindauer 1st pt. In: Schweiz Meteorol Zentralanst, Zürich (ed) Klimatologie der Schweiz, 311

Sieferle RP (1982) Der unterirdische Wald. Energiekrise und industrielle Revolution. Beck, München

Slicher van Bath BH (1963) The agrarian history of Western Europe AD 500–1850. Translated by O Ordish. Arnold, London

Stampfli M, Frey W (1987) 'Lieber schmale Bissen essen, als bei strenger Arbeit wohl zu sein'. Die demographische, ökonomische und soziale Entwicklung in den bernischen Ämtern Büren und Konolfingen zwischen den 1760er und den 1880er Jahren. Lizenziatsarbeit. Hist Inst Univ Bern

Stebler FG, Schröter C (1887) Beiträge zur Kenntniss der Matten und Weiden der Schweiz. Landwirtsch Jahrb Schweiz 1:77–190

Svedin U (1985) Economic and ecological theory: differences and similarities. In: Hall DO, Myers N, Margaris NS (eds) Economics of ecosystem management. Junk, Dordrecht, pp 31–40

Thober H, Lieth H, Fabrewitz S (1985) Modelling of the socio-economical and ecological consequences of high animal waste application (MOSEC). In: Hall DO, Myers N, Margaris NS (eds) Economics of ecosystem management. Junk, Dordrecht, pp 71–84

Ulbricht O (1980) Englische Landwirtschaft in Kurhannover in der zweiten Hälfte des 18. Jahrhunderts. Ansätze zu historischer Diffusionsforschung. Dunker & Humblot, Berlin

Vester F (1976) Urban systems in crisis. Understanding and planning of human living space. The biocybernetic approach. Deutscher Verlag, Stuttgart

Waddington VH (1975) A catastrophe theory of evolution. The evolution of an evolutionist. Cornell Univ Press, Ithaca, New York

White L (1949) The science of culture. Farrer, Straus & Giroux, New York

Williams M (1990) Global deforestation. In: Kates R, Turner W (eds) The world as transformed by human action. Univ Press, Cambridge (in press)

Zucker H (1983) Steigerung der Erzeugung und naturwissenschaftliche Forschung in der Tierhaltung. In: Ehlers E (ed) Ernährung und Gesellschaft, Bevölkerungswachstum – agrare Tragfähigkeit der Erde. Wissenschaftliche. Verlagsgesellschaft, Stuttgart, pp 183–199

An Anthropological Perspective of Environment, Population, and Social Structure in the Alps

P.P. Viazzo[1]

It is widely acknowledged that some of the most fruitful applications of ecological approaches in the social sciences have been associated with anthropology (Ellen 1985, p. 219), and much of the credit for this goes to the work and inspiration of Julian Steward, who was the first to advocate the explicit use of ecological concepts in anthropology, back in the 1930s. Besides being highly influential in anthropology, Steward's method (generally known as "cultural ecology") has found many admirers in other disciplines, particularly in geography and archaeology.

It is worth noting that the central aim of Steward's approach was to determine the extent to which modes of subsistence and, ultimately, forms of social organization are shaped by certain characteristics of the physical environment. He was therefore particularly interested in finding what he called "regularities", or similarities between cultures that recur in historically separate areas, and may thus be explained as a result of similar environmental features (Steward 1955). In so doing, "cultural ecology" has directed attention to a number of functional relations between environmental conditions and social organization, which other social scientists, including historians, may find very relevant to their own work (Ehmer and Mitterauer 1986, p. 10–11). But it should be stressed that Steward's approach was deliberately ahistorical, or even anti-historical, and that in several cases this methodological stand has led Steward and his followers to rather debatable conclusions smacking of environmental determinism (Viazzo 1989, p. 31–34).

Another noteworthy feature of this approach is that cultural-ecological models are conspicuous because of the virtual absence of demographic variables. This may at first sight appear surprising, but is a logical consequence of the fact that Steward and his followers were essentially concerned with those aspects of social structure and social organization that are most closely related to subsistence activities. Such a focus on subsistence patterns and their social-organizational correlates discouraged a serious exploration of those sectors of social structure which are related not so much to production strategies, as to the preservation of a balance between population and resources. Things have changed dramatically since the 1960s because of the growing popularity which the concept of ecosystem has enjoyed among anthropologists, geographers and other social scientists. In

[1]Cambridge Group for the History of Population and Social Structure, 27 Trumpington Street, Cambridge CB2 1QA, Great Britain

Brimblecombe, Pfister (Eds.)
The Silent Countdown

anthropology, the adoption of an ecosystemic approach has entailed a major shift from a concern with the relations between environmental features and modes of subsistence to the analysis of the relations between population and resources. In particular, many anthropological proponents of this approach have borrowed from biological models of the ecosystem the notion that the successful local population is one which adjusts its numbers in such a way as to maintain local resource stability. The study of population regulation has therefore acquired a crucial importance, and in the past 20 years a large number of anthropological investigations have tried to determine whether social practices ranging from marriage and infanticide to ritual and warfare could be related to the end of maintaining population equilibrium (Bates and Lees 1979; Ellen 1982; Harris and Ross 1987).

Yet "it is paradoxical," as the American anthropologist Emilio Moran has recently remarked, "that ecological anthropological studies have only rarely explored the changing population variable over time given the importance of demographics in population ecology" (1984, p. 17). This paradoxical situation is largely explained by the fact that most ecological anthropologists, like their colleagues in other sectors of the discipline, have generally relied on ethnographic data collected during their fieldwork and covering a very limited time-span. Admittedly, the notion of ecosystem has also been very popular in archaeological anthropology. However, for archaeologists it has mainly proved a useful conceptual device which has encouraged them to think in terms of systemic interrelationships. As a concrete unit of analysis, however, the ecosystem has had little role in archaeological research, not least because knowledge of prehistoric populations can be at best circumstantial (Jochim 1984).

Recent debate among ecological anthropologists indicates that it is increasingly being recognized that ecosystemic models are bound to remain mere "explanatory sketches" unless population dynamics are studied in the long term. Therefore, the efforts of ecological anthropologists are now concentrated on the development of a "processual" approach capable of overcoming what Benjamin Orlove (1980, p. 245) has called "the split between the excessively short and long time scales". There is, in particular, a growing consensus that hypotheses concerning ecosystemic regulation should ideally be tested over periods of two or three centuries, and that in order to do so it is necessary to borrow the conceptual and methodological tools of historical demography (Adams and Kasakoff 1984; Moran 1984, p. 16–19).

Perhaps the best example of this new approach is represented by the work of the American anthropologist Robert Netting on Törbel, a village in the upper part of Canton Valais in the Swiss Alps (Netting 1981; cf. also Netting 1972, 1976, 1984). Although Netting had made extensive use of conventional ethnographic techniques, he has also been one of the first anthropologists to realize that if ecological models are not to remain "distressingly hypothetical", it is necessary to go to the archives and to reconstruct the demography of the population living in the ecosystem under study. The choice of an Alpine village has not been accidental. For one thing, mountain areas still provide a classic terrain to evaluate the impact of environmental features on human social organization. Secondly, communities in the Alps, and particularly in Switzerland, have often been clearly bounded for centuries, which favours the application of an ecosystemic framework (Netting 1984). But perhaps the most important advantage, as Netting (1972, p. 132) has remarked, is the existence of detailed historical source materials which are not available for other major mountain areas like the Himalayas or the Andes.

Although Netting's study is exemplary both theoretically and methodologically, it would obviously be unwarranted to accept and generalize all its conclusions. We may wonder, in particular, whether and to what extent Törbel can be taken as representative of the Alpine area. Indeed, a parallel study I have conducted on a community which happens to be quite close to Törbel both geographically and ethnically (the village of Alagna Valsesia, on the southern side of Monte Rosa) has revealed a number of striking demographic differences, particularly as far as migration and fertility patterns are concerned (Viazzo 1983, 1989; Viazzo and Albera 1986, 1987, in press). In this chapter, however, I would like not to talk, for once, of "my" community. I would rather like to discuss briefly in a broader comparative perspective some general issues concerning the ways in which Alpine populations managed to keep a viable balance with local resources and to assess the validity of the model of ecological change and adjustment proposed by Netting.

If we assume that the population of an Alpine community is growing and its resources cannot be expanded any further, the only manner in which this community can respond to the pressures of overpopulation consists in limiting its numbers. This can be achieved in more than one way, but students of the Alps have mostly paid attention only to permanent emigration, which is invariably described as the vital "safety valve" allowing Alpine populations to escape the rigours of Malthusian positive checks. Impressed with Braudel's incisive characterization of mountains as "factories" producing migrant labour for other people's usage – "fabriques d'hommes à l'usage d'autrui" (Braudel 1966, p. 46) – many historians and geographers have indeed taken for granted that massive permanent emigration must have been an endemic feature of Alpine demography, thereby also accepting the notion that in the Alps the number of births tended to exceed the number of deaths by a large margin (e.g., Poitrineau 1983).

However, some leading authorities in the field have expressed diverging views. In spite of some similarities, the demographic model underlying Braudel's discussion of overpopulation and emigration in mountain areas is significantly different, for instance, from the model proposed in 1975 by Paul Guichonnet. Like Braudel, Guichonnet believes that fertility was high; but while Braudel's model implies the existence of a high rate of natural increase, Guichonnet maintains that fertility was nearly offset by very high levels of mortality. Natural increase was therefore restricted, and the size of permanent emigration relatively modest. Indeed, according to Guichonnet one of the main features of Alpine demography in what he calls the "age of autarky" is the lack of massive permanent emigration. Until 1850, he states, "les abandons définitifs sont l'exception" (Guichonnet 1975, p. 157–61).

In this respect, Guichonnet's position is similar to that of Elizabeth Lichtenberger, one of Austria's most influential Alpine geographers. Reacting against what she regards as extreme and preconceived views about the Alpine world, she has argued that the Alps can hardly have constituted an important reservoir of people for urbanization or industrialization because both overall fertility and the rate of natural increase have always been moderate, the main reason being that "the average age at marriage has always been high, and the percentage of married people has remained low" (Lichtenberger 1975, p. 7). Lichtenberger suggests, in other words, that the Alps displayed not so much a "high pressure" demographic regime characterized by high birth and death rates as a "low pressure" regime hanging on nuptiality. This may well reflect her greater familiarity with the rather distinctive social-structural and demographic features of the Austrian Alps, where a custom of impartible inheritance severely restricting marriage is known to have prevailed. But it is remarkable that the system she outlines fits very closely with the models proposed by a number of anthropologists who have worked on various mountain areas in the world and, nearly two centuries earlier, by Malthus himself.

As is well known, in the second edition of his *Essay on the principle of population*, Malthus discussed at length the case of Switzerland, whose demographic evolution, he reckoned, provided the clearest illustration of the general principles of his work. In particular, he used data from the Alpine parish of Leysin (Canton Vaud) to demonstrate that since "there is no land so little capable of providing for an increasing population as mountainous pastures", it is in upland regions that the preventive checks of high celibacy rates and late marriage should be expected to prevail to an unusual extent (Malthus 1803). Although phrased in a rather heavier jargon, the models advanced by ecological anthropologists are strikingly similar. Like Malthus's model, they tend to conceptualize upland communities as (closed) ecosystems with fixed resources. Also, they predict that mountain populations must rapidly reach a stationary state resting on a low-pressure demographic regime and expect marriage to function as a crucial regulatory mechanism (Friedl and Ellis 1976; Goldstein 1981; Viazzo 1989, p. 26–48).

To test these alternative views is clearly one of the most urgent empirical and theoretical tasks of any ecologically-oriented historical study of Alpine populations. It is obvious, in particular, that unless fertility and mortality levels are known, and the rate of natural increase is determined, the importance of the role played by permanent emigration can scarcely be assessed. But it is relevant to stress another point which is still seldom made when the relationships between population and the environment are discussed, namely, that although the same rate of natural increase may result from both high-pressure and low-pressure regimes, the latter allow for a more favourable ecological balance. For one thing, replacement is more efficient, because the number of people who live for a short time consuming resources and then die without realizing their full potential is smaller. Secondly, as Malthus had noticed for Leysin, the proportion of people under 16 years of age is a function of fertility levels. As a consequence, late-marrying populations with low fertility will have a more advantageous "dependency ratio" of children to the working population (Schofield 1976; Wrigley 1987).

Only a few years ago, broad comparative investigations of these topics would have been impossible. In 1975, Guichonnet had to acknowledge that the historical demography of the Alps was still in its infancy, a statement echoed by Jean-François Bergier in 1980 (Guichonnet 1975, p. 143; Bergier 1980, p. 166). As to anthropology, some interesting studies had already been completed, but they hardly contained any quantitative evidence. In the last few years, however, the situation has changed considerably. The results of a spate of recent studies show that for at least two centuries, from the mid-18th century to the Second World War, the levels of fertility and mortality were remarkably similar all over the Alpine areas, and significantly lower than in the surrounding flat regions. Crude birth rates rarely exceeded the level of 30 per 1000, while death rates typically ranged between 22 and 28 per 1000. The contrast with the high-pressure regimes of many adjacent lowland areas, where crude birth and death rates could easily reach 35 or even 40 per 1000, is striking. And there is little doubt that apart from the very interesting exception represented by a number of regions in the Western Alps (where low marital fertility appears to have coexisted with relatively high nuptiality well before the inception of birth control), such a balance of low levels of both fertility and mortality was achieved mainly through the check of nuptiality (Viazzo 1989, p. 83–93, 178–223).

This is enough to conclude that by and large Alpine demographic systems were not as heavily dependent on permanent emigration as implied by what we might call the Braudel model. Nor were these systems as "wasteful" as implied by the Guichonnet model. One of Malthus's main contentions appears therefore to be vindicated. However, Malthus's model, and here is a significant difference from the rather more static picture suggested by Lichtenberger as well as by some anthropologists, more generally predicted that the demography of Alpine populations would tend to work like a finely balanced homeostatic system, in which fertility was adjusted to changing mortality and fixed resources through variations in nuptiality. This hypothesis is as relevant and stimulating today as it was in the early 19th century, and a particularly useful testing ground is provided by a comparison between the demographic evolution of the Swiss Alps (which Malthus used as a crucial case-study to demonstrate the validity of his principles) and that of the neighbouring Austrian Alps.

A first important outcome of this comparison is the finding that in spite of different systems of inheritance, the Swiss and the Austrian Alps displayed very similar patterns of nuptiality. In both areas marriage was very late for both men and women, and rates of permanent celibacy exceedingly high (Viazzo and Albera in press). This would seem to support Malthus's contention that it is in mountain areas that the prudential check of marriage should be expected to prevail to a greater degree, and also the related view (much favoured by ecological anthropologists) that in marginal areas like the Alps marriage patterns are shaped more decisively by environmental constraints than by the formal properties of inheritance and succession rules. The decline in *both* nuptiality *and* mortality recorded in Switzerland and Austria alike in the course of the 18th century also would seem to corroborate Malthus's other contention, that since in mountain environments, resources tend to be fixed and cannot sustain population growth, then improvements in mortality necessarily entail a decline of nuptiality.

What Malthus had not forecast was the substantial expansion of resources made possible primarily by the introduction of the potato, which in most Alpine districts took place towards the end of the 18th century. In these critically changed circumstances, one would expect a "relaxation" of nuptiality, allowing population to increase so as to adjust to resources. In the Austrian Alps, however, the introduction of the potato appears to have had very modest effects. The population of its provinces either increased very little or stagnated. The population of the Swiss Alps, on the other hand, grew at a fast pace. But nuptiality, instead of increasing, continued to decline (Viazzo and Albera 1986).

Although detailed comparative studies of these puzzling questions are still lacking, there is reason to believe that the key to solve them consists on the one hand in considering marriage patterns more dynamically than has generally been the case in Alpine studies, and on the other in examining nuptiality in the context of the sharply different systems of social stratification found in Swiss and Austrian communities.

Apart from marriage age, the demographic indicators which have usually been employed in the study of Alpine nuptiality, be they celibacy rates, the Princeton index of proportion married, or simply crude marriage rates all provide *relative* measures. This may easily lead to paradoxical conclusions. Thus, the marked increase in the proportion of celibates recorded in Switzerland between 1750 and the late 19th century (Ruesch 1979) would seem to suggest that prospects for marriage were worsening. In the village of Törbel studied by Netting, the proportion of individuals reaching 20 years of age who never married increased from 20% in the second half of the 18th century to 33% one century later (Netting 1981, p. 132–135). However, the emphasis which demographers place on individual chances should not lead one to lose sight of the fact that proportions of people never marrying may refer to cohorts of very different size. As it happens, between the middle of the 18th century and the end of the 19th the population of Törbel more than doubled. The number of people born in Törbel in the period 1750–1799 was in fact just over 50% of those born in the years 1850–1899, and the number of those who eventually got married just over 60%. Thus, if we consider the *absolute* number of people getting married, and this is what matters most from an ecological rather than from a strictly demographic point of view, we find that in this period it increased considerably, and so did the number of estates, for the rule was that a new estate was formed with each marriage. It is important to note that since the total amount of land remained essentially the same, the average amount of land worked by each individual (or household) steadily declined. Yet it has been suggested that the balance between population and resources did not necessarily deteriorate, owing to the beneficial effects of the introduction of the potato (Wiegandt 1977; Netting 1981).

This is in marked contrast with what was apparently happening in the Austrian Alps, where population remained almost stationary (Klein 1971) and fragmentation was (apart from rather exceptional cases) very limited. It has sometimes been argued that in the second half of the 18th century the population of the Austrian uplands had virtually reached its ceiling. This is, however, hard to accept. As Michael Mitterauer has remarked, the "agrarian revolution" and in particular the introduction of the potato made it possible to feed a much greater number of people from the same acreage of land (Mitterauer 1982, p. 243–244). This was often the case in the lowlands of Austria. But the new potential was exploited only to a limited extent in the mountains. It is worth stressing that in Switzerland the cultivation of the potato was started by people belonging to the poorer stratum, who had been granted usufruct of tiny plots carved out of the communal land. In Alpine Austria, by contrast, the formation of cottages was strongly discouraged both by external power-holders and by the local farming élite, and the number of what Mitterauer (1986, p. 254) calls "potato-people" consequently remained quite small. While in Swiss villages, like Törbel, social-structural factors were opposing minimum resistance to the adjustment of population and nuptiality to the increasing carrying capacity of local ecosystems, in Austria marriage was sternly restricted in order to permit the perpetuation of the existing social structure, thereby preserving a situation of "static equilibrium" in which population size was blocked well below carrying capacity.

The social-structural factors examined in the previous paragraphs help make sense of the broad contrast between the demographic history of the Swiss and Austrian Alps in the 18th and 19th centuries. But if we look at the map of population growth from a shorter distance, it becomes apparent that significant differences existed within both the Swiss and the Austrian Alps (as, indeed, within all the other major sectors of the Alpine crescent). The problem is complex, and I have dealt with it at some length elsewhere (Viazzo 1989). But I would like at least to refer to the enlightening results of Werner Thut and Christian Pfister's comparative study of two districts in the Bernese Oberland: the eastern district of the Oberhasli, whose population increased between 1764 and 1846 at the fast annual rate of 1.1%, and the western district of Saanen, which only displayed a growth rate of 0.2%. Perhaps the most startling feature of the demography of the Oberhasli in this period was that the substantial surplus of births generated by a sharp decline in mortality and possibly by a certain increase in marital fertility was mostly absorbed, as in Törbel, by the local ecosystem: the annual rate of net migration amounted to only 0.15%, without any apparent sign of serious deterioration in the diet and living standards of those who remained. By contrast, Saanen displayed lower fertility, a rather heavier mortality and, most interestingly, a net migration rate which was twice as high as in the Oberhasli (Pfister 1986, p. 374–382).

These demographic differences are regarded by Pfister to be due to marked differences in the economy. The geographical features of the two districts were, he notes, fundamentally similar. But the Oberhasli had retained what Pfister calls a "dynamic balance" between the agricultural and the pastoral sectors, whereas Saanen had turned into a classic instance of commercial pastoral area – a *Hirtenland*, as Swiss economic historians say. As in many other Swiss upland districts, in the period of the boom in the cheese market in the 18th century hay-fields had been converted to pastures, and the capacity to support cattle over the winter had consequently been reduced in order to increase capacity in the summer. This was clearly a major economic and ecological transformation. But it is essential to note that such a transformation was closely interwoven with significant changes in the social structure of pastoral communities, for the new circumstances favoured those cattle-farmers who could engage in speculative land purchases, which in turn deprived the lower strata of their basis for subsistence. Thus, in the course of the 18th century pastoral communities like those of the Saanen district became far more inegalitarian than they had previously been. This system of stratification was still very different from the one which was typical of Austrian upland communities. Nevertheless, the new social and economic order of the *Hirtenland* apparently played a similar role in preventing population from adjusting to the new potential brought about by the introduction of the potato.

But there is still another intriguing question, namely why some Swiss communities became pastoral while others retained a more balanced and autarkic economy. Echoing the arguments already advanced in the 1940s by Richard Weiss (1946, p. 105), most scholars simply tend to attribute the spatial distribution of these two forms of upland economy to environmental differences and to single out the amount of precipitation as the critical determinant (Budmiger 1982; Mathieu 1985). Although it would be ill-advised to underestimate the importance of the "comparative advantages" offered by the various climatic zones, this is unlikely to be the whole story. Indeed, Rudolf Braun (1984) has recently argued that the degree of penetration of the commercial pastoral economy largely depended on a set of social-structural preconditions – inheritance patterns, once again, but above all the presence or absence of strong corporate community structures, whose importance he had already emphasized in his seminal works on the uneven development of cottage industry in the highlands of Canton Zurich (Braun 1960, 1978).

The main aim of this paper has been to show that the Alpine area provides a useful setting to test a number of hypotheses concerning the ability of nuptiality to adjust population to resources. This is clearly a question of major significance not only for anthropologists and historical demographers but also, as the continuing controversy over the causes of the Irish famine demonstrates (Mokyr 1983), for environmental historians. One of the merits of Robert Netting's work on Törbel (which nicely reconciles Malthusian and Boserupian perspectives) consists in having produced a carefully analyzed case-study which departs, in spite of significant similarities, from the classic Irish model and invites us to reconsider the question of the "relaxation" of nuptiality in a context of increased resources. The fascinating pattern of ecological change and adjustment suggested by Netting's work as well as by another anthropological study, also focussed on a village in the Swiss Alps (Wiegandt 1977), is that in spite of population growth, the agricultural and pastoral output per capita tended to remain virtually constant. This, however, was achieved through an increase in the age at marriage and in the proportion of celibates, which slowed down the fragmentation of village land and the rate of population growth itself. If future research should confirm that in some parts of Switzerland the output per capita remained almost constant, that would be an indication that local ecosystems could be, to use Netting's phrase, "as finely calibrated and dependable as an old Swiss watch" (Netting 1981, p. 89), and that nuptiality could work as a surprisingly precise regulatory mechanism.

The question is still open. And so are, indeed, even more basic questions concerning the causes of the introduction of the potato and the magnitude of its effects. Nevertheless, the analysis of the agricultural census of Canton Berne in 1847 has recently enabled Christian Pfister to estimate that towards the middle of the 19th century food production was sufficient to satisfy the nutritive requirements of the inhabitants of the upper zones, and that the potato had actually brought about a critical quantitative improvement in local peasant diets without undermining the nutritional balance. Also, the fact that in the late 1840s and in the 1850s no acute mortality crisis is recorded in the Bernese Oberland would seem to suggest that in this region local ecosystems were rather less vulnerable to crop failures than in Ireland (Pfister 1986).

But it should not be forgotten that in the same years other parts of upland Switzerland were struck far more severely by the potato blight, and Pfister himself has spoken of the progressive "Irelandization" of sizable sectors of the Swiss Alpine population, which apparently tended to exceed the carrying capacity of their territories (Head-König et al. 1987, p. 258–259). What is more, the evidence discussed in this paper suggests that social-structural constraints could be very effective in checking the tendency for a population to expand to the limit of its potential food supply and could actually keep population *below* carrying capacity. In this connection, it is relevant to note that in anthropology the adoption of an ecosystemic approach has resulted in a more sophisticated description of ecological interactions, involving computations of carrying capacity as well as estimates of energy intake, output and efficiency for different groups and activities. But it has also encouraged a view of human behaviour in many respects functionally equivalent to that of other animals (Ellen 1985). This has led many an anthropologist to overrate the role of environmental constraints and opportunities and to treat both population and, even more, social structure as mere dependent variables. The sharp contrast between the demographic and ecological evolution of a perhaps unusually egalitarian community like Törbel and that of the Swiss *Hirtenland* or of Alpine Austria indicates that the significance of social-structural forces in affecting the relations between environment and population should not be underestimated.

As a final note, I would like to suggest that it would probably be very rewarding to consider more carefully than has been the case so far the role played by "local" social-structural factors in the more recent ecological history of the Alpine area. To be sure, a considerable amount of attention has been paid to the role of different inheritance and succession systems in accelerating or "braking" agricultural exodus and mountain depopulation. However, surprisingly little has been done to assess the ability of local social-structural factors to affect other processes of major ecological significance such as, for instance, the origins and growth of tourism.

The main ecological implications of tourist development in the Alpine region are of course quite well known. Particularly the French experience of the so-called integrated resorts has brought out the risks of a strategy of tourist development guided by an exploitative logic reminiscent of the mining industry, and both the public and governments are now aware that in addition to an "environmental" carrying capacity there also exists a more subtle "perceptual" carrying capacity which has to be taken into account. This increasing awareness has stimulated a considerable amount of research projects, the best-known being perhaps the interdisciplinary study of the Tyrolese resort of Obergurgl, a locality which has been selected as an exemplary research site for the UNESCO project "Man and Biosphere". But as one of the members of the Obergurgl research team, Ursula Walther, has recently acknowledged, most of these studies have suffered from several shortcomings. Although it is understandable that these projects are primarily focussed on recent environmental and economic transformations, the lack of historical background may easily lead to a superficial and possibly misleading interpretation of change based on a poor understanding of an essentially unexplored past. Secondly, because of the emphasis placed on environmental and economic transformations, very little attention tends to be paid to the changes in local social structure brought about by growth of tourism, which have also been extensive and well worth investigating (Walther 1984).

In this connection, I would probably go a step further. If only few studies have examined the impact of tourism on social structure, even fewer have given analytical primacy to social-structural variables in order to gauge their impact on tourist development and, indirectly, on environment and population. Yet, the existence of a marked contrast between tourist development in the Swiss and Austrian Alps on the one hand and in the French and Italian Alps on the other has been a commonplace for geographers and economists for many years (Béteille 1968; Herbin 1969; Veyret-Verner 1971, p. 15; Veyret 1972, p. 120; Rota 1975, p. 472–480). Once popular explanations in terms of differential "mental attitudes" or at most in terms of differential "demographic vitality" can no longer be regarded as satisfactory. The much greater "demographic vitality" of the Austrian and, to a lesser extent, of the Swiss Alps is of course a factor of the utmost importance, but it can scarcely be regarded as a given requiring no explanation. There is in fact little doubt that to a large extent it depends on the differential ability to keep peasant families in the mountains displayed by the inheritance patterns found in the various sectors of the Alps. Moreover, the results of a number of anthropological studies of tourist development in several parts of the world strongly suggest that the amount of control which can be exercised by local bodies is a crucial variable to be taken into account if the dynamics of tourist development are to be properly understood (Smith 1981). It seems no accident that it is in the Swiss and Austrian Alps that institutions of long standing (reinforced by an unbroken line of protective legislation in the specific field of tourist development) define unambiguously the boundaries of the "local community" and its extensive prerogatives. This is not to say that the "Austrian" mode of tourist development is necessarily better than the "French" one (Herbin 1980; Dorfmann 1983). But the contrast suggests that strong institutional boundaries at the level of the local community can affect – today no less than in the 18th century – the degree of penetration of economic forces from outside, thereby playing a sometimes decisive role in both demographic and ecological history.

References

Adams JW, Kasakoff AB (1984) Ecosystems over time: the study of migration in long run perspective. In: Moran EF (ed) The ecosystem concept in anthropology. Westview, Boulder, pp 205–223

Bates DG, Lees SH (1979) The myth of population regulation. In: Chagnon NA, Irons W (eds) Evolutionary biology and human social behavior: an anthropological perspective. Duxbury, North Scituate, Mass, pp 273–289

Bergier J-F (1980) Le cycle médiéval: des sociétés féodales aux états territoriaux. In: Guichonnet P (ed) Histoire et civilisations des Alpes. Privat Payot, Toulouse Lausanne, pp 163–264

Béteille R (1968) Tourisme et milieu rural montagnard: l'example du Pitztal (Tirol autrichien). Rev Geogr Alp 56:367–76

Braudel F (1966) La Méditerranée et le monde méditerranéen à l'époque de Philippe II, 2nd edn. Colin, Paris

Braun R (1960) Industrialisierung und Volksleben. Rentsch, Erlenbach Zürich

Braun R (1978) Protoindustrialization and demographic changes in the Canton of Zurich. In: Tilly C (ed) Historical studies of changing fertility. Univ Press, Princeton, pp 289–334

Braun R (1984) Das ausgehende Ancien Régime in der Schweiz. Vanderhoeck & Ruprecht, Göttingen Zürich

Budmiger G (1982) Das Land der Walser. In: Budmiger G (ed) Die Walser. Huber, Frauenfeld Stuttgart, pp 24–38

Dorfmann M (1983) Régions de montagne: de la dépendance à l'auto-développement? Rev Geogr Alp 71:5–34

Ehmer J, Mitterauer M (1986) Zur Einführung. In: Ehmer J, Mitterauer M (eds) Familienstruktur und Arbeitsorganisation in ländlichen Gesellschaften. Böhlau, Wien Köln Graz, pp 7–30

Ellen R (1982) Environment, subsistence and system: the ecology of small-scale social formations. Univ Press, Cambridge

Ellen R (1985) Ecology. In: Kuper A, Kuper J (eds) The social science encyclopedia. Routledge & Kegan Paul, London, pp 218–220

Friedl J, Ellis WS (1976) Celibacy, late marriage and potential mates in a Swiss isolate. Human Biol 48:23–35

Goldstein MC (1981) The transformation of the social matrix of Tibetan populations in the High Himalayas. In: Baker P, Jest C (eds) Environmental and human population problems at high altitude. CNRS, Paris, pp 101–105

Guichonnet P (1975) Le développement démographique et économique des régions alpines. In: Guichonnet P (ed) Le Alpi e l'Europa, vol 2. Laterza, Bari, pp 138–196

Harris M, Ross EB (1987) Death, sex and fertility: population regulation in preindustrial and developing societies. Columbia Univ Press, New York

Head-König A-L, Hubler L, Pfister C (1987) Évolution agraire et démographique en Suisse (XVIIe-XIXe siécles). In: Fauve-Chamoux A (ed) Évolution agraire et croissance démographique. Ordina, Liège, pp 233–261

Herbin J (1969) L'insertion du tourisme dans la haute montagne: l'exemple de Tux dans les Alpes du Zillertal (Tirol autrichien). Rev Geogr Alp 57:665–705

Herbin J (1980) Le tourisme au Tyrol: une valeur d'exemple pour les Alpes françaises? Rev Geogr Alp 68:83–96

Jochim M (1984) The ecosystem concept in archaeology. In: Moran EF (ed) The ecosystem concept in anthropology. Westview Boulder, pp 87–102

Klein K (1971) Österreichs Bevölkerung 1754–1869. Mitt Österr Geogr Ges 113:34–62

Lichtenberger E (1975) The Eastern Alps. Univ Press, Oxford

Malthus TR (1803) An essay on the principle of population, 2nd edn. Murray, London

Mathieu J (1985) Ein Land von Hirten und Sennen? Inneralpine Wirtschaftsformen im 17. und 18. Jahrhundert am Beispiel des Unterengadins. In: Soc Suisse Hist Econ Soc (ed) Das Gebirge: Wirtschaft und Gesellschaft. Société Suisse d'Histoire Économique et Sociale, Lausanne, pp 1–15
McGuire R, Netting RM (1982) Leveling peasants? The maintenance of equality in a Swiss Alpine community. Am Ethnol 9:269–290
Mitterauer M (1982) Auswirkungen der Agrarrevolution auf die bäuerliche Familienstruktur in Österreich. In: Mitterauer M, Sieder R (eds) Historische Familienforschung. Suhrkamp, Frankfurt, pp 241–270
Mitterauer M (1986) Formen ländlicher Familienwirtschaft. Historische Ökotypen und familiale Arbeitsorganisation im österreichischen Raum. In: Ehmer J, Mitterauer M (eds) Familienstruktur und Arbeitsorganisation in ländlichen Gesellschaften. Böhlau, Wien Köln Graz, pp 185–323
Mokyr J (1983) Why Ireland starved: a quantitative and analytical history of the Irish economy, 1800–1850. Allen & Unwin, London
Moran EF (1984) Limitations and advances in ecosystems research. In: Moran EF (ed) The ecosystem concept in anthropology. Westview, Boulder, pp 3–32
Netting RM (1972) Of men and meadows: strategies of Alpine land use. Anthropol Q 45:132–144
Netting RM (1976) What Alpine peasants have in common: observations on communal tenure in a Swiss village. Human Ecol 4:135–146
Netting RM (1981) Balancing on an alp: ecological change and continuity in a Swiss mountain community. Univ Press, Cambridge
Netting RM (1984) Reflections on an alpine village as ecosystem. In: Moran EF (ed) The ecosystem concept in anthropology. Westview, Boulder, pp 225–235
Orlove BS (1980) Ecological anthropology. Annu Rev Anthropol 9:235–273
Pfister C (1986) Bevölkerung, Wirtschaft und Ernährung in den Berg- und Talgebieten des Kantons Bern 1760–1860. Itinera 5/6:361–391
Poitrineau A (1983) Remues d'hommes: essai sur les migrations montagnardes aux XVIIe et XVIIIe siècles. Aubier Montaigne, Paris
Rota G (1975) Il turismo alpino. In: Guichonnet P (ed) Le Alpi e l'Europa, vol 2. Laterza, Bari, pp 457–499
Ruesch H (1979) Die Demographie der Alpen zwischen 1650 and 1850. Schweiz Z Gesch 29:159–180
Schofield R (1976) The relationship between demographic structure and environment in pre-industrial Western Europe. In: Conze W (ed) Sozialgeschichte der Familie in der Neuzeit Europas. Klett, Stuttgart, pp 147–160
Smith VL (1981) Controlled vs. uncontrolled tourism. R Anthropol Inst Newsl 46:4–6
Steward J (1955) Theory of culture change: the methodology of multilinear evolution. Univ Ill Press, Urbana
Veyret P (1972) Les Alpes. Univ Press France, Paris
Veyret-Verner G (1971) Aménager les Alpes: mythes et réalités. Rev Geogr Alp 59:5–62
Viazzo PP (1983) Ethnic change in a Walser community in the Italian Alps. Thesis, Univ College, London
Viazzo PP (1989) Upland communities: environment, population and social structure in the Alps since the sixteenth century. Univ Press, Cambridge
Viazzo PP, Albera D (1986) Population, resources and homeostatic regulation in the Alps: the role of nuptiality. Itinera 5/6:182–231
Viazzo PP, Albera D (1987) Nupcialidad, fecundidad y estructura familiar: el caso de los Alpes Occidentales. Bol Asoc Demogr Hist 5, 3:3–40
Viazzo PP, Albera D (in press) Alpine marriage patterns in perspective. In: Smith RM (ed) Regional and spatial demographic patterns in the past. Blackwell, Oxford
Walther U (1984) Die Entwicklung Obergurgls vom Bergbauerndorf zum Tourismuszentrum. In: Ehalt HC (ed) Geschichte von unten. Böhlau, Wien Köln Graz, pp 183–200
Weiss R (1946) Volkskunde der Schweiz. Rentsch, Erlenbach Zürich
Wiegandt E (1977) Inheritance and demography in the Swiss Alps. Ethnohistory 24:133–148
Wrigley EA (1987) No death without birth: the implications of English mortality in the early modern period. In: Porter R, Wear A (eds) Problems and methods in the history of medicine. Croom Helm, London New York Sidney, pp 133–150

Environmental History: Some Questions for a New Subdiscipline of History

K. BLASCHKE[1]

1 Necessity of Environmental History

D. Meadows' *Limits to Growth* (1972) sounded an alarm. Preservation of environment is today a major task for mankind. Mastering environmental problems is an urgent one in the present age. There is an old adage: present concerns throw back a shadow into the past; so they are considered in the mirror of history and initiate inquiry into the roots of topical problems. The consciousness of history becomes enlarged by new material. The consequence is: environmental problems initiate the development of environmental history.

The present stage of development of society is characterized by a general feedback. Mankind has recognized that the hitherto existing development cannot be continued in the present way. Detrimental consequences of industrial technology have been detected, therefore, it is necessary to avoid, limit or eliminate the problems. Exuberant development is to be discouraged. On the time table of technical development it has become necessary to plan a "change of course" (Karl Steinbuch). Today we see a situation similar to the criticism of culture at the end of the 19th century. But here there is no cultural pessimism, no flight into romanticism. Likewise with the criticism of industrial society, it seems reasonable to argue for the end of exponential growth to prevent the imminent collapse of the world system.

Historians are able to identify parallel situations in the history of mankind: even in phylogenesis, it happened that certain evolutionary phases were interrupted, many projects such as the saurians did not continue. In the modern world this observation may be applied to technical-industrial development. This process of identifying parallels is a product of reviewing history, and has resulted in a new subdiscipline.

The name for this new subdiscipline, "environmental history" or "history of environment" seems to be fitting. The term "ecological history" would be too limiting. Here, we are not only concerned with nature, but with the whole system, consisting of nature, society, and culture.

[1]Am Park, 8101 Friedewald, GDR

Brimblecombe, Pfister (Eds.)
The Silent Countdown

2 The Competence of History

The present problem of environment is the result of historical development. The world system has fallen ill, any therapy demands a diagnosis. A profound analysis is only possible with knowledge of historical presuppositions. The man-environment relationship also has a temporal dimension. History is a science in the dimension of time, hence, it allows participation in the solution of environmental problems. There is a twofold interest of history in regard to environmental questions:

1. The exploration of environmental behaviour in history; and
2. The discovery of the roots of the present environmental problem in history.

The second question is the more important and more urgent one. From here there follows the question of a starting point; a rigid starting point does not seem appropriate, but environmental history must begin when the natural landscape changed to a cultural one, i.e., when human actions began to transform the environment. Palaeontology and all other palaeo-sciences are not included in the field of interest of environmental history.

3 Tasks

Since environmental problems have become prominent, historians have researched environmental questions in history, particularly case studies and a great number of isolated observations. Further works in this field are welcome, however, environmental history must not be purely descriptive; such a method would make the problem a normal phenomenon and, perhaps by implication, a harmless one. We must begin to dig deeper, we need a broad conception of environmental history as a guide to master the problems. It must be clear: the environmental problem today has a totally new quality, because population growth has placed new strains on the ecological system. Additionally, the possibilities of today's technology place much more pressure on the natural environment. The environmental problem of today has no historical comparison.

The new environmental history must hold in readiness a comprehensive supply of information about the environmental problems in history; it must become a useful and efficient auxiliary science for all types of environmental research.

4 Forming Periods

For a history of the environment, it is necessary to assign special periods regarding the relations between man and environment. The following three periods have been distinguished: (1) Over a long period of time man inhabited a natural landscape, and was dominated by it. It was one that he was not able to transform. (2) Only very recently did the intellectual and technical abilities of man give proof of his ability to change the environment and to transform it into a cultural landscape. This happened through artificial irrigation, clearing of woods, planning of towns, establishing traffic ways, agriculture, and horticulture. In this period the ecological system was preserved, in principle, but it was subjected to the goals of man. (3) Finally, man achieved such enormous means through industrial techniques that now he has gained the power to change his environment into an industrial landscape. This involved the destruction of the environment on a large scale; the ecological balance was extinguished. Gigantic slag heaps, open-cast working of brown coal, large, modern towns, expanded industrial areas, and broad asphalt or concrete highways are parts of this industrial landscape of today. The temporal demarcations between these three periods can be fixed differently according to the historical development in the different regions. At present, the natural landscape has been permanently transformed into a cultural landscape and this, in turn, into an industrial landscape. Therefore, the relationship of the dimensions is disarranged in favour of the younger systems to the disadvantage of the older systems. Today, all three kinds of landscape exist side by side.

5 The Scientific Surroundings

Environmental history appears at a time when a perfected system of science already exists. Its subject is partly included in adjacent sciences; their special experiences and methods complement environmental history. There are already "auxiliary sciences" which serve as a basis for this new environmental history. Among these are: historical geography, historical studies of territories (*Historische Landeskunde*), the history of settlements (*Siedlungsgeschichte*), agrarian history, history of climate, historical demography, economic and social history, mining history, industrial history, history of architecture, history of technology, historical statistics, history of law, and history of medicine. To discuss and clarify theoretical questions, interrelationships to sociology, philosophy, and theology are also important.

6 Sources of Environmental History

A wide variety of historical sources may be useful for research on environmental history. With regard to the Middle Ages, especially important are the customs and the codes of law. In modern times, reports about legal proceedings on environmental questions should be added, for example, cases of smoke damage, mining damage, and water pollution. Very important are land registers, field maps, and topographic maps, which provide information on surface changes. Names of places and field parcels often provide hints on the ecological conditions of the past. Statistical materials, including tax registers, give information on the amount of houses, men, animals, and production, etc. These examples illustrate the quantities within the environmental problem.

7 Mentalities

People of early epochs, living within a natural landscape, automatically had a natural, naive, usually harmonious environmental consciousness, because they were totally dependent on the environment. When human abilities increased, man became more independent, his consideration of environment decreased, and finally a break with earlier thinking and doing occurred, which depended on environmental consciousness. It seems that at some time or other an "ecological sin-fall" occurred, the fixed temporal point of which has yet to be more exactly defined (e.g. Renaissance or Industrial Revolution?).

In this context we have to consider the importance of money. It was introduced in between the producer and the consumer as a medium and dissolved the immediate relations of man to his elementary conditions of life. Ecological deliberations were replaced by economic deliberations. The ability to dispose of money became decisive for the extent of consumption, and no longer the individual's rate of work. The unrestrained struggle for the desired money destroyed the limits of an ecologically reasonable consideration of the environment.

Considering the fact that the environmental problem has its roots in Christian Europe, the supposition is obvious that the understanding of God and the theory of life and man given in the Bible have been responsible to a great extent for the formation of a mentality which gave man the right of domination and exploitation of the earth. This question must be pursued.

The present seriousness of the environmental problem demands the return of environmental consciousness, which should result from intellectual reflections and the understanding of the necessities.

8 Crucial Points of Research

Special attention should be given to exemplary, careful, and sensitive behaviour patterns in former times. The kind of environmental damage, its origin and the mental attitude of the originators must be explored. On the other hand, the resistance and useful measures against such damage also need to be examined. The arguments which were put forward at that time for and against environmental protection must be worked out. Indications of a well-developed environmental consciousness, for example, in customs or in positive law, should be made evident.

The development of the organization of labour should be examined in this direction regardless of whether it includes elements which are detrimental to the environment. In this regard we have to consider the importance of the division of labour, which caused mankind to lose the total view on all conditions of life, and to break up the immediate relations between natural resources and consumption. Contradictions between ecological and economic interests should be determined.

9 Goals and Effects

As part of the science of history, the history of environment is not able to grasp the environmental problem in a practical sense. But it must have a voice in shaping the consciousness of mankind today, providing a sense of responsibility for the environment. To that extent, environmental history has a pedagogic task. Historical knowledge can set standards for behaviour at the present time. It offers alternative solutions for technical tasks, since it helps to reduce industrial techniques to "tender techniques" and "human standards" (E. Schumacher). Mistakes of the past must be discovered, self-criticism by progress-credulous mankind must be stimulated. We should refer to environmental history concerning past experience and patterns of action, which can be made applicable to the present. Knowledge of faulty decisions in the past may help one to avoid mistakes in the future. The situation of the whole world in regard to environmental questions is so grave that environmental historians must also have the courage to express unpopular ideas.

10 Consequences

The seriousness of the environmental problems demands a contribution from many sciences. The science of history should offer broad possibilities to environmental solutions. Knowledge of the past makes the historian sensitive to interests in the future, thus leading to solidarity with our forefathers in favour of our grandchildren.

Reflections on a Typology of Historical Pollution: Complementary Conceptions

I. Mieck[1]

1 Introduction

In order to compare older and newer forms of historical pollution it seems reasonable to distinguish at least six types. These types are to be presented and illustrated in the six sections of the following article.

Except the last, all of them are based on two important observations: Through the increasing quantity of the different forms of pollution since the late 19th century, pollution in general took on a new quality, because it makes a great difference if the polluted water of 50 000 or of 500 000 inhabitants of a town is directed into the lakes and rivers of the environs, or if there are 2 or 200 chimneys blowing their smoke into the atmosphere of an urban district. This change of quality occurs at different times, depending above all on the concentration of people. In agglomerations like London and Paris with their 300 000–400 000 persons pollution reached dangerous dimensions much earlier than in little towns or villages. Therefore it has to be pointed out that the remarkable growth of European population since the 19th century is not to be seen as a cause but as an accelerating factor in the multiplication of environmental problems linked to the industrial revolution.

French historians have distinguished two types of pollution, the *pollution microbienne* and the *pollution industrielle* (Le Roy Ladurie 1973, p. 627–696; Goubert 1974, p. 59–69), but it seems rather clear that this typology is not sufficient because it neglects some important factors and does not cover the whole of the phenomenon. Taking into consideration the sources of pollution and their effects as well, there are at least six types of pollution. To characterize these types of historical pollution it is preferable to use the French expressions because they are perhaps the most descriptive.

2 *Pollution Microbienne*

The *pollution microbienne* or *pollution bactérielle* is the most natural of all historical and actual pollution. It is caused by bacteria living and developing in decaying and putrefying materials and stagnant water. Especially in the towns there were three focuses of this kind of environmental infection:

[1]Freie Universität Berlin, Fachbereich Geschichtswissenschaften, Habelschwerdter Allee 45, 1000 Berlin 33, FRG

Brimblecombe, Pfister (Eds.)
The Silent Countdown

1. The first problem arose from the streets that were rarely paved, always dirty, and very often muddy. In Paris, the local authorities had specialized tipcarts to transport the muds out of the town. These *tombereaux* were not very large, and carried by the middle of the 18th century about 1.5 m^3. In 1775 the mud transport was maintained by 74 carts in service 9 h daily. But the problem remained unsolved, even after the installation of the first gullies, which were plugged up by mud and refuse in a very short time. Calculations suggest that 750 m^3 of mud were transported out of Paris to special places of deposit each day. This is equivalent to 270 000 m^3 annually. Terrible odors and "an unbearable infection caused by the dead animals which are thrown in continuously" (Boudriot 1986, p. 517–519) spread from these areas. In 1758 the king ordered new places of deposit to be established at a greater distance from Paris and its suburbs.

A comparable situation was to be found in Berlin about 1800. "A pestilential odor" mixed with the odors from "the excrements of Berlin and the cadavers of its perished pets" arose from the refuse collections, situated not far from the eastern suburbs (Mieck 1973, p. 9).

2. The second cause of the microbial pollution was the dirty and polluted water coming from households and workshops. It was collected in the gutters on both sides of the streets. In Berlin, these little channels with a depth of several feet were open, often blocked by refuse or other nonfluid materials and sometimes decorated by the offal and the blood from the butchers' shops. "Even in the best quarters of the town the pestilential odor coming from the stagnant gutters was hardly tolerable during the summer months" (Mieck 1988, p. 510).

3. The third factor for the traditional microbial pollution consisted of the problem of human feces, collected in cesspools or transported daily to several discharging points on the rivers of the town. Those who could afford it, participated in the removal service: night by night, closed vehicles passed through the streets of Berlin, taking filled buckets out of the houses and changing them for empty ones. All the collected buckets were transported to certain points and contents thrown into the rivers or lakes – in the 1820s more than 200 000 buckets of feces annually at Jungfernbrücke alone. The inhabitants of this quarter were obliged to have their windows shut after 10 or 11 o'clock p.m. The odors were comparable when the cesspools were emptied, mostly by the peasants of the neighbourhood, once or twice a year (Mieck 1988, p. 510).

There was a different situation in Paris, where the houseowners *sont tenus avoir latrines et privéz suffisans en leurs maisons*. This medieval prescription was fixed in the article 193 of the *Coutume de Paris*. The cesspools were expensive, firstly because they were stone-built and had a volume of 34 m^3 on an average, secondly because they had to be emptied by specialists, the 36 *Maitres vidangeurs*, who carried out this dangerous work with some assistants, equipped only with shovels, buckets and bins. The price for this work, contaminating the whole house for about a week, was 55 livres for 7.4 m^3. Inexperienced workers were cheaper, but paid for their ignorance, often with their lives.

In 1776 the cleaning of the cesspools was confided to the *Compagnie du ventilateur*, which possessed a system to exhaust the poisonous gases before emptying. This procedure made the work less dangerous, but the costs rose from 55 to 90 livres. The house owners had four possibilities to lower the costs of the cesspools: they could make a hole in the wall of the cesspool, near the bottom, or empty them only partially, or empty their buckets in the gutters, or relieve themselves in public buildings like the Palais de Justice – surely a very bad habit which was *non seulement contre le respect dû à la Cour, mais aussi cause d'une infection extraordinaire* (Boudriot 1986, p. 521–523).

To fight against this special form of microbiological pollution, there were several developments during the 19th century. Firstly, canalization, which offered only a partial solution. The problem was "covered", because all the polluted water was collected in underground channels to be conducted to the rivers of the town – as before, but less visible. A second step in the right direction consisted in the creation of enormous areas far from the towns, reserved for accepting the polluted water in order to allow a settling or clearing process. Nevertheless, one problem remained unsolved, partially until the present time: because the underground channels of the canalization receive also rain water by the street gullies, it happens that after heavy rainfalls, the canalization net is filled up and threatens to flow over and inundate streets and cellars. In those cases the emergency exists are opened – with the result that all the contents of the canalization find its way into the rivers of the town. There are many examples for this kind of secondary pollution – one of the most recent in Cologne, where the clearing station stopped working in spring 1988 because of the high tide and let all the uncleared water flow into the Rhine. The example shows that a really effective fight against this kind of pollution demands a combination of several measures: two separate canalizations, clearing stations and settling areas are the most important elements of this concerted action.

3 *Pollution Artisanale*

The second type of historical pollution is closely connected with human activities in economics. Tanneries, potteries, smelting furnaces, paper mills, hammerworks and other pre-industrial workshops polluted air and water, and produced noise and waste. The situation did not really change from antiquity to the 18th century. This type of pollution emanated from rather small workshops of artisans or craftsmen and should be denominated as *pollution artisanale*.

There are many examples for this kind of pollution during the centuries. It was produced in towns and villages. Most incidences of pollution are to be deduced from laws and edicts, beginning with some interesting sentences included in the Corpus Iuris (Mieck 1981, p. 1141–42). Later on, in the Middle Ages and Early Modern periods, local or regional authorities issued many decrees to restrict the disturbances of the neighbours by smoke, stench, and noise. Some of the polluting professions were particularly disagreeable: string-makers, soap manufactures, bone-burners, glue makers etc. At any case, the pollution was restrained, touching only a relative small area, namely the immediate neighbourhood of the molesting workshop. Therefore, the measures were in general casuistic, orientated to a specific case, trying to diminish or to stop the environmental damages caused by this or that workshop. Only under extreme weather conditions could the *pollution artisanale* spread wider and touch a greater region. Normally the pollution remained restricted to a rather limited area, and the natural force of air and water to clean itself seems to have been sufficient in most of those cases.

4 *Pollution Industrielle*

The third form of historical pollution should be denominated *pollution industrielle*. It is characterized by the wide extension of environmental damage. Quarters, towns and even regions or provinces were (and are!) affected by this kind of pollution, which appeared about the beginning of the 19th century. The first objects of popular opposition were not the ironworks, but chemical factories. In 1829 70 Berliners complained about the exhalations of a chemical work causing a pollution by smoke and stench at a distance of several hundred meters (Mieck 1981, p. 1146–47), and in Menil-Montant, near Paris, 86 citizens protested against the toxic gases of a chemical factory damaging the health of the whole town in 1820 (Mieck 1982, p. 359).

With the progress of industrialization, this form of pollution took on enormous dimensions. In Berlin about 1845, nearly a dozen ironworks like Borsig were concentrated in a northwestern suburb, which people called *Feuerland* (land of fire). Another contemporary spoke of the "Birmingham of the Mark (Brandenburg)" – reminding by this association the center of the notorious Black Country in England. More dangerous than smoke were the sulfuric exhalations emanating from the soda factories, which ruined the vegetation and the agriculture of whole regions. The most terrifying example was perhaps to be found near Liverpool, where several soda factories were operated. In 1862 a correspondent of The Times resumed, that "whole tracts of country, once as fertile as the fields of Devonshire, have been swept by deadly blights till they are as barren as the shores of the Dead Sea or the banks of the Great Salt Lake". All the district around St Helens was "one scene of desolation. You might look around for a mile and not see a tree with any foliage on whatever". Houses, buildings and other artefacts of brick, stone, wood and iron were turned all to a "slaty-bluish colour" (quoted by Dingle 1982, p. 532).

The problem of industrial pollution persists. It is true that *der blaue Himmel über der Ruhr* is visible again and that the famous smog of London nowadays is less frequent than 20 years ago – but what about the terribly polluted regions around Krakow or between Halle and Merseburg? One of the greatest environmental catastrophes caused by industrial pollution in our days takes place actually in Central Siberia, near the town of Atschinsk. A concentration of several gigantic electric stations, all fired with brown coal, and an enormous aluminium-work with more than 20 chimneys transform the country into a land of ashes. Statistically, each inhabitant of Atschinsk receives 20 kg of ashes annually: "Streets, houses, trees, grass – all is covered by ashes. You can even taste it, it's crunching on the teeth" (Der Tagesspiegel, 13.2.88).

To end this section under another perspective, it must be noted that the industrial pollution of the 19th century had a consequence of great importance: the beginnings of national anti-pollution legislations, no longer aiming at isolated factories or workshops, but at certain branches of industry as a whole. The history of these national legislations is opened by a Napoleonian decree of 15 October 1810, which was completed and modified in the following years, with the result that in 1845 in France, no less than 307 branches of industry were classified as *établissements insalubres, incommodes ou dangereux*, needing for their settlement and for their economic activities an authorization by the local or regional police (Mieck 1982, passim).

In Prussia, there existed numerous edicts and orders against polluting workshops. They were compensated by the Paragraphs 26 and following ones of the Allgemeine Gewerbeordnung of 1845, which formed the first national anti-pollution law in Prussia (Mieck 1967, passim). There was a number of 19th century smoke abatement laws in England, but it was not until 1863 that Parliament voted in favour of the "Alkali etc. Works Regulation Act", a late but important edict (cf. Dingle 1982, p. 544–548).

5 *Pollution Fondamentale*

The extension of the *pollution industrielle* can still be surpassed when human activities begin to transform, not only regions, but whole countries or even parts of continents. Examples for this type of historical pollution, which may be called *pollution fondamentale*, are the deforestation of the mediterranean area since antiquity, or the famous soil erosion in some parts of the USA, a consequence of monocultural exploitation. There are also recent examples: the diversion of two rivers, realized for economic reasons, has dried up nearly half of the Lake Aral and has submitted the surrounding province of Usbekistan up to Buchara and Samarkand to a process of salination (Der Tagesspiegel, 25.5.87). There is no doubt that comparable consequences will result from the enormous destructions in the tropical forests, where millions of hectares disappear each year, for instance in Thailand, in Indonesia, and especially in Brasil. The projected Lake of Cararao, destined to aliment a gigantic electric station on the Rio Xingu, will cover a forest area greater than Hamburg and Bremen together (Der Tagesspiegel, 22.2.1989).

An important element of the fundamental pollution consists in the fact that the consequences of this type of pollution may touch the climatic conditions of the country. This is known from the Asswan Dam which brought great difficulties for the Egyptian agriculture.

Legislation against this *pollution fondamentale* seems to be very difficult, firstly because the State very often itself appears as promoter, organizer or protector of these projects, for instance in the USSR or in Brazil; secondly, because this type of pollution is mostly an international one. Therefore national legislation is not sufficient and it is the place where private organizations such as "Greenpeace" find their best fields of action.

While the names of the first three types of historical pollution – *microbienne, artisanale, industrielle* – are based on the main causes of these pollutions – *microbes, artisans, industrie* – the term *pollution fondamentale* is based on the consequences, because this type of pollution no longer allows the selection of isolated factors. It affects the whole, consequently it means that in the grades of historical pollution, the *pollution fondamentale* was certainly a qualitatively new step.

6 *Pollution Foncière*

The fifth category of historical pollution dates, it is true, from the 19th century, but has reached its maximal effects only in the second half of the 20th. It concerns the devastation of soil by the enormously intensified agriculture: the vast quantities of nitrogen, phosphates, and potash, the masses of herbicides and pesticides brings to the soil a chemical over-burden of unknown dimensions. Besides, through the employment of bigger and bigger agricultural machines the soil is more and more compacted. To characterize this phenomenon that has only just begun to interest the public, one could speak of *pollution agricole*, but there are other factors to take into consideration: the soil also receives quantities of acid rain, a result of the industrial pollution. The acid rain waters very often attack the plants and trees at the roots and make them die. Another element of soil pollution consists in toxic waste products, left behind by gas factories and other chemical enterprises working in former times and utilizing the soil as refuse deposit. In other cases the manufacturers did not pay attention and spoiled the soil by simple negligence; in Berlin, much soil pollution of this kind is known.

In 1986 the European Community and the Senate of Berlin organized a three-day symposium treating these problems. The results were summarized by a correspondent with the formula: "Europe's soil: contaminated, sealed, compacted." "The destructions of the soil", precised a participant, "are not to be seen and not to be smelt. Here you find a time bomb similar to forest die-back" (Der Tagesspiegel, 18.10.86). To express the importance of this kind of pollution, it seems appropriate not to consider the different causes, but instead, to pay regard to the results (as in the case of the fundamental pollution) and to speak of soil pollution, *pollution foncière*.

7 *Pollution Accidentelle*

It is not only the events of Tschernobyl that show that the five categories of historical pollution are not sufficient to cover all environmental evils today imagineable. The specific element of the last one, the sixth, lies in its uncalculability based on the circumstance that the normal situation has been completely changed by an accident. While all the other forms of pollution are to be foreseen, calculable and, in consequence, attackable, this last type avoids influence and becomes uncontrollable. Several of these accidents have taken place in the last few years, in the USA, Italy, India, USSR, and in Switzerland. The environmental catastrophes following those accidents have their own, extremely negative, quality and cannot be compared with the other forms of pollution. Even the *pollution fondamentale* is quite different, because its consequences are often not to be remarked at once but after 5 or 10 years, or still later, while the catastrophes of Harrisburg, Seveso, Tschernobyl, or Basle are to be felt immediately. Therefore it seems justified to introduce a new category for the environmental damages caused by these events, and to call it *pollution accidentelle*.

8 Conclusion

The six categories of historical pollution, described in a very concentrated form in this article, appeared in human history approximately in the same order as they have been presented here. Each new form of pollution produced an additional effect, because the older ones persisted, not only in the countries of the Third World. In Europe and in North America, the *pollution microbienne* has not yet disappeared. The actual problem consists in the fact that the negative consequences of all possible forms of pollution are leading the world to an environmental crisis, and nobody knows if mankind can succeed to manage it. But management is, perhaps, the only road to survival.

References

Boudriot PD (1986) Essai sur l'ordure en milieu urbain à l'époque pré-industrielle. Boues, immondices et gadoue à Paris au XVIIIe siècle. Hist Econ Soc 5:515–528

Dingle AE (1982) The monster nuisance of all. Landowners, alkali manufacturers, and air pollution 1828–1864. Econ Hist Rev 35:529–548

Goubert JP (1974) Environnement et urbanisation: rapport général. In: 6e Congr Int d'Histoire économique, Copenhague 19–23 août 1974:59–69

Le Roy Ladurie E (1973) Un concept: l'unification microbienne du monde XIVe-XVIIe siècles. Schweiz Z Gesch 23:627–696

Mieck I (1967) Aerem corrumpere non licet. Luftverschmutzung und Immissionsschutz in Preu*ß*en bis zur Gewerbeordnung 1869. Technikgeschichte 34:36–78

Mieck I (1973) Umweltschutz in Alt-Berlin. Luftverunreinigung und Lärmbelästigung zur Zeit der frühen Industrialisierung. In: Verein für die Geschichte Berlins (ed) Der Bär von Berlin. Jahrb Ver Gesch Berlins 22:7–25

Mieck I (1981) Umweltschutz in Preu*ß*en zur Zeit der Frühindustrialisierung. In: Büsch O, Neugebauer W (eds) Moderne preu*ß*ische Geschichte 1648–1947. Eine Anthologie (= Veröffentlichungen der Historischen Kommission zu Berlin, vol 52), vol 2:1141–1167

Mieck I (1982) Die Anfänge der Umweltschutzgesetzgebung in Frankreich. Francia 9:331–367

Mieck I (1988) Von der Reformzeit zur Revolution (1806–1847). In: Ribbe W (ed) Geschichte Berlins,

2 vols, 2nd edn, 1: Von der Frühgeschichte bis zur Industrialisierung. CH Beck München: 405–602

Mieck I (1990) Berliner Umweltprobleme im 19. Jhdt. In: Lamprecht I (ed) Umweltprobleme einer Großstadt. Das Beispiel Berlin. Colloquium Berlin: 1–26

Schramm E (1984) Soda-Industrie und Umwelt im 19. Jahrhundert. Technikgeschichte 51:190–216

Der Tagesspiegel (Berlin newspaper):

Marion Kern: Europas Boden: verseucht, versiegelt, verdichtet, in: Der Tagesspiegel vom 18.10.1986, Nr. 12485, p 19

dpa: Die Apfelbäume wachsen nur noch unter Zellophan, in: Der Tagesspiegel vom 13.02.1988, Nr. 12885, p 24

Carl Goerdeler: In Altamira ist die Luft voller Blei, in: Der Tagesspiegel vom 22.02.1989, Nr. 13199, p 3

Agricultural and Sylvicultural Impacts

Aspects of Historical Soil Erosion in Western Europe

J. Vogt[1]

Organizing common work with convergent aims by both naturalists and scientists, on the one side, and archivists and historians on the other, is a fruitful long-term undertaking. Of course, studies of historical soil erosion are typically interdisciplinary; knowledge in geology, foremost on surficial formations, often neglected by geological maps, and pedology, familiarity with economic and social history and an ability to trace sources hidden in records offices of all denominations are needed. This chapter is illustrated with examples, mostly from France, with an emphasis on Champagne (which is still waiting for a specific paper written by me, unlike Burgundy, Lorraine, the Palatinate etc.). An overall view of the authors' research on Western and also Central Europe, with a bibliography, is given by Blaikie and Brookfield (1987). For both methodology and specific features of Paris Basin see Vogt (1986a).

Actually this whole paper is rather down-to-earth, far removed from the philosophical and epistemological problems with which the author is busily dealing in other fields of research (geomorphology, historical seismology, etc.). However it should be emphasized that he always had the good fortune to be at crossroads, doing field work as well as digging into archives, all over the world. Nowadays, "environment", a strange and ambiguous word, is on everybody's lips, so that scientists and, with some delay historians, feel compelled to accommodate *à la sauce de l'environment*. Much research work undertaken for a long time without such a magical word is now opening doors, bringing money for research and opportunities for publications. Sometimes it is rather far from a strict definition of "environment", yet finds an audience.

Of course, studies of historical soil erosion fit well in the narrower definition of "environment", but it should be emphasized that intensive research work on western european plains, plateaux, and lower mountains had been undertaken from the beginning of the 1950s onwards, before the word "environment" began its glorious career. Typically undertaken for some years by independent agrarian geographers (at the crossroads, once more!) both in France and Germany, historical soil erosion studies were widely known from areas with extreme climatic and/or human conditions. Their emergence (or rather reemergence), with subtle settings (geological, pedological, economic, social, etc.), in the plains, plateaux, and lower mountains of western Europe was a surprise for many people, astonished above all by the wealth of thousands of sources, and fine forgotten

[1]Rue du Docteur Woehrlin, 6700 Strasbourg, France

Brimblecombe, Pfister (Eds.)
The Silent Countdown

appraisals, mostly from the 18th century. While "interdisciplinary", a word as popular as "environment" is in fact too often limited to plain juxtaposition[1] without the least imbrication, historical soil erosion is an outstandingly genuine field for interdisciplinary work.

Strangely enough, pioneering work of the 1950's has been forgotten to a large extent. There is a huge bibliography on soil erosion in western Europe, but mostly on present processes, with many strictly "scientific" papers on experimental plots, sophisticated measures of rain splash, etc., leading to abstract overall rates (erodibility factors, losses of tons per acre, etc.) (Auzet 1987; Sallenave 1987) in an often haphazard way. A typical example with a remote setting: an overall appraisal on soil erosion in West Africa's savannahs has been deduced from the silt load of rivers, forgetting widespread reworking of former huge silt fillings. Clearly, the background of quaternary geology has been forgotten by some "narrow-minded science" seemingly despising fieldwork. While former and present implications (soil profiles, valley fillings, economy etc.) of historical soil erosion have been mastered in some far-off, extreme areas (Mexico, Tunisia, etc., Cook 1986) they are seldom considered in western and central Europe. However, there is outstanding knowledge, thanks to archaeologists, of North Germany's *Auelehm* fillings, from Neolithic times on (Bork and Ricken 1983; de Ploey and Paulissen 1988), and the fieldwork on *Wüstungen* gives some information for the Middle Ages. Later centuries are seldom considered by specialists of soil erosion, a paradox considering pioneering work and a wealth of examples. Above all, overall abstract appraisals forget the need of analysis of the whereabouts and consequences of former catastrophic events, often with deep gullying, described in a precise manner by numerous writers. Three kinds of sources might be defined at first glance.

1. Outstanding are of course *systematic enquiries* on rural conditions, a number of which were undertaken during the 18th century, mostly in France, but also elsewhere. Typically, hundreds of *Cahiers de Doléances* deal with soil erosion as a major hazard, surprisingly to a wide extent in areas with, on the whole, slight relief, such as Champagne and Lorraine. At the same time, agricultural enquiries in some *departments* (Haut-Rhin, Yonne) also contribute a wealth of information, often as an answer to a specific well-formulated question. Some provinces even have a long-standing tradition in this field. Inquiries for taxation purposes might be biased, but often include very precise information which should be considered after corroboration with other sources. Additionally, an inquiry on the income of Lorraine's priests in the middle of the 18th century gives a fine illustration of erosion problems.

2. A second category is formed by thousands of reports on specific events, requesting tax or rent relief. Certainly, some caution is needed with these, but descriptions are often precise (depth of gullies, etc.) and confirmed by an

[1]In an astonishing way such a juxtaposition is shown by the preparation for a meeting on "History of environment" organized by CNRS (1985).

administration's or landlord's own enquiries (a widespread practice). As an example, such a control was organized in a systematic way by the States of Moravia in 1674: "wo . . . durch Wasser mit völliger Verlierung des Bodens eine ganze Ruine des Bodens praetendiert wird, das sollen die Herren Kreishauptleute mit Zuziehung anderer benachbarten solche Örter persönlich in Augenschein nehmen. . ."[2]. These reports often fit into a mosaic allowing an understanding of effects of specific events, e.g., heavy rains or snow-melt, over wide areas.

3. The third source is information provided by enlightened men, discussing the problem in a more general way. So, for example, a Champagne landlord complains in the 18th century about the gradual disappearance of a fertile loess cover. Consciousness of erosion hazard leads to protests against 18th century *édits de défrichement* of French "Physiocrats" in the same way as 20th century ecologists protest against the decisions of technocrats. Soil erosion is considered an important factor by agricultural reformers of the late 18th and early 20th centuries, for instance in Champagne where one wrote in 1819, in a general way, in addition to proposals of practical measures: "C'est un des devoirs de l'art géoponique . . . de s'emparer des eaux pluviales, dès l'instant où elles arrivent sur terre, de s'opposer à leurs dévastations, soit en dirigeant, divisant ou ralentissant leur cours dès l'origine, soit en couvrant le sol de végétaux qui résistent à leurs actions destructrices et surtout en ne la provoquant pas par d'imprudentes cultures"[3]. At the same time, some travellers make most interesting comments. One of them gives a wonderful overall view of the problem in the Bar valley (eastern Champagne): ". . . il y a quelques collines fort roides; on y trouve beaucoup de *vestiges d'anciens fossés* qui avaient été ouverts pour amortir l'action des eaux de pluie. Tous ces fossés sont comblés et je n'en ai vu nulle part que l'on s'occupât de rétablir; les collines dont la plus grande partie était en friches il y a 30 ans sont presque toutes cultivées aujourd'hui et comme il n'y existe aucun fossé il est plus que probable que la terre qui les couvre sera entrainée dans la plaine par les orages. . .". Crossing Clermontois region, the same concern arises once more: ". . . il faut pour conserver la terre végétale sur les coteaux rouvrir les fossés qui avaient été pratiqués jadis pour amortir l'action des eaux pluviales et qui sont comblés aujourd'hui"[4]. However, the background is not discussed.

There are two aspects to this background: (1) physical and (2) human. In the author's opinion first global climatic change is to be discarded because beside gradual processes leading in the long term to large losses of soil and correlative colluvial accumulation, heavy rains and sudden snow-melt are responsible for catastrophic erosion with, for long years, spirals of processes sometimes triggered by minor climatic events or vice-versa. Peasants are frightened by the quick succession of erosion that follows melting and rain, so in Champagne, at Marson, in 1784: "Dèjà les neiges de l'hiver dernier en avaient déterminé lors de leur fonte,

[2] State Archives of Moravia F 11/I/L 62 (carton 98).
[3] Essai sur l'état actuel de l'agriculture dans le département de l'Aube, Troyes, 1819.
[4] Voyage agronomique dans les départements du Nord de la France, Bibl. mun. Amiens Ms. B 1928.

mais ces eaux dernières étant tombées avec tant de force que sur ce terroir . . . de toutes parts s'offrent des ravins, de toutes parts terres enlevées, notament des héritages que le cultivateur réservait pour les ensemencer en seigle . . .. terres qui étaient les meilleures . . . il n'y reste . . . que le crayon qui était couvert de bonnes terres lesquelles venaient d'être labourées et se trouvent par cette pluie considérable totalement enlevées et portées dans le fond de ce terroir, mélangé avec grèves et crayons qui couvrent une petite prairie"[5]. Gullying by rain seems to have been prepared by snow-melting. In a significant way, the priest of Racines writes in 1788, after mentioning abundant meltwater: "Que Dieu veuille que les pluies ne soient pas trop abondantes, car toutes les terres qui sont en couture seraient absolument dégradées"[6]. Simple geological explanations, for instance from some rather abstract geological map, that neglect surficial formations should be avoided. In large areas of the mountains of central France, former weathering must be considered, while elsewhere widespread periglacial formations, covering slopes and filling so-called dry valleys are of relevance. Erosion in Champagne generally destroys shallow soils developed on frost-fragmented chalk, leading to modern fillings sometimes themselves considered periglacial (a supposition sometimes made elsewhere). Broadly speaking, genuine periglacial fillings are cut by gullies feeding modern fillings downward: the problem is not so much one of erosion of soil as one of surficial formations bearing a soil.

In a given physical setting, a set of human factors also triggers slow or catastrophic erosion. Generally, *surficial ploughing* is a most important factor, linked to such things as lack of manure, lack of fodder, and state of horses or oxen. These factors are often in themselves linked to tenure problems. Surficial ploughing leads to a discontinuity (*Pflugsohle*) near the surface, limiting infiltration, with widespread sheet-wash, rilling, and even gullying, with loss not only of soil but, moreover, of precious manure. Erosion is an essential part of a disastrous *spiral*, driven by both natural and human factors. In contrast, deeper ploughing limits erosion to a large extent; for example, at the end of the 19th century in Picardie, at Louvrechy, new ploughs mean that infiltration is quick: "On attribue ce phénomène à l'emploi des charrues nouveau systême qui permettent d'entamer davantage la couche arable, ce qui la rend plus perméable . . .". Formerly, limited infiltration led to numerous rills (at least): "La présence de fossés accusait autrefois une infiltration peu active."[7]

Erosion linked with surficial ploughing is, of course, particularly disastrous in fallow land prepared for winter crops. This is emphasized by numerous testimonies: from northern Switzerland, at Rorbis and Embrach (canton Zurich) we hear, June 1574: ". . . die Brach hat es dermassen verschwemmnt und . . . hinweg geführt, dass es ein Jammer zu sehen . . ."[8]; from Champagne etc., at Greux, June 1788: ". . . terres de versaines . . . situées sur des coteaux très rapides . . . enlevées par les torrents d'eau . . . qui ont causé beaucoup de ravins

[5]Archives départementales de la Marne C 946.

[6]Archives départementales de l'Aube. Notes historiques extraites des régistres paroissiaux.

[7]Archives départementales de la Somme. Monographies communales.

[8]Zentralbibliothek Zürich. Ms. F 23 (Wickiana).

qui rendent presque toutes ces terres méconnaissables . . ."[9]; at Sogny, May 1790: ". . . les terres nouvellement labourées et fumées ont été entrainées en plusieurs endroits à pure perte pour les propriétaires . . ."[10]; at Cuisy, June 1851: "Les terres préparées pour les semailles prochaines ont été tellement ravinées et enlevées que l'on peut à peine s'y reconnaitre . . ."[11]

Most important is the partition of land into "fields" (*soles, saisons, Zelgen, Seiten*), such that huge areas were ploughed at the same time, bore the same crops, or lay fallow. Water-flow became organized with widespread sheet-wash and/or deep gullying. Such a system, although sometimes considered a technical achievement of the early Middle Ages, is actually a key contributor to catastrophic erosion. It is one of the main causes for many rural problems that persisted in some localities to the very end of the 19th century.

Even from vast expanses of slight relief there are often striking descriptions of local or widespread disasters. While sheetwash is seldom clearly defined, the gullies are often accurately described. In Champagne, at Chauchigny, an inquiry from 1789 complains about numerous gullies cutting across narrow fields: ". . . coteaux dont le sol est . . . sujet aux ravines, de façon qu'il y a de certaines piéces . . . qui n'ont pas trente cordes de largeur qui sont traversées quelquefois par trois, quatre ou cinq ravines . . ."[12]. From Picardie, at Epaumesnil, we hear, June 1859, of meter-deep gullies: ". . . Des fossés d'une profondeur de deux mètres en moyenne ont été pratiqués par l'écoulement des eaux; des terres récemment labourées ont eu toute leur partie arable emportée de sorte qu'il ne reste plus que le sous-sol".[13] Sometimes gullies were so deep that access to fields was hindered, so we find in Champagne, at Soulaines, from a 1788 enquiry: "Le finage se trouve extrêment gâté par les eaux qui creusé des ravines, ce qui empêche que plusieurs champs ne peuvent recevoir la culture dont ils seraient susceptible, ce que cause aussi beaucoup d'embarras et de retard pour les voituriers qui sont obligés de faire de très grands tours pour enlever les emblaves . . ."[14] In addition to the immediate loss of crops, vast expanses of land could not be used for years. From Champagne, at Vanault-le Châtel, we hear, June 1792: "crues d'eau si rapides lesquelles ont enlevé partie des terres sur les côtes . . . qui ont dégradé tellement lesdites terres qu'il impossible avant dix ans de les remettre en même nature qu'elles étaient auparavant . . ."[15] Here and there eighteenth century Champagne villagers remember formerly cultivated fields, now destroyed by erosion, so at Avant for some 50 acres: ". . . montagnes . . . autrefois cultivées mais dont les grosses pluies ont entrainé les terres dans les vallons de sorte qu'il n'y reste que de la craie ou grève, ce qui les rend incapables de culture"[16] or at Courdemanche: ". . . terres . . . il ya longtemps abandonnées à cause que ces . . .

[9]Archives départementales de la Meuse, C 390.
[10]Archives départementales de la Marne, I L 423.
[11]Courrier de Verdun 2/8/1851.
[12]Archives départementales de l'Aube, C 1318.
[13]Archives départementales de la Somme, Mm 107365.
[14]Archives départementales de l'Aube, C 849.
[15]Archives départementales de la Marne, 7 L 28.
[16]Archives départementales d l'Aube, C 1213.

terres ont été entrainées par les grosses nuées . . . ce n'est rien . . . que de la pierre"[17]. Further, crops and grass in the valleys were often covered with thick debris, a loss often more important than from erosion on slopes, considering high yields on colluvials and the shortage of good grass. One example from Burgundy, at Chanay, June 1810: ". . . Les terres des montagnes ont été entrainées dans le vallon, les vignes déracinées . . . , les prairies couvertes de pierres et de vase, les blés couchés et ensevelis sous la boue . . ."[18]

What is going on, often very later, in the so-called *Révolution agricole*? The most important reasons for the reduction of the erosion hazard were the disappearance of "fields" and fallow, and deeper ploughing. However, the role of new crops was ambiguous. Cover crops were used and quickly emphasized by enlightened people as one of the best ways to stop erosion. Finally, an 18th century gentleman farmer was convinced that cultivation of peas, a traditional, but limited crop, on fallows would quickly limit erosion in the alsatian Sundgau (Vogt 1968). The merits of new forage crops were discovered, at the very beginning of the ambiguous *Révolution agricole*. From Champagne, we are lucky enough to have a fine testimony from . . . 1762, on "sainfoin": "On en sema avec de l'orge au mois d'avril . . ., au mois de mai . . ., il survint un orage furieux, les eaux prirent leur écoulement au milieu du champ et la violence du courant dégrada le sol, enleva les plantes d'orges, celles de sainfoin résistèrent"[19]. Later pamphlets emphasize its merits: "Il est fort utile de semer en sainfoin des côtes en pente, surtout celles qu'il est difficile à cultiver. Le sainfoin empêche la terre végétale d'être entrainée par les pluies. . ."[20]

On the other hand, crops like potatoes and beets do not protect soil. Actually, the danger of erosion is limited in long narrow fields alternating different crops, and with step-building processes at the limit of protected and unprotected fields. However, the danger grows with both the "clearing" of the country and the formation of huge fields. The first process is found in Champagne at the beginning of the 19th century and could partially explain a quick succession of erosion disasters. Agronomists tried to stop the frenzy of "clearing": ". . . la plus grande erreur serait de croire qu'il faut tout labourer. Au lieu donc de culbuter, comme on le fait depuis plusieurs années dans les plaines de la Champagne . . . ces terrasses naturelles, connues . . . sous le nom de *haules* et qu'on formerait ailleurs grands frais, il faudrait y favoriser la plantation des bois . . . le bien que l'on retire de ces défrichements n'est que passager; . . . la bonne terre gazonnée que l'on arrache des haules sera bientôt . . . entrainée dans les ravins, avec celle des champs qu'ils soutenaient. . ." (Pauffin-Tiercelet, An XI). The second process, more or less, reconstitutes the traditional system of "fields" with, however, the compensation of deep ploughing and hence better infiltration. The erosion hazard was still widespread, but it seems that the resistance to intense storms was

[17]Archives départementales de la Marne, C 435.
[18]Journal de la Côte d'Or 15/7/1810.
[19]Mémoire sur la culture du sainfoin et ses avantages dans la Haute Champagne (1764). Bibl. Nat. S 14685.
[20]Réflexions sur l'utilité des prairies artificielles dans le Département de la Marne. Almanach Champenois pour l'année 1809.

higher. Less gullies, less debris, more mud? We will set aside the specific problem of erosion on ground frozen at some depth, by meltwater or rain. Whatever the reason, the erosion hazard disappeared in some areas of marginal and formerly sensitive soils by abandonment of cultivation, the *Sozialbrache* of German geography, a method now reinforced by the E.E.C.

Problems on the lower mountains, for instance of central France, should be discussed for their own peculiar features of climate, soil and agriculture. A book could be written spanning three centuries of erosion disasters in Auvergne, with seemingly huge losses of land and possibly one of the reasons for important emigration. Once again, general comments are made by enlightened people, for instance by Arthur Young for the region of Brioude: "The mountains . . . too much cultivated; the earth is, by such means, washed away by storms and torrents drive away everything". Such a disaster was described near Issoire in July 1788: ". . . l'eau a entrainé le terrain ainsi que les pierres détachées et n'a laissé que le noyau de roches à nu, ce qui emporte, pour bien des années encore, les espérances du cultivateur"[21]. Years are needed for weathering and desegregation: "Il faut pour ensemencer la campagne attendre que par le concours de l'eau, de la pluie, du gel et du dégel la montagne soit assez exfoliée pour que la charrue puisse y pénétrer". Saint Babel's *Cahier de Doléances* – let's quote at least one! – considers erosion and its consequences a central problem: "Depuis près de vingt ans les inondations ont enlevé le terrain de plus de 500 septérées de terre, n'y ont laissé que le rocher, de sorte qu'il est impossible d'en tirer la moindre récolte. D'autres septérées, au nombre de plus de 400; ont été . . . lavées par les inondations et couvertes de pierres. . .". One case, quartier de Las Perras, at Saint-Dier, 1792, is of peculiar interest for a better understanding of processes: "Une grosse pluie . . . a entrainé le terrain des champs que l'honnête cultivateur préparait à la semance, a formé . . . des cavités énormes, enlevé la plus grande partie des *pommes de terre*, récolte de la plus grande ressource dans ce pays et n'a laissé à ces malheureux citoyens que la misère et le désespoir"[22]. Indeed it seems that after a traditional, more or less shifting cultivation, adapted to harsh conditions, potato cultivation leads to an intensification of agriculture in the Auvergne as in Vosges, with a sudden aggravation of soil erosion. To this we should always add correlative accumulation. This is an explanation for the accumulation of alluvium in the Lauter and Queich valleys which invaded the ditches of Landau fortress, to the dismay of French army engineers (Vogt 1986b). In southwestern France, after the earlier adoption of corn cultivation, such a process is also suspected; once again, the human factor should be emphasized.

Of course protective devices are widely known and sometimes put into practice in a clever way. However, their use is to a large extent often prevented for social reasons: tenure (above all), short lease, and, worse, share-cropping. At first glance, it seems that France's vast areas of share-cropping are "rewarded" with widespread and unhindered soil erosion (Vogt 1958).

Vineyards, often on steep slopes and sensitive surficial formations, should be considered separately. It would lead us astray to consider the details of changing

---

[21]Archives de l'Academie de Médecine, Paris, 141/40.

[22]Archives départementales du Puy-de-Dôme, L 1173.

techniques (working by hand or ploughing, size of vineyards, presence or absence of grasscover . . .), so a few random testimonies should be considered. Significantly, newly worked vineyards were eroded in Champagne, at Varennes, April 1842: "Les vignes, nouvellement houées, ont été ravagées par les eaux qui ont entraîné beaucoup de terre végétale dans les plaines adjacentes . . ."[23] Everywhere, rills and gullies are described, as at Goussaincourt, June 1788: ". . . la grande affluence d'eau avait fait des saignées dans plusieurs . . . vignes, en avait emporté une partie des terres, détérioré les ceps, ce qui ne peut se réparer qu'avec beaucoup de peines et après de grands travaux"[24]. Huge gullies were cut in May 1724 in vineyards of northern Switzerland, for instance at Glattfelden (canton Zurich): ". . . in den Weinbergen . . . entsetzliche Riss gemacht und teils Orten eines hohen Hauses tief . . ."[25] From Bôle (canton Neuchâtel) we hear in August 1733: ". . . l'eau . . . a fait de grands fossés et emmené la terre et les ceps"[26]. Indeed whole plantations are sometimes lost, so in Champagne, at Arcy, in 1774: ". . . vignes . . . comme anéanties, il a fallu les arracher, les autres ne peuvent subsister qu'autant que les propriétaires pourront y faire reporter les terres, ce qui nécessitera uned dépense énorme . . ."[27]

While soil erosion seems to have been globally reduced elsewhere, it ranks as a crucial problem in many wine-growing regions even to the present day. Thanks to modern techniques it may be worse than ever. Among others, the following reasons are given for the major disaster occurring in the summer of 1905 in the famous Epernay vineyards, following two heavy rains, in June and August: ". . . le coteau . . . il y a quelque trente ans . . . était encore parsemé de trous, fossés, broussailles qui retenaient en partie les eaux . . . Mais . . . il a été procédé à des défrichements importants, la plus grande partie des trous et des fosses ont été comblés . . . La surface du coteau . . . formée d'une terre rocailleuse, peu meuble, a été . . . mélangée de sable et de composts, de sorte que la couche végétale . . . est devenue très mobile et . . . est facilement entraînée par les eaux"[28]. Once more, landscape cleaning is a major factor! Changing techniques could also partly explain the disaster in Baden, at Durbach, in the summer of 1906: ". . . ein Wolkenbruch, der tiefe Rinnen . . . riss und wagenweise den besten Grund aus den Reben fortschwemmt; solche Güsse wiederholten sich in den kommenden Wochen mehrfach und in solcher Heftigkeit, dass in einzelnen Rebstücken die Erde bis auf die Wurzeln fortgeschwemmt ist"[29]. A succession of rains seems an important factor in both cases. Nowadays, streets of alsatian wine-growing villages are often invaded by flows of debris or mud, so that a need for cover-crops is still felt.

---

[23] Revue de l'Est 9/4/1842.
[24] Archives départementales de la Meuse, C 390.
[25] Zentralbibliothek Zürich, Ms. Z I/396.
[26] Bibliothèque Cantonale Neuchâtel, Ms. A670.
[27] Archives départementales de la Meuse, C 954.
[28] Etude des mesures à prendre pour mettre la ville d'Epernay et la route No. 51 à l'abri des inconvênients et des dégradations occasionnés par les orages. Archives départementales de la Marne, I S 566.
[29] Stadtarchiv Offenburg, Archiv von Neveu 1484.

In conclusion, for centuries, human factors have largely explained the rather surprising spectacle of sheet-wash on a large scale, and of numerous, sometimes house-deep gullies (still to be seen today). Such spectacles are now mostly known from areas with extreme physical and/or human conditions. In other words, human factors, as well as the mode of tenure, allowed an extension of a "brand" of erosion usually considered mediterranean to the far north. Actually, the proportions of some erosion features are such that they attain a geological scale. Deep gullies of the Bray region (eastern Normandy), developing in former centuries along agrarian discontinuities were thought to be neotectonic features, a testimony of recent upheaval of the Bray anticline by geologists who failed to consider more down-to-earth factors.

Research in the vast field of historical soil erosion should occupy a unique position in the same way as studies of historical landslides, rockfalls, etc. Although catastrophes often occur, the now widespread "catastrophism" should however be avoided, to allow patient, interdisciplinary analysis. Indeed such research needs a wide documentary basis, far from "tit-bit work" sometimes seen in economic and social history, with its harvest of hasty and unreliable conclusions. Actually, such methods are a major danger for "environmental history", likely to become a "lunch ticket" for opportunists.

Besides publishing some papers (for bibliographies see Vogt 1986; Blaikie and Brookfield 1987), mostly in regional and internationally unknown periodicals, the author has accumulated vast amounts of notes over the last 35 years that await exploitation. He would be glad to share his knowledge with young interdisciplinary people eager to deal with such a challenging and stimulating problem, for joint publication. It is a pity that CNRS' specialists did not show the least interest for a proposal to exploit these materials for a book.

References

Auzet V (1987) L'èrosion des sols cultivés en France. Ann Geogr 96:529–556

Blaikie P, Brookfield H (1987) Land degradation and society. Methuen, London

Bork HR, Ricken W (1983) Soil erosion, holocene and pleistocene soil development. Catena Suppl 3

Cook AG (1986) El control de la erosion en Tlaxcala: un problem secular. Erdkunde 40, pp 251–262

de Ploey J, Paulissen E (1988) Processus géomorphologiques, formes du relief et interprétations des paléo-environnements quaternaires. Bull Assoc Geogr Fr, 65:1, pp 65–78

Pauffin-Tiercelet (An XI) Réflexions sur les inondations, leurs effets et les moyens de les prévenir dans le Département des Ardennes. In: Mémoires choisis parmi ceux lus ou adressés à la société libre d'agriculture, arts et commerce du Département des Ardennes, 4me partie, Soc d'Agriculture

Sallenave M (1987) Approche économique de l'érosion dans le vignoble champenois. Génie Rural, No 5–6, pp 29–32

Vogt J (1958) Protection des sols et modes de tenure dans l'ancienne agriculture. In: Comité des Travaux Historiques et Scientifiques (ed) Congr 1957. Bull Sect Geogr, pp 117–129

Vogt J (1968) Un texte remarquable sur l'érosion historique des sols au Sundgau. Rev Géogr Est, No 3–4, 8

Vogt J (1986a) Un problème méconnu: l'érosion historique des formations superficielles et des sols, l'example du Bassin de Paris. In: Association des Ruralistes Français, Groupe 2 (ed) Colloque National, Strasbourg, 14 pp

Vogt J (1986b) Aspects de l'érosion anthropique des sols en Alsace et dans le Sud du Palatinat. Les Vosges, 65:3, pp 3–6

Natural Environment and Human Settlement over the Central European Lowland in the 13th Century

T. Dunin-Wąsowicz[1]

1 Introduction

The research carried out by English and American geophysicists (UNESCO 1963) into the changes which took place in the natural environment of Europe during the Middle Ages, is corroborated by the work of Polish researchers on climatic change and on the effect of these changes upon the patterns of human settlement in Europe during the same period (Dunin-Wąsowicz 1974). Their research enables us to put forward certain hypotheses. I would like to draw special attention to two: (1) the effect of human activity in the field of irrigation in forested lowland areas; closely related to this is the exploitation of natural resources. (2) The other pertains to the general climatic changes which have taken place irrespective of man (Wigley et al. 1985).

The comprehensive research of geographers, historians and archaeologists carried out during the last 20 years charted the pattern and evolution of early medieval settlement in the lowlands of central and eastern Europe. The works of Hilczer (1967), Kurnatowski (1963) and Herrmann (1959) provide evidence that human settlement during the early Middle Ages gradually moved from river valleys and marshy areas onto higher land.

This trend was significantly intensified during the middle of the 13th century, although an earlier pattern of migration to areas of higher land can be detected (Fig. 1). At one time this pattern was thought to have been induced by dramatic floods, recorded in the annals and chronicles of the period. It seems more likely, however, that it was brought about by changes in the system of irrigation. These in turn reflected a more intense settlement in mountain regions, which had led to large-scale cutting down of forests. According to some geographers and climatologists, these extensive high-altitude forests had regulated the degree of rainfall and atmospheric humidity. It has recently been demonstrated, however, that the most extensive utilization of forests (if this can be measured statistically) took place during the 16th and 19th centuries, although these centuries show no major climatic changes. The cutting down of forests, if it exceeded a certain level, which could have been quite low in the context of a millenium, had nonetheless an adverse effect on the European hydrographical system. However, there may have been other factors at work as well.

[1]Institute of History of Material Culture, Polish Academy of Science, Swierczewskiego 105, 00-140 Warsaw, Poland

Brimblecombe, Pfister (Eds.)
The Silent Countdown

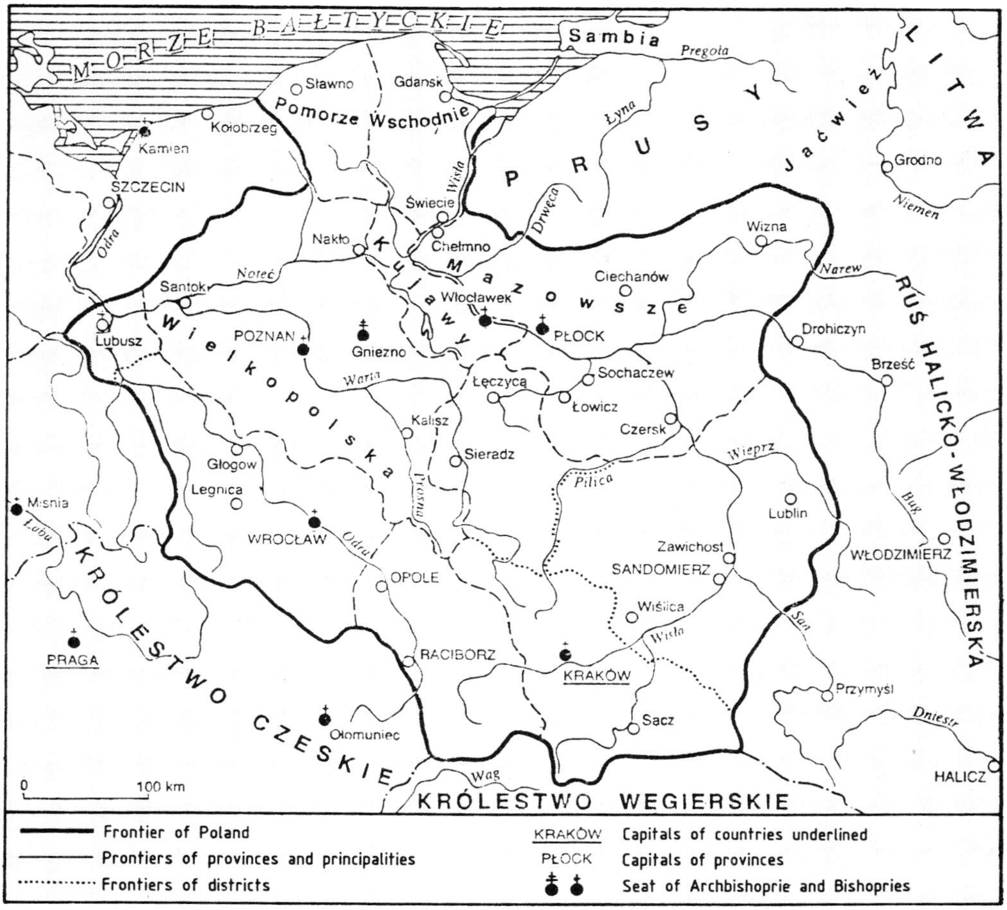

Fig. 1. Poland in the year 1206 prepared by Kazimierz Pacuski Institute of History of Polish Academy of Sciences

The theory of Schove (1955), i.e., the theory of the spectrum of time, estimating the maximum degree of solar activity and ascribing to it climatic changes, in conjunction with the findings of the American scholars Douglas (1919) and Huntington (1924), have made possible the construction of a graph depicting solar activity over a long period of time. According to this interpretation, the years of reduced solar activity are associated with natural disasters and, possibly, with periods of social upheaval. For our purposes, it is especially relevant that solar activity is found to have been at its lowest level three times during the course of the Middle Ages: for a short time in the middle of the 11th century; for almost the whole of the 13th century and the beginning of the 14th century, and throughout the 15th century. For Poland, as well as for some other parts of Europe, these were periods of successive natural disasters.

The large-scale migration of people from the lowlands to higher-lying regions would seem to indicate not only a high degree of atmospheric humidity and flooding in the river valleys, but also a general improvement of agricultural conditions in the highlands. In the southern regions of Poland, these migrations also coincided with the first cultivation of loess soils, extremely fertile ones depending on favourable water conditions. Since this was the case from the 13th century on, it acted as a great stimulus to the development of agriculture in these regions, while the valleys became utilized more for stock-breeding.

Since for the period from the 9th to the 15th centuries, the findings of Polish geographers, historians, and archaeologists (Radwański 1972) impressively corroborate the theoretical graph of Schove, the records of solar activity must be treated with respect as a serious method for testing changes in the geographical environment of Europe. This approach might also be extended to the whole Central European Plain (Alexandre 1985; Dunin-Wąsowicz 1968, 1973; Podwińska 1971, 1972; Tyszkiewicz 1986). The necessity of having an agricultural area to draw on was almost always an important consideration, conditioning the possibility of growth into larger settlements. Since during the 13th century water resources in Central Europe were lower than today, such conditions existed only in the river valleys of the area, i.e. (from west to east) the basins of the rivers Loire, Rhine, Danube, Oder, and Vistula. It was the increase in water resources in the catchment basins of smaller rivers (Seine, Moselle, Spree, Warta, Pilica) that made possible their development into agricultural areas.

In the higher lands, however, it is only from the 13th century that an agricultural base existed for extensive settlement. In Poland, the areas of high land in Little Poland and Silesia, especially those of fertile loesses – hitherto less productive – rose in value at this time, and from the 13th century on, they came to be some of the most fertile and productive regions of Poland.

At the same time, the fertile silt valleys of the larger rivers and the depressional basins surrounding lakes became unsuitable for permanent settlement as a result of the marked rise of the groundwater table; violent and unpredicted floods, intensified in the mid-13th century, destroyed the crops in the valleys. In view of this, new settlements avoided these areas. A safer location on a raised bank or terrace was chosen instead.

2 Changes in Natural Environment as a Result of Human Activity

2.1 Forests

The cutting down of forests in the Central European plain had important consequences for the natural environment. Geographical research has emphasized the fundamental relationship between diminishing stands of forests and the hydrography of an area. The progress of settlement in the higher regions of Central Poland, involving the cutting down of forests at the sources of mountain rivers, caused changes in the amount and type of rainfall and also affected melting of snow and ice. All these factors contributed towards causing sudden floods, recorded in historical sources from the 13th century. A cessation in the destruction of forests, however, was not possible, since timber was the basic building material for the inhabitants of the area. On the contrary, the demand for timber steadily increased with the advance of settlement and the erection of fortifications ensuing from the formation of states. It would be interesting to estimate the acreage of oak forests needed for the construction of the early medieval settlements of Gniezno, Kruszwica, Lad, Konin, Giecz, and Poznań. A close look at the map of the area between the rivers Vistula and Pilica is very instructive in this regard. One can easily detect the blank spaces once covered by forests around medieval settlements; the region of Radom, Skrzynno, and Sieciechów are good examples. Even the more important routes of communication in the 13th century led from one forest clearing to another, and these were inhabited by much scattered settlement. Prior to the 13th century these routes had led through forests, most often over watershed ridges. Water transport of heavy materials such as timber was relatively easy; even in winter time, when the rivers were frozen, timber could still be moved more easily over the ice than overland. The importance of forests in early medieval transport and communication was also of another kind. By extensive archaeological research, carried out over large areas of Poland, Germany, Russia, and Czechoslovakia, it has been established that timber was a basic material in road-building. It is almost impossible to estimate the amount of wood utilized for such work. Stone was rarely used for housing in those areas. It was only used for building jetties in rivers and lakes and at river crossings to keep the fords passable. Elsewhere, archaeology has revealed the exclusive use of timber structures for bridges, dams, and roads, some of them quite small, others stretching for miles along important routes, including roads leading as causeways to human settlements on islands. Classic examples of the latter, often mentioned in relevant literature, are the bridges in Teterov in Mecklenburg (Schuldt 1968) and in Ostrów Lednicki in Poland. As settlements grew, they also required more and more timber. Archaeological research in Gdańsk and Opole has revealed that the timbered roadbeds needed replacing every 25 years, excluding normal, current repairs. Timbered street subgrades had to be replaced every 50 years.

Apart from fortifications (Fig. 2) and road-building, there was a further use of timber in this period as the material out of which almost all everyday utensils were made. Apart from possibly peat, it also served as the main source of fuel.

The destruction of deciduous forests in the post-glacial European plain, particularly in densely populated river valleys, is further confirmed by diagrams compiled by Firbas (1949–1950) and Flohn (1968) revealing a dramatic decline in the amount of alder pollen and an increase in the amount of grass pollen. This reflects not only a considerable change of the climate (which all of Europe underwent at that time), but also the development of settlements in Europe north of the Alps (Boháč 1988).

Fig. 2. Mound of the stronghold of Poznan from the 10th century (article of T. Dunin-Wąsowicz)

2.2 Hydrography

The problems of hydrography, conceived in their widest sense, are closely interconnected with those of forests. Here we are particularly concerned with the importance of water for changes in the natural environment. Our task is to try to establish the degree of man's interference in hydrographic conditions during the early Middle Ages. Apart from flowing and stagnant water, there are two other important categories: groundwater and springs giving drinking water. It is difficult to establish the degree of interference in the first two cases. The other two have tended to be neglected by historians studying the development of settlements.

Historians have always agreed about the importance of river valleys and the basins of channel-valley lakes as primary locations for settlement. It is the archaeological evidence of settlements in low-lying areas, sometimes even in floodland and marshland, which has been more controversial. On the basis of the assumption accepted by historical geographers that the climate was generally milder (particularly due to higher humidity) and that water levels were generally higher than they are today, it has been difficult to explain this pattern of settlement. The present writer has arrived at different conclusions.

We know relatively little about climatic changes in the early Middle Ages. From some specialized research it can be deduced that there were at least two very dramatic changes during this period. Schove's graph indicates that they took place in the 11th and in the 13th centuries. These changes coincided with topographical shifts in settlement patterns. They can also be deduced from the studies of water conditions in the areas under discussion at the same time. The changes in micro-regions and micro-climates affected the general climate to a large degree. We must therefore allow for local variations which were dependent on the lie of the land and its morphology, the development of settlement, and the degree of interference with hydrographic conditions. Here we touch upon a further controversial point, which is whether or not the construction of weirs and water-mills altered the drainage patterns in medieval Poland.

This must have been the case, but to what a degree is difficult to judge. Generally speaking, these constructions interfered with the natural water flow in the area by diverting some of the water to irrigate the surrounding land. They also affected the speed of water run-off at certain periods.

Closely connected with hydrography is the problem of waterways and their use during the early Middle Ages. We see more and more clearly how closely interconnected were land routes and waterways. They frequently complemented each other over long distances (Dunin-Wąsowicz 1959). In the northern European plain they were certainly fully used for transport purposes. The capitals of early medieval states often lie at points of convergence of natural routes of communication, i.e., roads and waterways. Waterways were used for the transport of heavy goods, such as timber and rock salt. It may seem that in the higher regions, particularly in the mountains, lakes and streams played a lesser role. West European specialists, however, have supplied evidence that mountain streams and high-lying lakes were used for purposes of transport (Lopez 1956). This greatly reduced the cost of transport. For some years now Swiss geographers (Bergier 1963) have been providing evidence that an extensive communication network existed between the alpine lakes, with relatively short stretches of roads. It seems, however, that an abrupt change took place in the network of waterways in the northern European plain during the 13th century. This was probably due to human interference in the natural course of rivers, with the development of transport by land. The latter was facilitated by the more widespread use of modern horse-drawn vehicles. This also meant an increase in the total amount of transported goods. A further factor was the economic development of cities, which were often without natural waterways.

The role of the sources of drinking water and the importance of suitable springs remain to be discussed in the advance of settlements. The gradual shifting of the city centre of medieval Warsaw from Bródno to Kamion and Solec, and then to Jazdów and what eventually became the old town, provides an excellent example of a shift which took place probably because of the need of new sources of drinking water (Gieysztor 1967, 1986). Some of the springs discovered at that time are used to this day.

2.3 Agriculture and Stock Breeding

Issues connected with agriculture and stock-breeding have long been considered of great importance by European historians and archaeologists. Geological and pedological research has been fully utilized, but there still remains a great deal to be investigated. The studies of E. Le Roy-Ladurie (1968), mainly dealing with France and the alpine countries, have proved that the yield of harvest depended much more on adequate water supply to the crop, than on intensive methods of cultivation. It is therefore the question of climatic changes which provides a key to the problems of agriculture and stock breeding, and these changes can be deduced from hydrographical records. There has been controversy amongst specialists for many years about whether in the early Middle Ages lands less fertile but easier to cultivate were utilized in preference to richer loess or humus soils, requiring much more labour; the results have so far been inconclusive. We can assume that in Poland the key factor was also the amount of soil humidity because, as has already been suggested, this is often more important to the fertility of the soil than the quality of the soil itself. Soil fertility was critically affected by the cutting down of forests. The next important factor was the introduction of cultivation and the increase in stock breeding. The topography of certain small settlements (such as Sieciechów, Pułtusk and Lad) seems to confirm this theory. It is relevant here to mention the current debate about the scale and degree of innovation exhibited prior to the 13th century by the agriculture practised by monastic estates.

Horticulture, and especially viticulture, of early medieval Poland provides one more argument in support of the thesis that the climate used to be considerably milder, and that a climate suitable for viticulture existed. Viticulture was not only practised in the region of the upper Oder and the valley of the Nida rivers, but in the whole of southern and central Poland, as proved by local place-names, e.g., Winiary, Winna Góra, Winnica (Modzelewski 1964). To this day, viticulture continues in the Zielona Góra region on account of its relatively mild climate.

Pastoral farming had better conditions in the low-lying areas of central Poland than in the higher lands to the south. The marshy areas covering a large proportion of available land in the central plains presented conditions particularly favourable for pastoral farming. Topographers have associated the name of Lad with "meadow", meaning an area of dry land situated amidst watermeadows and marshes. Topographical names of early medieval settlements very often betray their early geographical location, e.g., Krosno on the Oder, Lad on Warta Kopanica, (Berlin), Gniezno.

2.4 Saline Waters and Bog-Ironstones

Salt, its location, and the methods used to obtain it, provide a key to understanding the patterns of settlement. There has been much recent research on the subject of rock salt. At present work is being concentrated on the early medieval distribution of saline waters, their extent and the ways in which they were utilized. Let us examine how these questions relate to the problem of changes in the natural environment. As in the case of the exploitation of deposits of bog-ironstone in early medieval Poland, it is extremely difficult to date these happenings. Despite extensive research, it is often impossible to decide with any degree of certainty whether a particular source of saline water, or a smelting furnace for bog-iron, was in the use during the first millenium or only during the second millenium A.D. The profusion of sites of these ancient smelting furnaces in Poland testifies today to the fact that they were in common use. A similar case can be made with respect to saline waters. Geological maps, and maps of the distribution of natural resources, are as useful to the researcher by their toponymy. The line of deep mine-working, following traces of bitumen and salt deposits, runs diagonally across Poland from Kołobrzeg to the range of the Świętokrzyskie Mountains. Apart from this, we find deep deposits of rock salt in the middle reaches of the Warta river and again in the Warta river valley north of Poznań. Areas rich in bog ironstone are situated particularly north of Radom, towards the valley of the Pilica river and in southern Masovia, in the basins of the Utrata and the Bzura rivers; next, also in the large region of Konin-Kalisz and Sieradz; and yet another, situated in the area north of Poznań in the basins of the Warta and Wełna rivers.

This distribution leads to far-reaching conclusions. The concentration of relatively easily accessible basic minerals like salt and bog-ironstone was an essential prerequisite for the development of the political and economic centre of northern Poland.

From toponymy we also gain some helpful hints. The names of localities obviously deriving from salt can be clearly divided into two groups: the first imply trading in salt, the other (e.g. Solnica, Solanka, Słona, etc.) refers almost certainly to actual exploitation of saline waters. There is no doubt that the Kołobrzeg salines and the trade in Ruthenian salt in Poland was established in the period before the 13th century. These have been the subject of many publications. It is tentatively suggested that saline waters were widely utilized in the areas of Great and Central Poland. This, however, has yet to be verified on both local and regional scales. It seems that, apart from the saline waters in the Kujawy region, salt for local consumption was also produced in other low-lying areas. The effects of this large-scale exploitation of both bog ironstone and saline waters must have had a considerable effect on the natural landscape as the result of exploitation of local deposits of bog ironstone. In addition to the utilization of local timber stands, smelters affected the level of groundwaters after smelting had ceased. The same phenomenon can be observed in the regions of Radom and Konin.

Full research into the impact of the exploitation of saline waters on the natural landscape has not yet taken place, apart from studies of the coastal saline deposits, which vary widely between the Baltic, the Atlantic, and the Mediterranean. Otherwise the exploitation of saline waters has almost certainly resulted in depressions, both in the region of the middle stretch of the Warta river and in the Kujawy region. Salt deposits on neighbouring fields may well have adversely affected the vegetation. The effect it had on the productivity of pastoral land is not clear. It seems likely that the vegetation was stunted, but the grazing pasture might well have been richer in minerals, important in stockbreeding.

To conclude, we would suggest that all the factors which we have mentioned above as constituting a natural environment, i.e., timber, water, soil, salt, and bog-ironstone, were there for men to exploit to the point of creating devastation through exhaustion, since they were requisites for the very existence and development of the early medieval settlements. All these resources were available in the low-lying lands, and thus it was in these areas that there was a concentration of settlements resulting in the devastation of this natural environment – a process which reached a threshold in the 13th century. After this date, settlement shifted increasingly to other, often high-lying areas.

The evidence of increasing devastation – both that caused by man and as a result of general climatic change in Europe – is expressed in the increasing number of references to floods found in documentary sources from the 13th century. Floods had been recorded previously, especially at the turn of the 10th and 11th centuries in the territories outside the Polish heartland, but then they were noted as extraordinary phenomena. By contrast, references from the 13th century describe floods in terms of a common natural disaster, such as military invasion or epidemics. This is also certainly how floods were viewed by the populace at that time. Since from this time the threat of floods was a primary consideration in the minds of settlers when founding their homes on the banks of great rivers, what they had to reckon with were not the regular, yearly Spring and Autumn inundations – which irrigated and enriched the soil – but flood disasters, which threatened life and livelihood.

The topographical pattern of settlement must therefore have been affected by these considerations.

3 Changes in the Natural Environment of Central Europe as a Result of Climatic Changes

In medieval Europe, and elsewhere in the world, interest in research into climatic and hydrographic changes has been growing in recent years. It has brought new hypotheses and controversies. Overall generalization has been replaced by an increased awareness of the need to assess each case individually. The results of regional studies ought to be compared; often these comparisons will be drawn between very disparate areas. The published results of the Rome conference of 1963 (UNESCO 1963), which was devoted to problems of climatic changes highlighted the dangers inherent in heedless comparisons of changes taking place simultaneously in regions as distant from each other as Japan and Canada. The London Congress of 1966 (World climate from 8000 to 0 B.C. 1966) reached similar conclusions. At a symposium held in Aspen in 1962 (Le Roy-Ladurie 1968), in which many European historians and geographers took part, attention was drawn to some other pitfalls, e.g., neglect of the importance of microclimates. The investigation into the role they play is a necessary prerequisite for understanding the changes of climate and the shifting patterns of settlement in eary medieval

Europe. This is the view held by the French historian E. Le Roy-Ladurie (1968), who, in his numerous papers, stresses the importance of small alpine glaciers on the quality and quantity of agricultural crops in the late Middle Ages and in the early 16th century (Pfister 1985a,b,c). The subject of European microclimates has also been considered in numerous papers and in the doctoral thesis of J. Schove, who compiled an interesting graph of maximum solar activity throughout several centuries. These two scientists have provided the essential inspiration for the present study, which is an attempt to consider these problems within the strictly defined chronological and topographical limits of early medieval Poland.

In the more recent literature on the subject, two works deserve particular attention. The first is the extensive and multivolume work edited by C. Weikinn and published 1958–63, which utilizes and systematizes information on the subject of climatic changes in Europe, obtained from chronicles and narrative sources (Weikinn 1958). The other is the work of Titow (1970), published in 1960 and 1970, in which this author analyzes climatic changes in the Bishopric of Winchester between 1209 and 1450 A.D. This latter attempt is particularly interesting, because the author utilizes very detailed sources from this period.

To discuss these problems in the context of eastern Europe, which is rather poorer in sources for this period than medieval England, presents greater difficulties. The archaeological evidence from the vast plain between Warsaw and Berlin supports the scanty written evidence on hand. Publications such as those of Herrmann (1959) and Kahl (1964) deal with problems related to the climatic and hydrographic changes in the region of greater Berlin, i.e., the early medieval settlements around Spandau, Kopanik (today Köpenick), Brema, and Brandenburg. It seems that their conclusions corroborate those of Polish researchers on this subject.

In an attempt at defining the nature of the connection between hydrographical and climatic changes in Europe as a whole, all the regions of this continent must be included. What so far has been done in this field on a global scale was presented at the Aspen Congress and later published in the works of Le Roy-Ladurie (1968). These changes are more readily demonstrable for the prehistoric period. For the modern period, one can only consider them beginning with the 16th century. At the Aspen Congress, a working diagram was presented; it compared some information about climatic changes in Europe, America and Asia in the 11th century. For Africa, only data about the Nile floods were included. The very attempt at a comprehensive approach to the problem in this working diagram reveals the great lacunae existing in our present knowledge. All information is available only on a purely local scale, and even then it is incomplete; while the changes in the natural environment are so complex that climatic changes represent only one element in the process. At some future conference, concerned with climatic changes in historical times, it would be interesting to see a diagram illustrating the state of knowledge on this subject for the 13th century. At present the lack of coordination of research on the subject makes it very difficult to compare even the data concerning changes within the same hemisphere. Japanese research (Arakawa 1956) into climatic changes which have occurred in the patterns of cherry-tree flowering, leads to the conclusion that, even in this instance, a major change took place during the 13th century, a change which seems clearly related to interference with the natural environment. In the case of Canada during this time, however, the effects of human activity must be discounted and the climatic disturbances charted by Schove must therefore have resulted from elemental changes. The question arises whether, in the present state of knowledge, the graph of Schove is sufficiently supported by historical evidence to warrant wider application to the progress of human settlement. Only further systematic investigation with regard to the northern hemisphere can test the validity of Schove's hypothesis with respect to all climatic changes during the course of the last millenium. At present his work provides encouragement for further investigations along these lines. These will need to be checked, wherever possible, against concrete historical, archaeological, and geographical evidence. In the case of Poland, such an attempt has been made only in regard to Cracow. Radwański's research (1972) on the region of Okole tallies with the diagram of Schove but it remains only an isolated case and therefore one cannot generalize from it on a world-wide, European, or even a Polish scale.

This support for Schove's diagram has found a further striking confirmation in the work of Biernacki concerning the whole of the valley of the Vistula. This seems to supply additional evidence in support of the hypothesis that about this time a dramatic climatic change, affecting other areas of the world, took place. This coincided with a time of considerable economic development in central Europe, which by then had already adversely affected the natural environment.

The combination of these two factors (the small-scale destruction brought about by human settlement, and the sudden deterioration of the climate) brought about great changes in the hydrographic balance, which reached proportions of a natural disaster during the course of the 13th century.

It would be interesting to compare the situation in regions morphologically different, but geographically close, e.g., in the loess areas of the basin of the Black Sea and in the drainage area for the Dniester river, especially if one could find examples in areas less intensively economically developed.

Geophysical research, especially that of Schove, has drawn the attention of historians to periods of intense climatic changes which, it now seems, took place all over the northern hemisphere. These effects on the natural environment were different in different parts of the world, e.g., in Japan compared with Canada and Europe. Why should this be so? Climatic changes used to be only one amongst several factors which affected the natural environment of a given region. Regional studies made in the area of Central Europe and particularly in the area of the Central European plain have proved that, besides climate, man was the most instrumental factor in bringing about permanent changes in the microclimates and microregions. Only by merging these different types of study will the full picture emerge.

Increases in population, the advance of settlement into forested areas, the progress of economic activities in terms of agriculture, pastoral farming and the utilization of water resources, exploitation of salt deposits and mining of bog ironstone and iron ore – all these factors contributed towards upsetting the balance in the hydrography of central Europe. They also interfered with local microclimates, which were among the greatest natural resources of the area.

The ability properly to utilize a specific microclimate was an attribute of the ancient Greeks and of the other ancient societies displaying a high level of civilization. It seems that the population of early medieval Europe lacked this ability. One can assume that it was the new settlers arriving in central Europe from the west, south and east who introduced certain improvements and skills into this area. After the invasions by nomadic peoples who, as we know, brought considerable contributions to the material culture of Europe with them, but rather less towards the management of a settled economy – the time came when the specific skills of the settled indigenous peoples could be utilized. Ancient skills and values, which had come down from societies of a higher stage of development, were now revived. Knowledge of the proper utilization of natural resources had been preserved in various parts of the Byzantine empire and in the monasteries of Scotland and Ireland. It is therefore not surprising that we first discover the introduction of viticulture in these monastic estates, and that the first attempts at reclamation of boglands and irrigation of dry land in Central Europe were first started by learned monks. The clergy and the nobility followed examples given by the monastic estates. Horticulture and fruit farming, which can be well traced from archaeological evidence, are further examples of the proper utilization of a specific microclimate.

A microclimate is, however, very susceptible to changes brought about by human activity. The present-day cases of Venice and Rome can be cited here. Man's interference in the natural balance of forests and water systems is an obvious example. Irrational intervention by man has always resulted in the destruction of particular microclimates. It may seem that this is unlikely to affect world climate, yet it must be remembered that the world climate consists of a multitude of microclimates. It may be, therefore, that man's interference with his environment, even on a local scale, can bring about a total "climatic crisis". In certain areas of central Europe it seems that the 13th century has been a threshold in this respect.

References

Alexandre P (1974) Histoire du climat et sources narratives du Moyen Age. Le Moyen Age 1:101–116

Alexandre P (1977) Le climat du Moyen Age en Belgique, Rhénanie, Nord de la France. Annales Economies, Sociétés, Civilisation. Paris, pp 183–197

Alexandre P (1985) Le climat en Europe au Moyen Age. Contribution à l'étude des variations climatiques de 1000 à 1425 d'après les sources narratives de l'Europe occidentale. Paris

Arakawa H (1956) Climatic change as revealed by the blooming dates of the cherry blossom at Kyoto. J Meteor 13, 6:599–600

Bergier JF (1963) Les foires de Genève et l'économie de la Renaissance. Paris

Boháč Z (1988) Historical ecological aspects of the Bohemian feudal state economy. Hist Ekol 1:11–59

Douglas E (1919) Climatic change and tree growth. Washington

Dunin-Wąsowicz T (1959) Research on the medieval road system in Poland. Archaeol Pol 2:125–140

Dunin-Wąsowicz T (1968) Questions ouvertes sur le milieu géographique des états situés sur la Grande Plaine Européenne. Miedzynarodowy Kongress Archeologii Słowiańskiej. Ossolineum. Wroctaw. vol 4. pp 156–158
Dunin-Wąsowicz T (1973) Remarques sur les changements dans l'hydrographie et la topographie de la région ou se situe actuellement Berlin. Internationaler Kongress für Slavische Archeologie. Berlin. vol 3. pp 119–124
Dunin-Wąsowicz T (1974) Zmiany w topografii osadnictwa wielkich dolin na niżu środkowo-europejskim w XIII w. Changes in the topography of settlement in the great valleys over the central European lowland in the 13th century. Ossolineum Wrocław Warszawa
Dunin-Wąsowicz T (1975) Climate as a factor affecting the human environment in the middle ages. J Eur Econ Hist 4:3
Dunin-Wąsowicz T (1986) Il paessagio medioevale dell'Europa Centrale. ambiente e insediamente. In: L'uomo e l'ambiente nel Medioevo. Galatina. pp 11–25
Firbas F (1949–50) Spät- und nacheiszeitliche Waldgeschichte Mitteleuropas nördlich der Alpen. vols 1–2. Fischer. Jena
Flohn H (1968) Le temps et le climat. Paris
Gieysztor A (1967) Aux origines de Varsovie. In: Miscellanea mediaevalia in memoriam Jan Frederik Niermeyer. Groningen. pp 177–186
Gieysztor A (1986) La Vistola nel Medioevo: asse dell'insediamento umano e delle comunicazioni. In: L'uomo e l'ambiente nel Medioevo. Galatina. pp 27–51
Herrmann J (1959) Wasserstand und Siedlung im Spree-Havel-Gebiet in frühgeschichtlicher Zeit. Ausgrab Funde 4:90–106
Hilczer Z (1967) Dorzecze górnej i środkowej Obry od VI w. do początków XV w. Basin of the river Obra high and middle course from the 6th to the beginning of the 15th century. Ossolineum Wrocław
Histoire de Pologne (1971) PWN. Warsaw
Huntington E (1924) Civilization and climate. New Haven
Kahl HD (1964) Slawen und Deutsche in der Brandenburgischen Geschichte des zwölften Jahrhunderts. Die letzten Jahrzehnte des Landes Stodor. vols 1–2. Köln
Kurnatowski S (1963) Études sur l'habitat du haut Moyen Age dans la région de l'Obra inférieur. Archaeol Pol 6:261–278
Le paysage rural: réalités et representation (1980). Rev Nord 62:244
Le Roy-Ladurie E (1968) Histoire et climat. Paris
L'évolution de l'habitat et des paysages ruraux d'Europe (1978) Actes de la conférence tenue a Varsovie en septembre 1975. Geogr Pol 38
Lopez RS (1956) The evolution of land transport in the middle ages. Past Present 9:22–32
Modzelewski K (1964) La division autarchique du travail à l'échelle d'un État: l'organisation "ministeriale" en Pologne médiévale. Ann Economis Sociétés Civilisation 19. 6:1125–1138
Pfister Ch (1985a) Bevölkerung. Klima und Agrarmodernisierung 1525–1860. vol 2. Stuttgart
Pfister Ch (1985b) Klimageschichte der Schweiz 1525–1860. vol 1. Stuttgart
Pfister Ch (1985c) Veränderungen der Sommerwitterung im südlichen Mitteleuropa von 1270–1400 als Auftakt zum Gletscherhochstand der Neuzeit. Geogr Helv 4:186–195
Podwińska Z (1971) Zmiany form osadnictwa wiejskiego na ziemiach polskich we wcześniejszym średniowieczu. Zreb. wieś. opole. Changes of forms of rural settlement on the Polish territory during the first phases of Middle Ages. Zreb. village. opole. Warszawa
Podwińska Z (1972) Structure et formes de l'habitat rural sur le territoire polonais depuis le VIe siècle jusqu'au début du XIIIe. Kwart Hist Kult Mater 20. 3:415–436
Radwański K (1972) Stosunki wodne wczesnośredniowiecznego Okólu w Krakowie. ich wpływ na topografię osadnictwa i próby powiązania tych zjawisk ze zmianami klimatycznymi. Water conditions in the Early Medieval Okole of Kraków and their influence on the Topography of Settlement and Attempts at Connecting those Phenomena with Climatic Changes. Mater Archeol 13:5–40
Schove DJ (1955) The sunspot cycle. 649 B.C. to 2000 A.D. J Geophys Res 60:127–146
Schuldt E (1968) Slavischer Straßen- und Brückenbau in Mecklenburg im frühen Mittelalter. *Liber Iosepho Kostrzewski octogenario a venatoribus dicatus quem Konrad Jazdzewski redigentum curavit.* Wrocław. pp 355–362

Titow J (1970) Le climat à travers les rôles de comptabilité de l'éveché de Winchester (1350–1450). Ann Veon 23, 2

Tyszkiewicz J (1981) Człowiek w środowisku geograficznym Polski średniowiecznej. Man in the geographical environment of Medieval Poland. Warszawa

Tyszkiewicz J (1986) L'ambiente e le epidemie nelle Polonia del Medioevo. In: L'uomo e l'ambiente nel medioevo. Galatina, pp 53–73

UNESCO (ed) (1963) Les changements du climat. Actes Coll Rome. UNESCO Organisation Météréologique Mondiale. UNESCO, Paris

Weikinn C (1958) Quellenteste zur Witterungsgeschichte Europas von der Zeitwende bis zum Jahre 1850. Berlin

Wigley TM, Ingram MS, Farmer G (eds) (1985) Climate and history. Studies in part climates and their impact on man. Cambridge

World Climate from 8000 to 0 B.C. (1966) Proc Int Symp Imperial College, London, 18th and 29th April 1966, London

Sylviculture and Forest Administration in Hungary 11th–20th Centuries

I.N. Kiss[1]

1 Introduction

By the mid-20th century, the forests were quite desolate and exploited, and became even more endangered by industrialism and civilization. However, there also appeared, as a token of conscience, a conscious intention to start nature conservation.

A. Barthelmess book, *Wald, Umwelt des Menschen* (*The Woods as Human Environment*, 1972) dealt with the problem of forests from the historical point of view. He cited W. Sombart: "The forests were the cradle of the European culture both materially and spiritually."

According to Barthelmess, the great period of the sylvan history of Europe, involving a movement of settlements and the clearing of forests in the centuries of the great migrations, settled down as late as the 14th and 15th centuries. Actually, it consumed two-thirds of the forests, establishing the rates of husbandry and settlements. This process, however, had its own social history. In Germany, for example, the peasantry's demand for wood had been of a quite threatening nature from the Peasant's Revolt of 1525 through the French Revolution until 1848.

The existing and fatal features of forest economy, i.e., clear felling, forest grazing, and gathering forest litter, became particularly dominant on account of the extension of capitalism. They thwarted natural afforestation and reclamation.

Another form of danger was *speculation*. "Between 1800 and 1884 two-thirds of state and foundation forests were sold or put in pawn for wealthy industrialists and speculators in the Austro-Hungarian Monarchy." During the 19th and 20th centuries, the ever-increasing exploitation of forests, especially for industrial and mine timber as well as railway sleepers, turned the woods into tree farms. In fact, despite modern methods, the intensive use of forests was extremely harmful. This process can be exemplified by several instances in Germany or Hungary. As a result of intensive exploitation, the fertility of sylvan soil was spoiled; in addition, the forests lost their ability to resist severe weather conditions and diseases.

What are the forest reserves of this ill-fated economic policy? According to the FAO's survey from the late 1950s, approximately 4 billion ha of forest still cover one-third of the earth (Fig. 1). At least half of it, i.e., 55%, can hardly be

[1]Hungarian Academy of Sciences, Budai N.A. 5, 1137 Budapest XIV, Hungary

Brimblecombe, Pfister (Eds.)
The Silent Countdown

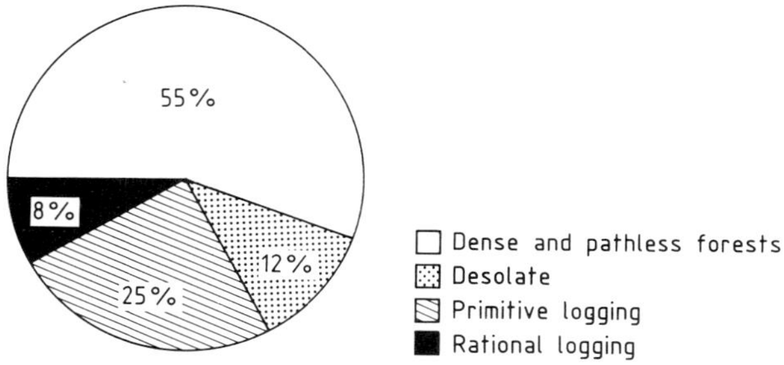

Fig. 1. Forest stand in the world (4 billion ha) around 1960

reached, whereas only two-thirds of the rest are exploitable, i.e., 33% of the whole. As few as 8% of the world's forest stand is exploited rationally, whereas 25% of it is used up rather primitively. The remaining 12% is regularly exploited even now.

2 The Case of Hungary

After the conquest of the Carpatian Basin, the Christian Hungarian Kingdom was founded around 1000 A.D. The large, almost intact woodlands were organized into crown forests. The forest became gradually populated and became part of the property of landlords and newly founded towns. According to general estimations, forests made up 75% of the country; most of them belonged to royal estates.

Contemporary forestry was based de facto on ruthless exploitation of forests. Reproduction was exclusively restricted to natural growth and letting the forests rest. The income from forests, mainly the rights for hunting, were controlled by "woodrangers". They were half-freemen, and many of them became members of the lesser nobility at the turn of the 13th and 14th centuries. Like the population of hunting villages, they lived on royal estates. The highly specialized royal hunters were also half-freemen and many of them rose to the status of the nobility.

The serfs' common of estovers (fire-bote and housebote) can be proved from the 12th century on. Clear-felling and natural growth led to a thickening (*silva caedua*) and deterioration of the forest stand itself. The appearance of prohibited forests was aimed at protecting sylvan growth.

Animal farming remained the most important branch of the Hungarian economy until the turn of the 18th and 19th centuries. This, however, was mainly reliant on grazing in forests. Pig-feeding was almost exclusively possible by leasing old oak groves. As the royal estates were gradually becoming the property of landlords between the 13th and 15th centuries, the devastation of forests was inevitable because of large-scale forest grazing.

Until the late 15th century (Fig. 2), the reproduction of forests proceeded in a traditional and natural way (regrowth). A charter from 1262 mentions acres of wood which "had grown from hand-sown seeds" (Lesenyei 1936), but even if the charter is authentic, it can be regarded as a national rarity. However, the

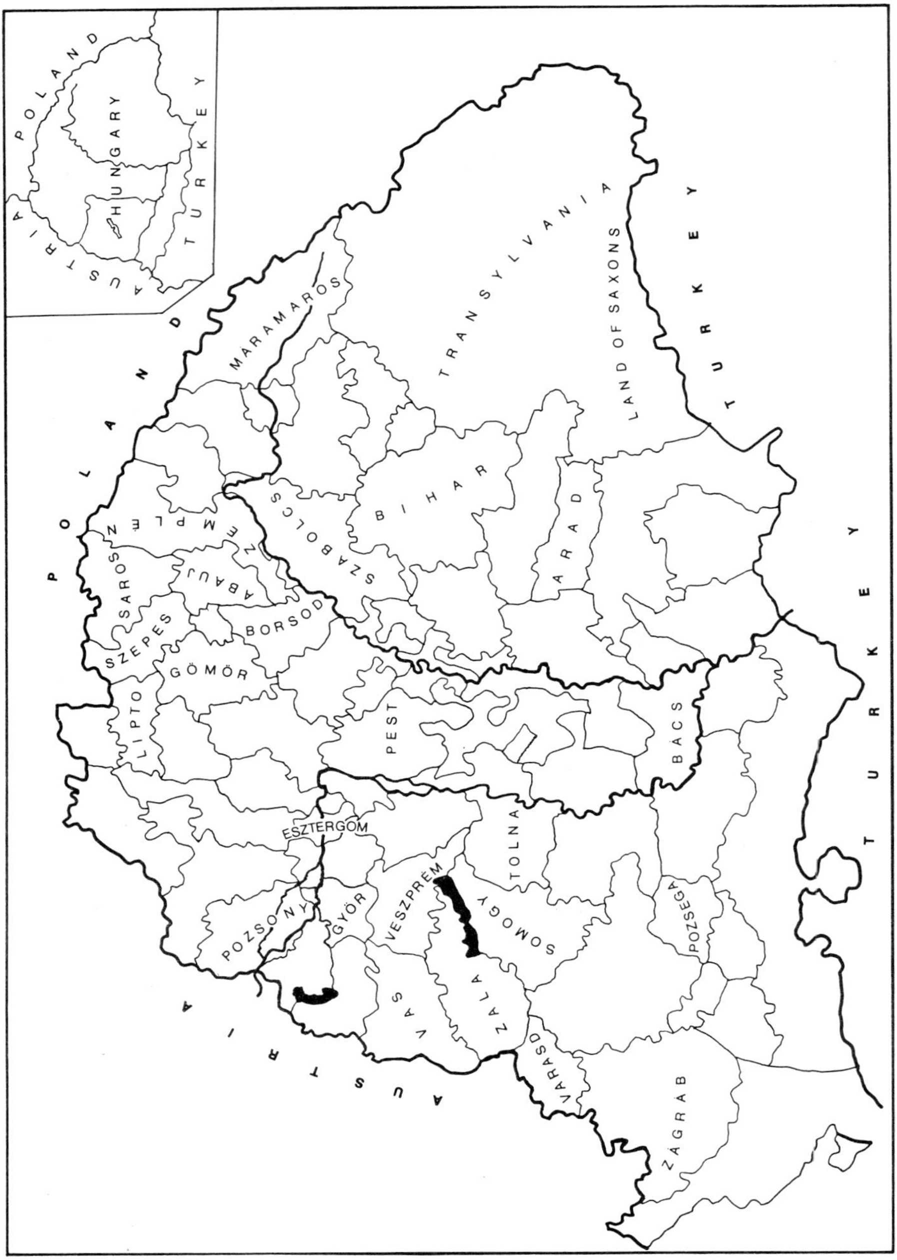

Fig. 2. Kingdom of Hungary XV. Century

Tannansaer ("pine-sower") movement of 14th century Nuremberg had its followers in Hungary, too. In 1485 a German town dweller in Buda imported 20 bushels of pine kernels for landscape gardening. According to Hungarian historiography, the first nurseries were developed at that time; nevertheless, they were all for fruit trees.

3 Extent of Forests

There is no concrete and comprehensive data available concerning the extent of 15th-century forests. However, István Szabó, a Hungarian medievalist, analyzed the fields of 66 Transdanubian parishes in the 15th century. His data cover 50 000 ha in nine of the counties of Transdanubia. His research stated that forests had made up 45.3% of cultivated areas:

Period	In million ha	Percentage of territories (%)
11th century	(estimated)	75.0
15th century	(50 000 ha; see Szabó 1975)	45.2
1855	4872	23.4
1920	1098	12.6
1980	1608	17.3

The decrease of forests in Hungary was not only striking, but, even by European standards, remarkable as well.

4 Manorial Forest Policy

The ruthless exploitation of forests reduced their extent; much of the woodlands of the Great Hungarian Plain were cleared away. Although this sometimes gave rise to lack of wood in certain areas, it could not be considered to be generally true everywhere.

Manorial tree farming was usually equivalent to unrestricted exploitation: the "common of estovers" was ensured, timber was sold, rents were drawn from grazing, whereas reproduction was absolutely neglected (enclosure, protection, etc.).

5 Treasury Forest Policy

From the 13th and 14th centuries on, the development of precious metal production and the royal privilege for mining made it necessary for the state to reserve certain forests supplying timber for the mines, subordinating them under the Treasury. This way, the Treasury became the controlling (administrative) body for both the Crown Forests and a greater part of the manorial forests.

This situation, which had developed by the 15th and 16th centuries, remained unchanged in Hungary for several centuries. Mining, smelting and other industrial activities were based on state forests as late as the 19th century. According to a partial summary, more than 3000 km^2 of state forests existed in 1817. The territory of certain big Crown Forests could occasionally be quite a few hundred square kilometres.

Treasury administration subordinated the forests to the interests of mining. The planning of clearing and resting, as well as the notion of "prohibited" forests were introduced, barring logging and grazing. Mines and furnaces were entitled to occupy forests from 1 to 6 miles.

Forests as supply bases played a particular role in Treasury administration. Apart from the woodcutting settlements, the 94 families employed by the mining chamber of Besztercebánya were active in the following five branches of industry in 1707:

1. 12 families in the official and trade center as woodrangers, wood reeves and forest clerks;
2. 10 carpenters and coopers of 16 skilled workers;
3. 21 melters and 26 charcoal burners for the furnaces;
4. In the group of other staff (11 families): 2 forest guards;
5. 8 more families whose function is not directly described.

We know the exact posts of 86 employees as follows: 6 woodrangers and forest guards; 10 carpenters and coopers; 26 charcoal burners; this sums up to 49% of the whole staff! Half of the constant staff of the mining chamber, or more explicitly, that of the center of iron, copper, and precious metal production, drew its living from forestry and the wood industry.

6 The Mining Towns and the Treasury

By the 15th century, two strikingly different branches of sylviculture had evolved in Hungary: (1) the private forest use of landlords and towns; and (2) the so-called treasury forest policy promoting mining itself.

At the turn of the 13th and 14th centuries, the mining towns, mostly inhabited by German immigrants, and founded to exploit precious metals, obtained considerable forest territories from the king. Dozens of serf villages belonged to all of the towns. Their population supplied manpower for mining. The serfs, doing the earthwork, charcoal-burning, wood cutting and transporting, worked as paid labourers.

The three big mining towns had approximately 60 000 ha of forests; however, 54% of the forests remained in the hands of the citizens, compared to 46% by the mid-16th century.

In a constant struggle with the towns and landlords, the Mining Chamber expropriated more and more forests, depriving the towns of the fundamental mining conditions.

The forest stand of *Bergrevier* consisted of pines, beeches, and oaks. Pines were the best for pitwood and timber, whereas the oaks were regarded as the most important construction timber, and the beeches served as raw material for charcoal-burning.

The Mining Chamber possessing the remains of the forests of former crown lands, proved to be a dangerous competitor for the mining towns. The Treasury, together with the Royal Minting House of Körmöcbánya, tried to have access to as many mines as possible and to seize the neighbouring forests by leasing, donation, or pledge. They collected not only the royal mining assessments (*urbura*), but arranged the compulsory metal delivery as well. Enormous central copper and precious metal furnaces were built, while the endeavour to get even more shares in mining became quite usual.

According to contemporary levies, a great many town properties were in the possession of the Treasury; in 1542 and 1559, the Treasury owned as many as 64.5 and 60 tax units (*porta*) in Besztercebánya and its neighbouring villages, compared to 59 and 52.5 of the town.

Between 1502 and 1546, the Thurzó-Fugger enterprise rented the Mining Chamber from the Hungarian king. During this period, for example, they obtained 20 000 ha of forest for their copper furnace. After the Fugger leasing, the Treasury of Lower Austria rented the Hungarian Mining Chamber (1548–1741). In fact, its activity promoted foreign interests, which led to the restriction of the freedom of the towns and hindered their economic progress.

The mining towns protected their forests with an accurate sylvan policy, for the interests of the wealthy *Ringbürger* and *Waldbürger* coincided with those of the less well-off town dwellers. The regulations of forest use, e.g., the supply of fuel, propwood, charcoal and timber, were centralized. The forests of the town council (*Bannwald*) were administered independently from the public forests (*Freywald*).

In fact, possessing quite large territories of forests, Körmöcbánya was only able to successfully resist the expansion of the mining chamber. After the citizens of Selmecbánya and Besztercebánya had been driven out of mining, they lost most of their forests and had to import charcoal and timber.

Enterprise and paid labour, as distinct from the general feudal tableau, were fairly widespread in the mining towns concerned with timber industry and commerce; moreover, society was highly stratified. In 1707 the number of citizens in Körmöcbánya was 853, whereas that of the chamber personnel totalled 598. Some 2502 persons, mainly dependent on the timber industry and mining, lived in the villages annexed to the town. According to the assessment, some 36% of the inhabitants belonged to the wealthy stratum, compared to the remainder, i.e., 64% who had no actual property. In Selmecbánya the number of citizens was 1899, including the servants, as opposed to that of the chamber personnel who made up a population of 309. In addition, 2567 poor and 2474 miners or metal workers (*metallici*) without any property were put on the list in the villages belonging to the town as well. The same year about 23% of the town dwellers were regarded as wealthy and 77% as pauper! The assessment of 1707 comprised 725 burgher families and 174 families of the chamber personnel. Forty-nine percent of the 819 families belonged to the citizens, whereas 51% was considered to be poor (*inquilini*).

By and large, an assessment of 1730 displayed a similar situation, totalling the number of workers of the neighbouring mines up to 5000. They traditionally came from the town *inquilini* and the peasantry of mining towns. However, a remarkable social tension had come up between the citizens and the poor, who were reliant on paid labour.

The gradual setback of the mining towns, which had originally been founded for mining, brought about very difficult economic conditions under which they had to fight for their former privileges between the 16th and 18th centuries. The forest regulations issued by the Treasury in 1565 sanctioned its own forest policy. It completely forbade contemporary manorial tree farming (*Waldfeldwirtschaft*), forest grazing, the maintenance of sawmills, and the marketing of timber and plants. The regulations did not even make mention of mining and forest-use by the citizens. Its main goal was to ensure the permanent supply of timber and charcoal for the chamber mines and furnaces. The royal mining statute of 1571 authorized the Treasury with exclusive mineral rights which might entail the expropriation of the forests of mining towns.

Following a series of protests, the *Waldamt* of the Treasury was set up which led to the end of town administration in the town councils' forests (1736).

The subordination of the Chamber of Hungarian Mining Towns to the Treasury of Lower-Austria resulted in the expropriation of town forests, abolishing former tree farming and absolutely submitting them to the interests of mining. This situation hindered the development of towns both legally and economically until the mid-18th century.

Foreign chamber administration deprived the mining towns of their various incomes, obstructed their timber supply, and encumbered their industrial progress. Apart from the resistance of Körmöcbánya, Selmecbánya and Besztercebánya were dependent on regular wood and charcoal imports whose prices were constantly growing. The citizens were ousted from mining, so commerce, in Besztercebánya, and farming and wine growing, in Selmecbánya, became the main pursuits of these towns.

On the other hand, the Treasury could not really follow its own forest policy in favour of mining, either. In spite of official prohibition it had to approve animal keeping and forest grazing by its personnel further more, opening forests for grazing. A higher inspection of 1635 found at the site of 17 charcoal burners, as many as 405 heads of cattle. Besides the manors of chamber officers, wood-cutters and charcoal burners cleared the forests, opening up new pastures and arable lands. In 1767 there were 24 settlements of wood-cutters in the territory of chamber forests.

The notion of mining actually was comprised not only of the skilled workers, i.e., pitmen, furnacemen, but of woodworkers too: wood cutters, carpenters, cartsmen, etc. The two groups were considered as the same in parliamentary acts.

Manorial and chamber sawmills were seldom profitable, but rural industry had always been effective even in remote regions. As a result of losing their previous influence in mining, the citizens of mining towns tried to invest their money in real estate and land. The manorial forests of the wealthy citizens had become of great importance by the 18th century.

7 The Use of 18th and 19th Century Forest Maps

The history of forestry and forest economy prior to the capitalism in Hungary cannot simply be described by laws and statutes. The analysis of forest maps, found in central archives, made it possible to outline the real plight of forests, the actual forms of use and forest cultivation, and, last but not least, the extent of sparing or environmental destruction.

Sopron, possessing 8 serf villages with 15 000 acres of forest, had been a chartered royal free town since 1277. The forests were administered by the mayor in association with the council. The town needed to import timber. Nevertheless, afforestation cannot be traced back to the 16th century in the documents. In accordance with Baron Geusau's Habsburg Imperial Forest Regulation of 1767, the Hungarian forest regulation was issued on December 22, 1769, which, however, was regarded as a summarizing collection of contemporary practical advice.

The territory of forest covered 144 km^2 in the salt-producing estate of the Treasury at Bocskó (Máramaros County) in 1781. There was no sign of afforestation, although storm damage had been recorded. Some parts of the forest were to be cut down to establish a common. In 1800 there was a reserved forest (*reservata silva*) of 1.4 km^2 within the actual extent of 5.7 km wood of Apátfalva in Trencsén County. In the 18th century, a reserved forest of 19 km^2 was registered in the large crown forest of Bács County (229 km^2). The village of Wallendorf (Szepes County) had a wood of 20 km^2 in its fields of 40 km^2. In 1787 12% of 3 km^2 of forest in the village of Salamon, Veszprém County, was already meadow, but in the forests of the adjacent village of Varsány (10.5 km^2) the meadows and pastures represented only 3%.

The crown oak woods of Vaszar, Veszprém County, had an extent of 7.2 km^2 in 1787. There remained only 51.5% of its previous extent, including its coppices and young growth. Half of its territory had been taken by other branches of farming. However, 99% of the Szentlélek Forest of Esztergom County was still intact in 1788. The woods of Debröd and Tapolcz (14.4 km^2), Abaúj County, could be described in 1790 as follows: 47% matured forest, 48% coppice and young growth, 5% desolate.

The Mindszent-Rudnok Foundation Forest was surveyed in 1790. The age and condition of trees were registered in every lot of 29 km^2 (Fig. 3).

However, the forest stand of the area to be cut down was either young or sparse. According to the map, about one-fifth of the woods of the given area was useless, whereas the amount of coppice and young growth was relatively low, i.e. 12.3%. Clearings and bushes were derived from neglected logging; the sketch contained the following legend: "felling that has become desolate", "bush" and "clearing" (*Aufflug*). A large part of the forest had been lost to logging and fuel

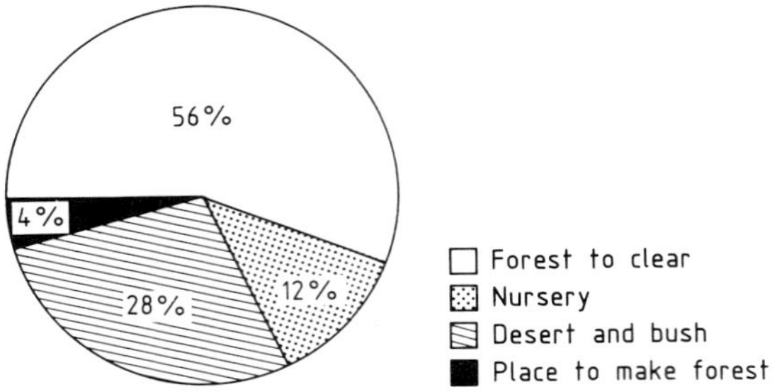

Fig. 3. The forest of Mindszent (29 km²) in Abauj county, 1790

gathering. Nevertheless, the extent of planned afforestation was highly restricted, too, i.e. 4.3%. The forests of the river flats in Krisalyán, Southern Hungary, were in Zala County. There, forests had been extremely dissected by the Drava, so the exploitable wooded areas had a great many strips of bushes; what is more, some signs of formerly cultivated lands (*vestigia agrorum*) can be seen even today.

Between 1795 and 1799 the amount of forests in Western, Northern, and Eastern Hungary can be exemplified by numerous data throwing some light on the rate of other branches of agrarian production. Thybe Forest, Szepes County, still had four-fifths standing, but 18% of it had been completely cleared. The forest of Kiszemlin, Bihar County, was reduced to 20% of its original extent, i.e., the rest had already become arable land and pasture. The amount of arable land, pasture and meadow was 39%, whereas the woods made out 61% in the forest of the market town of Tüskevár, Veszprém County. There remained 72% wood, including the coppice and young growth, compared to the 21% rate of other branches of agriculture in the forest of Sajólád, Borsod County. The woods covered 76% of a forest area around Köröm, Borsod County, in comparison with pastures and arable lands, which represented 24%.

Even the internal structure of a remnent forest can be observed in the previous woods in the case of the manorial forests of Patka, Györ County, in 1801 (Fig. 4).

This forest, in an extremely poor condition, covered only 72% of its former extent. Ten percent had been taken by arable lands, pasture and garden, whereas vineyards represented 19%. The internal structure of the still remaining woods was as follows: One-fifth of the forest had been devastated and some problems in the afforested areas seemed to arise too. At least, the following remark might imply it: "As this part of the wood has become rather sparse, it has to be continuously improved by its sparing use and sowing kernels".

In the case of the small forest of Karakó, Vas County, covering only 3 km², the following amounts could be observed: its forest stand with young growth, 82%; bush, 3%; arable fields in the wood (*agri inter sylvas situati*), 15%. One-fifth of its former extent as a forest had disappeared. However, the manorial forest of Kayata, Sáros County, showed an exaggerated rate of afforestation in 1806 (Fig. 5). The quantity of still exploitable forest stand of medium quality remained under 40%, while on the other hand, the rate of afforestation reached 50%. Bush and cleared areas made out approximately 10%.

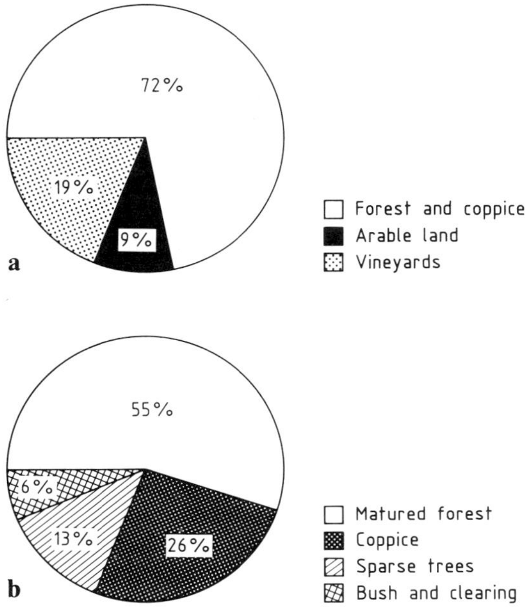

Fig. 4. a Forest stand at Patka (11 km²) Györ County, 1801. **b** Division of forest stand at Patka

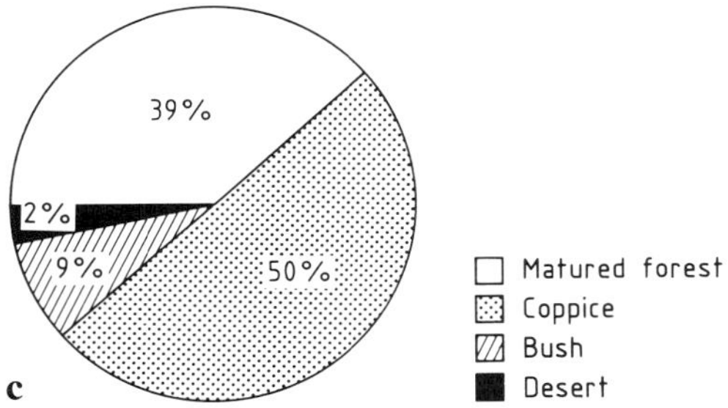

Fig. 5. Forest stand at Kayata (4 km²) Sáros County, 1806

The manorial forests of Csolnok, Esztergom County, were in quite a bad condition in 1806; apart from a cattle run of 5%, the so-called vacant spots (*vacuitas*) made out 23% of the woodlands. One can hypothetically draw the conclusion from this vacancy to be afforested (*vacuitas implatanda*) that a nursery was planted there; however, it has not yet been proved. According to the forest map of Matacs, Veszprém County, the amount of cleared woodlands was one-third of the total area in 1835.

Quite often the squandered areas became the obstacles to rational tree farming; for example, the forests of László Piller from Tizsite, Sáros County, were situated in several ports in an area of 2.5 km^2, 5 or 6 km apart.

It was very typical of contemporary forest stand that serfs received a woodland as cattle-run when common property was distributed in 1842.

Normally, it appears from numerous examples, forest statutes, the condition of forests and the problems of distribution were not necessarily in harmony with the regulations. Woodland areas were continuously decreasing in favour of other branches of farming. The actual forest stand was increasingly exploited for material reasons. It was cleared but left mostly desolate.

The number of glades can directly infer the extent and condition of forests on an exploitation map. One can certainly expect a complete change in forest economy if 72–80 clearings are planned. The crown forest of Kuttejovo, Pozsega County, with an area of 107 km^2, was dissected by 294 glades in 1804. Where the woods were already sparse or of low quality as few as 50–70 glades were allowed, with an approximate average of 60. A foundation forest of Bihar County was situated in the neighbourhood of four villages in 1795. It covered an area of more than 13 km^2. Not one clearing was planned in the first village, but five were intended in the second, thirteen in the third, and twenty-five in the fourth; their extent was naturally diverse in each village. The maps shows 20–30, or sometimes 40 glades, the forest stand was highly damaged or desolate. The main woods of the foundation forest of Legoglava, Varasd County, was dissected by 80 glades, whereas only 10 clearings were planned in advance in the small, scattered woods.

In the case of clear-fellings, the extent of lots was the same only in one forest. In spite of the various size of planned glades, the average size of lots to be cut down was suprisingly similar in the different parts of the country. The average extent of logging in the four great regions of the country is shown in the following:

Transdanubia	0.14 km^2
The Carpathian Mountains	0.19 km^2
The Border Lands	0.21 km^2
The Great Plain	0.19 km^2

Consequently, despite the above mentioned differences, the average extent of logging was quite similar from a regional point of view. The average is between 19 and 21 ha! This is only somewhat lower in Transdanubia, which might be due to widespread, smaller forest properties.

Considering the extent of analyzed forests, the following results can be obtained:

Regions	Number of maps	Forests in km^2
Transdanubia	39	252.2
The Carpathian Mountains	34	938
The Border Land	40	755
The Great Plain	15	559
	128	2504.2 km^2

As the average extent of a present day clearing totals up to 1 ha, whereas it was about 5 ha in the Great Plain and rarely covered 10 ha in the forests of river flats, we must not disregard the fact that the average extent of clearing in the 17th and

18th centuries surpassed that of present day clearing by four to twenty times, which implies a rather more intensive exploitation.

Unfortunately, there exist no comprehensive date to exemplify the process and extent of forest destruction from the 16th to 19th centuries.

Our examples and evidence mainly go back to forest exploitation by the Treasury in favour of mining. The director of the salt office in Sóvár reported in 1747 that serfs of the local Jesuits had "so much destroyed Schogen Forest that it cannot possibly be replanted even in a hundred years!" Between 1495 and 1526 20 000 m^3 of wood had to be annually delivered for a copper furnace rented from the Fuggers. At the same time, the forest rented from the estate of Likava Castle was cut down in 10 years; what is more, the forest stand of the so-called Királyerdö had been entirely exploited three times within 50 years. In 1792, the forests of a chamber furnace in Ó-Deszna, Arad County, was reported to have been absolutely devastated. Its 15 km^2 of exploitable area had been reduced by 12.5%. The above-mentioned forest policy of the Treasury prohibited forest grazing and scraping off tree bark, which was a common practice among the "vlah" shepherds. However, renting forests for grazing was extremely profitable for the mining towns. Some remarks appropriately characterizing the situation appear regularly in the records: "by and by the inhabitants have made the whole area desolate and they still possess it", or "the neighbourhood of a village has been fully devastated", and "one can still see withered trees in some glades."

There are some forest maps which were drawn by Treasury engineers; yet they do not refer to the forest policy of the Treasury supporting mining, but that of landlords. Deterioration of forests can also be observed here in various forms, as confirmed by the former examples.

Fifteen percent of the wood were already cattle-run in Taszar, Somogy County, in 1771. In 1817, 80% of the forests of six Transdanubian villages still covered an area of 31 km^2, whereas 20% of it had become bush, desolate, or cleared woodland. In 1799, however, 90% of Perbesháza woods, Gömör County, had been totally devastated. The forest map of Peklin, Sáros County, describes the situation in 1758, showing 62 glades in an area of 24 km^2. At least half of its forest stand was exploitable, while the other glades consisted of coppices and young growth. One glade was reserved for the salt industry and another one for a goldmine; however, most of the glades were labelled as follows: "without expecting serious annual income." In 1788, 10, 20 and 40% of three minor forests in Szepes County were regarded as pasture, meadow, or barren fields.

There remained quite a few maps of the counties of the Border Lands. Thirty-nine percent of the sylvan area (286 km^2) in five villages of Arad County had been wasted. Some parts were characterized as woods of debased quality, clearings, bushes, but it mostly had become cattle-runs. The maps of Nádas and Kresztu-Ménes, Arad County, contain the following remark: "owing to the border lines of the two villages forest belong to the serfs' cattle-run", i.e., 36.72 km^2.

We can have access to a lot of information about the sylvan conditions of the eastern counties of the Great Plain. Of the 44 km^2 of forests near Asszonyvására and Ránk, Bihar County, clearings and bushes made up to 56%. Forty-five percent of the oak wood of Barát, Szabolcs County, was declared good, but 55% of it proved to be of very bad quality in 1805. It was stated about the woods of Géres, Szatmár County, in 1805: "They used to be forest, but now they are pastures!" The largest forest of Bihar County could be found around Debrecen. It comprised the border fields of three villages; an area of 72.4 km^2, it contained only 45% exploitable woods in 1840, whereas the rate of clearings and sparse strips totaled 55%.

And now let us throw some light on forest economy in highly desolate woods at the turn of the 18th and 19th centuries.

In 1780 "the ideal sketch map of Körösmezö Forest Office of Máramaros County" contained a great number of records which referred to the general conditions of contemporary forest economy, for example, "bushy woods", "a mixed and desolate forest at high altitude", "poplar and spruce bushes." These three circumscribed ervirons made out more than half of the several square kilometres of woods. By the late 18th century, the forests in the Carpathian Mountains had been extremely exploited. A detailed survey was made about the woods around Modor, Pozsony County, at approximately the same time. According to this survey, 60% of the forest was very young growth. The chart of the table had an extra line with the following inscription: "present day damage of the woods." Typical tree diseases: "Sterile tops", "forest with dry tops!" The wood in question was mainly cut down and consisted of coppices, and had been suffering from diseases.

In 1798 a small Carpathian forest in Zemplén County was reported in the following condition: 26% exploitable; 34% young growth and coppice; 40% waste or pasture. The same year, however, half of the forest of the capital, Buda Kamaraerdö, was in the hands of 23 glade holders. The exploitation map of Marienthal Forest, Pozsony County in 1804, showed that 41% of it was ripe for clearing, 28% had been completely hewn down, whereas there was 31% of young growth.

By the rise of capitalism in the mid-19th century, most of the forest seemed extremely deteriorated. About 1850 the manorial forests of Földvár, Tolna County, were recorded with 21% of clearing, as compared to 72% of standing wood, although most of the trees were not older than 10 or 20 years. In the same county, the foundation forest of Németkér (7 km^2), having been fully exploited during the previous decades, contained only 51% of young growth on its original area in 1867, which was already obviously in need of afforestation and special care. Here the other branches of farming had taken 24% of the total area, whereas desolate clearings amounted to 25% of the total.

Besides, the notion of wood care was manifested by the extremely rare symbol of reserved forests, in so far as the reserved forests badly needed care and protection. In the 16th century woods of Késmárk, the so-called protective felling method was applied. "This method was highly advantageous, for the woods did not become sparse; setting a fully cut down glade between two intact strips." In addition, cattle grazing was prohibited: "the introduction of harmful grazing has been decidedly forbidden."

Peasants around Debrecen who were in charge of taking care of the woods were called *vakáncsos*. In the 17th and 18th centuries, they were given temporary permission to settle down and farm by the city of Debrecen, in return for the protection of coppices for a set time. The existence of a reserved forest near the castle of Hradek, Liptó County, was also recorded in the 18th century which was called "Hratska Hora, a reserved forest for particular cases." In 1789 a pine forest of 344 km^2 was recorded, which had originally been reserved for salt mining in the neighbourhood of Aknarahó, Máramaros County. A special forest "has been reserved for salt transportation by rafts." "Certain parts of the arables and pastures" were planned to be afforested in the Crown Forest of Baraczka Bács County. The intention of afforestation was obvious without referring to its actual method.

A rare example of forest care, as has been mentioned above, was the Royal Statute of Forest of 1565, which was mainly concerned with how to care for forests and how to establish reserved forests. However, the Royal Mining Chamber itself failed to make others observe the prohibitions. It could not forestall the damage of clearing and grazing, even in its own forests.

Wood shortages can be proved to have been a kind of contemporary energy crisis in 18th-century Hungary. Quite recently, Joachim Radkau (1986) has dealt with this theme, analyzing both the problem of shortages and consequences. Radkau, examining the sources, drew the conclusion that the so-called wood shortage derived from a general atmosphere of panic in the western and eastern parts of Europe. In fact, the rate and trend of devastation were exaggerated, just like the sporadic shortages. Actually, a liberal endeavour, which was aimed at breaking the state monopoly and ascending wood prices, was responsible for this atmosphere of panic. Radkau also stated that charcoal supply continued to be undisturbed even in times of wood shortages. There was "a general shortage of bacon and not wood" in the mining district of Styria. J. Hazzi (1986), a Bavarian Chief Councillor, stated in 1804 that "one has always been able to buy wood for money." The shortage of wood was only a trend, for the supply of wood for furnaces, mills, and other manufacturing activities "was roughly kept in balance as a result of wood reserves of preindustrialism." According to Radkau's ironical remark, the industrialists tried to obtain certain privileges by constantly drawing attention to the lack of wood," i.e., in order to have access to newly donated forests.

Contemporary shortages of wood could also happen in 17th- and 18th-century Hungary. The Queen's domain of Diósgyör, Borsod County, (20 000 acres) whose logs had been consumed by the local glassworks and iron furnaces, was highly in need of oak trees at the beginning of the 20th century; this was most certainly due to the neglect of afforestation. In Besztercebánya, around 1530, the copper furnaces ate up the surrounding forests. Most of the mining towns were dependent on wood importations, so the council of Selmecbánya, for example, had to close down the lime-burning furnaces of the town in 1630. The Forest Statute of Maria Theresia put a special emphasis on the extraordinary wood consumption of mining, charcoal burning melting, and potash making. Therefore, the resettling of furnaces was frequently ordered in areas of sufficient woods.

8 Afforestation and the Sparing Use of Forests

After examining the forest economy of the Treasury as well as of the domains and stating the extent of damage and devastation in the forests, let us have a closer look at the exact data of 18th- and 19th-century afforestations. The replacement of forest stand was mainly dependent on natural methods, so the new woods began sprouting on areas which were completely cut-down.

A forest of saplings grew again in this way in the foundation forest of Szetica Pratrowecz, Zágráb County. The methods of afforestation in Apatin, Bács County, can be characterized by the following remark: "to have a new forest of saplings from the seeds in a sparse wood." In 1789 a manorial forest of 12 km^2, in Fürje, Zala County, had 87% of its area still standing, although it was of very bad quality (*malae conditionis*), "therefore a preliminary confirmation is necessary, confirming every tenth felling in advance."

However, the neglect of woods and afforestation can be exemplified by the care of Asszony-vására, Bihar County, in 1799. As the surveying engineer stated: "it can hardly be called a forest." It is even more shocking when it was more specific: "ten or at least twenty trees can be seen in an acre *without* any kind of young growth or coppice." The areas, covered with bushes, can obviously be regarded as a sign of total a lack of care for the woods. In 1795 there were 60 glades in the manorial forest of Pecze-Szent-Márton, Bihar County. Fifty of them had been totally exploited, two of them were a coppice and a forest of saplings, and the remaining eight glades were already covered by bush. In 1790 the reasons for the devastation of Debröd Foundation Forest, Abaúj County, were enumerated as follows: charcoal burning (*carbonaria*), grazing (*pascua*) and tillage (*ager*). Seventy percent of the 1.5 km^2 forest was expoitable, 20% was coppice, 7% was desolate. The desolate area was designated for afforestation, without making mention of the intended methods. The care of Kosfariszko, Szepes County, was an extremely good example of an exploited forest: one-third of the forest of 1.5 km^2 was still exploitable, whereas two-thirds of it consisted of young growth.

From the late 18th century onwards, numerous cases of afforestation were recorded. As to the crown forest of Bujin, Bács County, for example, where the local serfs with the inhabitants of Tovarisov were given 7 km^2 of wood for farming, the map holds the following note: "desolate lands appointed for afforestation." However, "young-growth" was mentioned as a single form of afforestation in Puszta-Szent-Mihály, Veszprém County, even after 1870.

Consequently, the so-called natural afforestation must have been rather dominant in the country, but this method brought about losses which were largely due to indifference and inefficient care. The other method, artificial afforestation (seeding, nursery, etc.), was not really widespread. The following are some examples from the sporadic data: in 1806, the forest of Mogyorós, Esztergom County, was of extremely bad quality. What remained of it was virtually a coppice of 7–22-year-old trees. Five big clearings had been planned to be afforested (*vacuitas pro implantatione*).

In the case of the adjacent Karva forest, the process of afforestation was described as follows: "three glades are to be planted with trees where only grass has grown apart from some small trees." However, the glades appointed for felling could hardly survive, according to the forester.

Count Pál Festetich, Royal Commissioner, issued the forest regulations for the Chamber Office of Máramaros (salt mines) in 1778. He not only emphasized the regularity of felling and the prohibition of grazing, but underlined the clearing of glades and sowing pine kernels in the given areas too. According to their plans for exploitation, 9.5 km^2 from the Chamber Office's 221 km^2 of forest had been separated for the supply of serfs with wood and pasture. Some other parts were rented or appointed for reserved forests. The extent of glades, sown with pine-kernel, was only 1 km^2. The newly sown area had been established during the last 12 years, covering only 0.42% of the forest and making up 0.68% of the area appointed for exploitation.

Officially, the first signs of afforestation could be observed in Hungary in the 1760s. It appeared first in Bánát, in the south of Hungary, and then in the Great Plain and in the Carpathian Mountains at the turn of the century. The rate of planting saplings was rather low and at the same time it was quite precarious (damage caused by grazing animals). Between 1818 and 1842, 230 ha of land was annually afforested from about 87 500 ha in Bánát. This annual afforestation made up 0.26 percent.

In 1799, 90% of the foundation forest of Kemej, Borsod County, was normal wood or coppice, 3% under afforestation, 3% arable land, and 4% pasture. The latter was "appointed for afforestation". The map of Pusztaszentmihály, Veszprém County, showed a "nursery" too. At the beginning of the 19th century, afforestation was also referred to on forest maps. Around 1855, for example, the clearings of the manorial forests of Putnok, Gömör County, were afforested. Oak and pine kernels were regularly sown in Szerede Forest, Pozsony County. Count Károlyi had the deteriorated nursery of his domain of Aldebrö hoed out in 1855.

Early Hungarian capitalistic forestry and the condition of mid-19th century woods were characterized by Borsiczky, a contemporary expert, and Galgóczy, a statistician and economist. Károly Borsiczky, Forest Assessor of Bereg County, wrote about forest economy in 1852. According to his manuscript, sylviculture was concerned with taking care of woods, and the sparing use of forests in accordance with their normal exploitation. He promoted thinning out instead of clear-felling and definitely was against forest grazing and collecting litter from the woods. In addition, he preferred the natural ways of afforestation, protecting coppices. He wanted to apply artificial methods, i.e., planting nurseries, particularly against wind and quicksand. In 1855, 48 000 km^2 were covered by woods; 24.4% of the territory of the country at that time. According to Galgóczy (1855), "the woods of the Great Plain were cut down because of an unquenchable greed for profit from grain growing in lieu of forest economy. Another reason for the deterioration of forests was the peasants' common of estovers and grazing which constantly devastated coppices. Charcoal burning and lime burning were also to blame for reducing forest stand. Galgóczy gave a list of model farms at the end of his examination of Hungary's economic and geographic conditions, but it contained only three or four concerned with forest economy. In 1790 the feudal parliament passed an act on the protection of forests and left its enforcement to the counties. A few years later in 1807, the afforestation of drift-blown sand was ordered in Hungary. Clear-felling and forest grazing were prohibited at the same time. "Furthermore, it is rather advisable to avert the harm of building houses of wood, thus houses are to be built of some solid material in the future." (Markus 1901a). Later Hungarian forest Laws (1848) were mainly concerned with property problems and incorporated the Austrian Forest Code (1852). After Hungary regained its independence (1867), it was legally stipulated under the capitalist government in 1879 that 87.3% of a territory of 74 000 km² had to be continuously retained for forests. The new administrative organization was circumscribed, as well as an obligatory inspection of privately owned forests which could even entail the nationalization of misused forests. The parts of clear felling had to be afforested within six years.

He concluded with the recognition of a new capitalistic wood market which had come about on account of the abolishment of feudal bonds; "nevertheless, owing to its high prices it was able to produce the same amount of profit in tree farming" without respect to the quality of forest stand. During the capitalistic period (1848–1948), the extent of woods was 74 000 km², making up 26% of the whole territory in 1867. The condition of forests, as described by Galgóczy, had only been affected negatively; the average output was 3.1 m³/ha.

In 1920, Hungary lost two-thirds of her population and 71% of her territory, due to the Peace Treaty of Trianon. There remained only 10 000 km² of forest from the original 74 000 km². Formerly the population density was 2.8 persons/ha, after Trianon 7 persons/ha, and 7.9 persons/ha in 1980!

Between the two world wars only 6% of the forests belonged to the state. Forest economy was mainly based on hunting and logging. The gathering of sylvan litter and forest grazing, though officially prohibited, still remained dominant even if 1 ha of good cattle-run was equivalent to 6–8 ha of sylvan pasture. The signs of a more advanced forest policy could naturally be observed in state forests; between 1931 and 1934, 120 000 ha of forest was afforested, 43% with natural methods and 57% with artificial ones.

As with German, Swiss, and American examples, the new Hungarian forest laws condemned pollution and tried to protect the forests with punitive sanctions too.

After the nationalization of private property, 74% of the forests became state property, 19% was owned by cooperatives, and only 7% remained private property. Coppices had mostly been afforested until 1962. The system of nurseries particularly spread in the past decades, so 40% of the forest stand was well under 20 years in the 1970s. Clear felling has recently been replaced by the so-called afforesting felling. Its introduction is already a sign of the sparing use of forests.

According to the figures of 1964 (Danszky and Rott 1964; Danszky 1973), 15% of the Hungarian woods had become desolate, bush, or pasture, i.e. deteriorated forest, which was due to the ruthless exploitation throughout the centuries. This phenomenon, however, was quite common in Europe. The deteriorated forests covered with bushes, made up 23% of the forests in France, 24% in Bulgaria, 30% in Yugoslavia, 50% in Greece, and 80% in Spain. Apart from bad quality, which was a heritage of the past, new serious biological diseases could be observed. They were dated from the last period of industrialism. In the 1980s, oak plague has also become widespread in Hungary.

The harmful effects of air pollution can be measured in the whole of central Europe. "As a result of the ever-increasing industrial production, the forests of industrial regions are in extreme peril" said a scholar who worked in East Germany (Lampadius 1974). Relying on local examples, he stated that "according to a certain summary, forests polluted with smoke can produce 27.6% less spruce and 23.8% less pine.

Finally, Hungary lost 75% of her forests in the past six centuries. The actual use of forests meant a ruthless exploitation both under feudalism and capitalism. The method of afforestation had become stale and the woods were neglected. It was only after the nationalization in 1945 that forest economy became of better quality, which was mainly based on afforesting-felling and artificial afforestation. Nevertheless, forest economy continued to be subordinated to the interests of industrialism; besides, industrial damage became widespread by the second half of the 20th century.

References

Barthelmess A (1972) Wald, Umwelt des Menschen. Dokumente zu einer Problemgeschichte von Naturschutz, Landschaftspflege und Humanökologie. München, p 333

Lampadius F (1974) Beitrag zum Nachweis der Wertminderung des Waldes als Folge vom Immissionseinwirkung. Abh Säch Akad Wiss Leipzig, Math Naturwiss R 42.3:1–28

Lesenyei F (1936) A magyar erdögazdaság története és mai helyzete. Budapest, p 25

Márkus D (ed) (1901a) Corpus Juris Hungarici. Az 1740–1835. évi törvények. Budapest, p 510

Orisiczky K (1852) Az erdömüvelés elemei. Manuscript, Munkács paginae 150

Further Reading

Balogh I (1936) A vakáncsosok. Debrecen, p 16

Buis J (1983–1985) Historia Forestris. AAG Bijdragen 26/27:1058

Csapodi I (1975) Sopron város erdöbirtokának kialakulása és a középkori erdögazdálkodás nyomai. In: Kolossváry Sz (ed) Az erdögazdálkodás története Magyarországon. Akad K Budapest, pp 107–123

Csöre P (1980) A magyar erdögazdálkodás története. Középkor. Budapest, p 310

Danszky I (ed) (1973) Erdönevelés. Budapest, p 923

Danszky I, Rott F (eds) (1964) Általános irányelvek. Budapest, p 346

Galgóczy K (1855) Magyarország, a Szerb vajdaság és a Temesi bánság mazögazdasági statisztikája. Pest, p 430

Gleitmann RJ (1980) Rohstoffmangel und Lösungsstrategien. Das Problem vorindustrieller Holzknappheit. In: Duve F (ed) Technologie und Politik, pp 104–154

Gleitmann RJ (1984) Wege aus der Energiekrise: Holznot und Holzsparkünste im 18. Jahrhundert. In: Energie in der Geschichte. Düsseldorf, pp 401–407

Kolossváry Sz (ed) (1975a) A magyar erdögazdaság történeti fejlödése. In: Az erdögazdálkodás története Magyarországon. Akad K Budapest, pp 15–79

Kolossváry Sz (ed) (1975b) Az erdögazdálkodás története Magyarországon. Akad K Budapest, p 526

Magyar E (1983) A feudalizmus kori erdögazdálkodás az alsómagyarországi bányavárosokban. Budapest, p 222

Márkus D (ed) (1901b) Corpus Juris Hungarici. Az 1879–1890. évi törvenyek. Budapest, p 389

Paulinyi O (1971) Eigentum und Gesellschaft in den niederungarischen Bergstädten. In: Bog I (ed) Der Außenhandel Ostmitteleuropas, 1450–1650. Köln, pp 525–568

Péch A (1884, 1887, 1967) Alsó-Magyarország bányamüvelésének története. Tom I– III. pp 502–846–645

Public Archive of Budapest (ed) OL S 11 – Signatures of old maps and plans of forest from The Royal Mining Chamber in The Public Archive of Budapest (Hungary)

Public Archive of Budapest (ed) OL S 12 – Signatures of old maps and plans of forest from The Consilium Regni Locumtenetiale in The Public Archive of Budapest (Hungary)

Public Archive of Budapest (ed) OL S 13 – Signatures of old maps and plans of forest from The Great Foundations in the Public Archive of Budapest (Hungary)

Public Archive of Budapest (ed) OL G 28, V 2 d – Rákóczi Archive in The Public Archive of Budapest (Hungary)

Public Archive of Budapest (ed) OL E 158 – Conscriptiones portarum, Irregestrata IV, anno 1707 – in The Public Archive of Budapest (Hungary)

Radkau J (1986) Zur angeblichen Energiekrise des 18. Jahrhunderts: revisionistische Betrachtungen über die "Holznot". Vierteljahresschr Sozial Wirtschaftsgesch 73:1–37

Reisch I (1974) Waldschutz und Umwelt. Springer, Berlin Heidelberg New York, p 568

Schmidt FA (1834) Chronologisch-systhematische Sammlung der Berggesetze der österreichischen Monarchie. II. Ungarn

Szabó I (1975) A magyar mezögazdaság története a XIV. századtól az 1530-as évek végéig. Budapest, p 100

Szilas G, Kolossváry Sz (1975) A diósgyöri kincstári uradalom erdögazdálkodása. In: Kolossváry Sz (ed) Az erdögazdálkodás története Magyarországon. Akad K Budapest, pp 140–174

Szücs J (1955) Városok és kézmüvesség a XV. századi Magyarországon. Budapest, p 339

Tagányi K (1896) Magyar erdészeti oklevéltár. Tom I–III. Budapest, p 735, 900, 789

Troßbach W (1987) Bäuerlicher Widerstand in deutschen Kleinterritorien zwischen Bauernkrieg und Französischer Revolution. Einige Bemerkungen zu Formen und Gegenständen. Z Agrargesch Agrarsoziol 35:1–16

Varga F (1980) A tölgypusztulás Magyarországon. Erdészeti es Faipari Tudományos Közlemények 2:11–17

Wilstacke L (1986) Die voraussichtliche Resonanz von Flächenstillegungsangeboten und Aufforstungserleichterungen bei praktischen Landwirten. Ber Landwirtsch 64:528–557

Zakar J (1975) A máramarosi kamarai sóbánya-uradalom 1778. evi rendtartása. In: Kolossváry Sz (ed) Az erdögazdálkodás története Magyarországon. Akad K Budapest, pp 124–139

The Ecological Background of the Livelihood of Peasants in Kuusamo (NE Finland) During the Period 1670–1970

Y. VASARI[1]

1 Introduction

Kuusamo is a vast commune – the total area of present-day Kuusamo is 5805 km^2. It stretches across the great watershed of Finland which divides waters running west to the Baltic Sea, from those running eastwards to the White Sea. Close to the Arctic Circle and bordering on the Soviet Union, Kuusamo has until the present time been known as a wilderness area. Indeed, today forests still cover 55% of the total area; the percentage of mires (= peatlands) is 29% and that of lakes 14%. Only 2% of the total area is covered by fields.

Kuusamo is a part of the great "Sea of forests" covering the eastern and northern parts of Finland. It belongs to the northernmost subzone of the boreal coniferous forests (see Fig. 9 in Ahti et al. 1968).

The history of Kuusamo is not very lengthy. The permanent agricultural settlement of Finns dates back only to the 1670's; until then the region had been inhabited by Lapps, who lived by hunting, fishing, and keeping small herds of reindeer. Such a way of life was in complete harmony with the local conditions and left hardly a trace on nature. During the summer season, the Lapps lived scattered over the area as individual families, while in the winter they gathered together in winter villages, of which there were but two in the whole vast region. The total number of the Lapps probably did not exceed 200 persons.

The Finnish immigration to the region was a consequence of measures the King of Sweden had taken in 1673 when he gave the so-called Lapland decree. This was meant to encourage agricultural settlement of the vast wilderness of Lapland hitherto inhabited by only a few Lapps. The Finnish pioneers penetrating the Kuusamo region were mainly of East Finnish origin. This is important, because people in eastern Finland were prone to adopt the slash-and-burn or *swidden* cultivation, a very extensive use of forests (e.g., Sarmela 1987).

The Norway spruce (*Picea abies*) is the most important tree in the forests of Kuusamo; spruce-dominated forests make up more than half of the total area of forests (Figs. 1, 2). Spruce forests were also best suited for the slash-and-burn cultivation. They had, however, also been very important as hunting grounds, and within 30 years of the beginning of wide-scale slash-and-burn cultivation hunting, could no longer be considered the backbone of Lapp economy. "By 1720

[1] Department of Botany, University of Helsinki, Unioninkatu 44, 00170 Helsinki, Finland

Brimblecombe, Pfister (Eds.)
The Silent Countdown

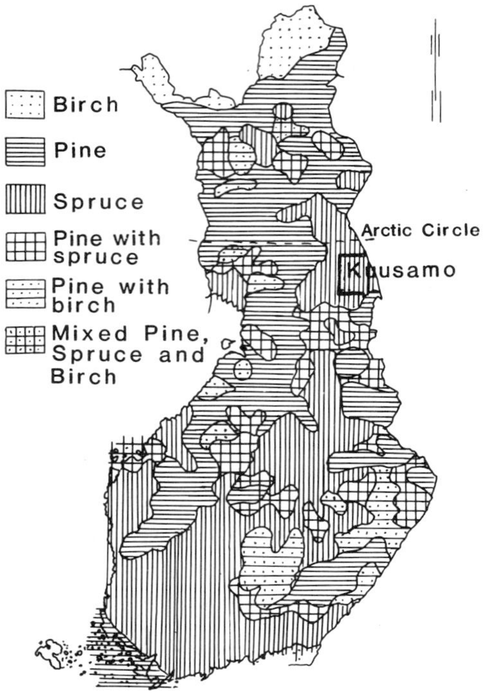

Fig. 1. Composition of forests in Finland and location of the Kuusamo region (Fig. 1 in Hicks 1985)

Fig. 2. Spruce forest of the thick-moss type in Kuusamo. Photo V Vasama

the Lapps had disappeared completely from South Kuusamo and by 1767 neither of the two Lapp winter villages remained" (Hicks 1985, p. 54; cf. also Ervasti 1978 and Kortesalmi 1975).

2 Forest Resources

2.1 Slash-and-Burn Cultivation

A pre-requisite for the slash-and-burn cultivation was an abundance of suitable primeval mesic forests. The procedure known in Finnish as *huuhtaviljely* meant that after a parcel of land to be cultivated had been cut in the first year it was burnt-over in the second year and burnt again and sown with rye in the third year. This meant that most of the forests in Kuusamo were burnt down at least once during the period in which this type of cultivation was practised (ca. 1680–1870), largely due to forest-fires originating from the "burning-over" (Ahti and Hämet-Ahti 1971 p. 7; cf. also Kujala 1921; Hicks 1976, p. 374). Land which had once been burned and cultivated could be used again after 40–50 years, and even then the yield was not as good as it had been the first time.

Pollen-analytical studies on the forest soils of the spruce forests in Kuusamo have shown that (1) bits of coal can be found almost regularly in the forest soil, often connected with occurence of cultural indicators and (2) that the primeval forests of the region when the first Finnish settlers came there were of a more mesic, luxuriant type than the present, generally quite unproductive, so-called thick-moss forests. This impoverishment must have partly resulted from the exploitation of the nutrient reserve of the soil, partly, however, from a climatic back-ground. During the entire era of slash-and-burn cultivation, a markedly cold and humid, Little Ice Age climate prevailed. This resulted in a rise of the groundwater level and thus a better availability of nutrients. The end of the Little Ice Age at the close of last century and the subsequent warmer and drier climate led to a general lowering of the groundwater and drier soils, which could not provide the vegetation with nutrients as well as before (Vasari 1965).

The poor spruce forests of today represent a greatly impoverished variant of the earlier primeval forests. The combined effect of human exploitation and climatic change immediately following it had led to this marked alteration.

During its heyday in 18th century, the slash-and-burn cultivation led to a surplus of rye in Kuusamo and its export, partly as grain and partly as spirits. During the 19th century, the slash-and-burn cultivation gradually dwindled away; the main reason for this was probably a lack of unexploited forests, but the state also began to oppose this practice which was regarded disastrous.

2.2 Use as Reindeer Fodder and Bark Bread

Before the beginning of this century, the forests had no commercial value and they were accordingly treated in many ways we may now find difficult to accept.

Reindeer farming, for example, often suffered from a lack of winter pastures. Reindeer, as we know, feed on lichens during the winter and when the ground lichens were no longer accessible lichens hanging in trees could be consumed. The reindeer herdsmen, knowing this, used to fell trees, mainly old, big spruces to feed their animals. The trees were cut in late spring at the snow level which meant that high stumps were left standing. From a modern point of view this was lavishness in the extreme. However it kept the reindeer alive. It is estimated that to feed one reindeer, two big trees had to be felled per day (Blomqvist 1959)!

One more, somewhat exotic, way to live directly from the resources of the forest was to make bread of the phloem of pine. This was a common practice everywhere in Finland and it survived in remote areas until the beginning of this century. For this purpose big pine trees, 60–100 years old, were felled. It was estimated that in order to feed one person, 100 pines per year were needed and for a whole family about 500 trees (Blomqvist 1959).

Thus, during more than two centuries, the pressure upon the forests in Kuusamo was very great – the spruce forests were used for cultivation and reindeer fodder, the pine forests, besides for timber and fuel, for human consumption, too. Therefore, the forests which one even to-day would be inclined to regard as primeval ones, in fact usually represent a secondary stage after quite heavy previous exploitation.

3 Cattle Breeding Based on Natural and Seminatural Meadows

For some time during the 18th century, the main source of livelihood in Kuusamo was the cultivation of rye and barley, supplemented by fishing, hunting, and reindeer and cattle herding. With the decline of slash-and-burn cultivation, cattle breeding gradually became more and more important. From 1870's onwards, Kuusamo surpassed all its neighbouring communes in dairy production.

Abandoned rye fields were used to some extent as pastures for cattle and sheep. At the same time, hay that was needed for winter fodder was collected from natural meadows, i.e., open mires which very often were not particularly productive. Alluvial meadows along some big streams were also used for this same purpose.

In 1770, Elias Lagus, the local priest, described the Kuusamo Parish (Ervasti 1975). According to him, the local meadows gave only poor yields; hay from them could be made only once in two years. The meadows were small and often far away from the farmsteads (Fig. 3). Leaves from trees were used as additional fodder. The meadows were not managed in any way except when they were occasionally flooded by means of dams. This was done in order to kill the mosses and to fertilize the soil. This is the first account we have from Kuusamo of artificial improvement of the natural meadows. According to Grotenfelt (1899), flooding was seldom practised in Kuusamo in the 18th century.

In 1784, the local chief of police, Henrik Planting, claimed that after several unsuccessful attempts at improving natural meadows by other means, he had begun flooding them with dams (Kortesalmi 1975, pp. 299–300). This might well

Fig. 3. The meadows were small and often far away from the farmsteads. Photo Tuomo J Niemelä 1977

have been an unfounded boasting; the truth is, however, that in the last years of the 18th century, the number of new farmsteads in the region had considerably increased (from 85 in 1754 to 159 in 1787), and flooding of meadows by means of dams had become a common practice.

A somewhat more recent invention than the use of dams was flooding meadows by means of catchworks (Kortesalmi 1975). The use of these dates only from the 19th century. In the early 19th century we find the first records of lakes whose levels had been lowered in order to get hay meadows (Sjögren 1828). Old-fashioned use of natural meadows was actively practiced until the 1950's and 1960's. Only then was this replaced by modern cultivation of hayfields (Kortesalmi 1975; Vasari and Väänänen 1986). In the early 20th century, however, yield of the natural meadows still furnished more than half of the local fodder (Pöyhönen 1910).

3.1 Management of Wet Meadows

In his papers dealing with irrigated and naturally flooded hay meadows in North Sweden, Elveland (1978, 1979) presents a clear classification and a good des-

Fig. 4. Alluvial meadows on the shore of River Oulanka in Kuusamo. Photo Tuomo J Niemelä 1977

cription of their management. According to Elveland, two natural and two seminatural types of meadows were used for haymaking. Among natural types, the first group included several more or less productive open mire types, and the second group, various strand meadows, both alluvial meadows along rivers (Fig. 4) and meadows on shores of lakes and the sea.

Seminatural flooded meadows include two groups, i.e., "flowing water-meadows" (*dammängar*) and "catchwork water-meadows" (*silängar*). The various types of natural and seminatural wet meadows are illustrated in Fig. 5.

The aim of treating seminatural meadows with water and ice was to level the surface of the meadows, to kill trees, bushes and mosses, and to favour the growth of plants suited for fodder, like grasses, sedges, and some herbs. Elveland gives an excellent, concise description of the treatment of both types of irrigated meadows, of flowing water meadows and catchwork water meadows (Elveland 1979, pp. 8–9). Customs in Kuusamo were rather similar, with only minor exceptions.

In Kuusamo the hay was usually made every second year, in northern Sweden generally every year. The following was the practice in Kuusamo: in the year in which the meadow had been lying fallow, the water was raised in late August or early September, after the growth of the grass had ceased. This was done by carefully closing the dam. The water had to be raised to a height to completely cover the meadow vegetation up to the grass tops and yet never freeze down to the bottom during the winter. In the autumn, a well-kept meadow thus

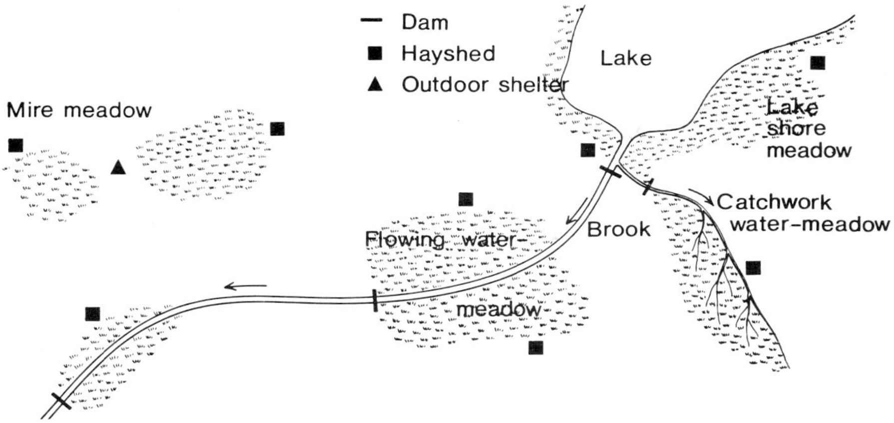

Fig. 5. Various types of natural and seminatural wet meadows in Kuusamo. Redrawn by Marja Koistinen after an original figure of Elveland (1979)

resembled a small lake. The continuous flow and the snow-cover prevented the water from freezing solid during the winter.

During the spring thaw, the dam was opened and the water allowed to drain away until the growth of the plants began. The dam was then closed again, this time not quite so tightly as in the previous autumn, and the water was allowed to rise to keep pace with the growth of the plants. It was regarded as important that the water level was continuously high enough to allow only the grass tops to reach above the surface. The meadow remained flooded until almost the beginning of the haymaking season in early or middle July. Then the water was either drained away one to two weeks before the haymaking began or immediately before. After the hay-making, the meadow was usually left on its own until the autumn of the next year.

If the meadow surface needed levelling, either when new meadow land was cleared or when the meadow needed "a make-up", the flooding could be made to last two to three years on end, killing all the existing vegetation. The ice could be used to level the surface if the unfrozen water was drained away from under the ice during the early spring. The ice then sunk down to the surface, pressing it level; even tree stumps were said to become pressed down into the peat (Grotenfelt 1899).

By means of flooding, the vegetational succession could be directed as desired. The water also brought to the flooded areas suspended fine material, including mineral particles which acted as fertilizers.

Catchwork systems were usually made on sloping ground further away from rivers and brooks. The previous vegetation could be killed by the means already described. When, through the action of water and ice and then the clearing of the trees and bushes, the surface had been levelled, the management of the new, fairly hard meadow could begin. Irrigation was begun in the spring; the meadow was covered with shallow water and irrigation was continued almost until hay-making began. Usually catchwork water-meadows were cut every second year; when fallow they did not need irrigation.

The use of meadows formed through lowering the lakes was rather simple compared to both flowing water-meadows and catchwork water-meadows. At first they gave quite high yields – later on, however, growth had to be encouraged, and often this was done by flooding.

3.2 Significance of Natural and Seminatural Meadows

The period of active management and use of natural and seminatural meadows in Kuusamo lasted from the late 18th century until about the 1950's, a little less than 200 years. During this time, such meadows and the hay collected from them formed the basis for stock and horse raising in the area. Pöyhönen (1910, p. 550) cites the farmstead of Mäkelä as an example of the importance of the hay yield thus obtained: though earlier the farm could support only 7 cows and 1 horse, the active management of water-meadows allowed the people a little later to keep 30 cows and 5 horses!

Exact information about hay yields from water meadows is rather scanty, but Pöyhönen (1910, pp. 552–553) gives some figures, noting that water-meadows of good quality could produce about 550–726 kg hay ha^{-1} a^{-1}.

The bulk of the winter fodder fed to cattle consisted of various sedge species (like *Carex acuta*, *C. aquatilis*, *C. lasiocarpa*, *C. rostrata*), water horse-tail (*Equisetum fluviatile*) and *Calamagrostis* spp. (cf. Pöyhönen 1910, p. 553; Elveland 1979, p. 8). Horses were given the little hay that could be made from "hard meadows" on mineral substrate (Pöyhönen 1910, p. 553). Pöyhönen points out that despite this diet, almost exclusively of sedges and the like, both total milk production and its fat percentage were almost equal to the values in southern Finland. Natural and seminatural meadows thus made the raising of cattle and horses possible in the Kuusamo region before the period of modern agriculture. However, an enormous amount of labour was needed to collect hay from small patches of meadows lying generally far away from the farmsteads. Distances of several kilometres, even several tens of kilometres, were usual, and the haymaking season usually lasted 5–6 weeks, often until the first snow fell (Kortesalmi 1975).

Despite the difficulties involved in collecting hay from natural and seminatural meadows and the great labour needed, the cattle breeding experienced quite a boom. Between 1767 and 1880's, in slightly more than 100 years, the number of cattle rose about four- to sixfold. The same was true for horses and sheep (Ervasti 1978; Hicks 1985).

As a result of the general practice of haymaking from natural and seminatural meadows, the role of open mires became very conspicuous in the landscape. They formed, in fact, a characteristic feature of the natural surroundings in Kuusamo. The cessation of this practice in 1950's led to profound changes in the landscape. The formerly open mires are now in the process of becoming grown over by bushes and trees, and without preventive measures will soon disappear or at least diminish greatly in number and area.

The widespread technique of flooding meadows can also be read in the stratigraphy of the mires, although the signs are not very clear (Hicks 1975, 1985; Vasari and Väänänen 1986).

4 Recent Time

The slash-and-burn cultivation came to an end at the turn of this century, while the use of natural and seminatural meadows lasted about 50 years longer, until the 1950's. The 1950's saw a fundamental change in the peasant life in Kuusamo. Due to special historical reasons, the general reparceling of land there was not finished until well after the Second World War; only then were individual farmers allotted their own farms. A special feature of this area is, that not all of the forest land was divided between the inhabitants, a very significant part of it being retained as commons. The great change in the proprietary relationship did, however, mean a great leap forward. The forest, i.e. timber, then gained commercial value and gave the inhabitants a chance to modernize their lifestyle. A short-lived boom was unfortunately followed by a depression which resulted in large-scale emigration. The peak in the population was reached in 1968 when it amounted to about 21 000 (Ervasti 1978); now it has fallen to about 18 000.

The present basis of livelihood in modern Kuusamo is forestry and agriculture. As a result, the landscape has experienced great changes. Tourism in this region, called "the Switzerland of Finland", has become very important, and Kuusamo is now one of the main tourist regions of our country. The local people are well aware of the economical importance of the tourist industry and this has led to a general appreciation of the resources of the natural beauty of Kuusamo.

There are several national and nature parks in the region and more and more tourists and lovers of the nature go there every year. That, too, is a way of exploiting natural resources.

References

Ahti T, Hämet-Ahti L (1971) Hemerophilous flora of the Kuusamo district, northeast Finland, and the adjacent part of Karelia, and its origin. Ann Bot Fenn 8:1–91

Ahti T, Hämet-Ahti L, Jalas J (1968) Vegetation zones and their sections in northwestern Europe. Ann Bot Fenn 5:169–211

Blomqvist AG (1959) Reseberättelser från åren 1867–1869. Silv Fenn 100:1–132

Elveland J (1978) Skötsel av norrländska rikkärr. Studier av vegetationsförändringar vid olika skötselåtgärder och annan påverkan. Management of rich fens in Northern Sweden. Studies of various factors influencing the vegetational dynamics. Rapp Stat Naturvårdsv 1978 PM 1007:1–96 (in Swedish with English summary)

Elveland J (1979) Dammängar, silängar och raningar – norrländska naturvårdsobjekt. Irrigated and naturally flooded hay meadows in North Sweden – a nature conservation problem. Rapp Stat Naturvårdsv PM 1174:1–124 (in Swedish with English summary)

Ervasti S (ed) (1975) Lagus E (1770) Försök til Beskrifning öfwer Kusamo-Sokn uti Kemi = Lappmark. Kuvausyritys Kemin Lapin Kuusamon pitäjästä. Acta Univ Ouluensis, Ser B5, Hist 3:1–74

Ervasti S (1978) Kuusamon historia I. Koillissanomat Oy, Kuusamo, pp 1–861

Grotenfelt G (1899) Det primitiva jordbrukets metoder i Finland under den historiska tiden. En redogörelse. Akad Afhandling, Helsingfors

Hicks S (1975) Variations in pollen frequency in a bog at Kangerjoki, N.E. Finland during the Flandrian. Commun Biol 80:1–28

Hicks S (1976) Pollen analysis and archaeology in Kuusamo, north-east Finland, an area of marginal human interference. Trans Inst Brit Geogr New Ser 1:362–384

Hicks S (1985) Problems and possibilities in correlating historical/archaeological and pollen-analytical evidence in a northern boreal environment: an example from Kuusamo, Finland. Fenn Archaeol II:51–84

Kortesalmi JJ (1975) Kuusamon talonpoikaiselämä 1670–1970. Helsingin Liikekirjapaino Oy, Helsinki (Kuusamon historia II)

Kujala V (1921) Havaintoja Kuusamon ja sen eteläpuolisten kuusimetsäalueiden metsä- ja suotyypeistä. Beobachtungen über die Wald- und Moortypen von Kuusamo und der südlich von dort gelegenen Fichtenwaldgebiete. Acta For Fenn 18:1–68

Pöyhönen M (1910) Niityn kastelu Kuusamossa. In: Grotenfelt G, Enckell K, Suuronen P, Nylander H, Cajander E (eds) Maahenki, Maataloudellinen tietokirja II. Otava, Helsinki, pp 533–556

Sarmela M (1987) Swidden cultivation in Finland as a cultural system. In: Raumolin J (ed) Special issue on swidden cultivation. J Finn Anthropol Soc 4/1987:241–262

Sjögren A (1828) Antekningarne om församlingarne i Kemi Lappmark. Helsingfors

Vasari Y (1965) Studies on the vegetational history of the Kuusamo district (North East Finland) during the late-Quaternary period. IV. The age and origin of some present-day vegetation types. Ann Bot Fenn 2:248–273

Vasari Y, Väänänen K (1986) Stratigraphical indications of the former use of wetlands. In: Behre KE (ed) Anthropogenic indicators in pollen diagrams. Balkema, Rotterdam Boston, pp 65–71

Urban and Industrial Impacts

Environmental History – The Environmental Evidence

F. OLDFIELD[1] and R.L. CLARK[2]

1 Introduction

Historians and palaeoecologists alike study the past, but usually rely on quite different kinds of evidence. Most palaeoecologists who study past environmental change are concerned with long time scales stretching beyond the reach of history proper. They also overwhelmingly use the present as the key to the past, in the classic Huttonian sense (Hutton 1788), rather than the record of the past to inform our understanding of the present and future. Historians may thus be forgiven for seeing few links between their approach to environmental history and that of palaeoecologists. Yet there is a significant and growing minority of the latter who have focused their expertise on the recent, "historical" past and tackled problems with direct bearing on present day issues (Oldfield 1983, 1987; Wasson and Clark 1985; Clark 1989). Our contention is that these studies have much in common with document-based historical research. There is a continuum of concern across the wide range of studies embracing "scientific" and "documentary" approaches to environmental history, and there are parallels between the two methodologies which often complement and reinforce each other in important and interesting ways.

Before amplifying these statements, we would like to illustrate the value of recent, problem-oriented palaeoecology by considering an example.

2 The History of Lake Acidification

Several causes have been proposed for the acidification of surface waters in Britain and Scandinavia. Table 1 summarizes these and gives a simple outline of how, using the record of acidification in the sediments, it is possible to test each proposal through historical-empirical study. Such an approach comprises, in essence, a series of experiments already conducted by the environment and readily accessible to our retrospective analysis and appraisal (Deevey 1969).

The first hypothesis, that acidification at the present day is the expression of a recently culminating long-term process of soil leaching and podsolization, is

[1] University of Liverpool, Department of Geography, Roxby Building, P.O. Box 147, Liverpool L69 3BX, Great Britain
[2] CSIRO Division of Water Resources, Canberra, Australia

Brimblecombe, Pfister (Eds.)
The Silent Countdown

Table 1. Proposed causes of current lake acidification and ways of testing each hypothesis

Cause	Test
1. Natural long-term changes	Establish timing of change: rapid and recent or ancient and gradual
2. Decline in farming	Compare lakes with or without recent land use change
3. Conifer afforestation	Compare lakes with or without recent conifer afforestation
4. Fossil fuel combustion emissions	(a) Compare documentary evidence for industrial emissions with sedimentary evidence of acidification (b) Compare sedimentary records of acidification and industrial deposition

attractive to the traditional palaeoecologist, for this trend has been well demonstrated through studies of vegetation and associated soil changes during the last two interglacials (e.g., Iversen 1964), both of which lasted for over 10 000 years, and the latest of which ended some 100 000 years ago. Jones (1987) has studied the record from the whole postglacial period (the last 10 000 years) preserved in the sediments of currently acidified lakes, where the catchment soils have become strongly acidic through peat formation or podsolization in preindustrial times. Her results confirm that there was no response in the lake system to the early, natural acidification of catchment soils. The fossil diatom record, which is our key to past acidity changes (Davis 1987), shows that the lake waters remained mildly acid throughout the period of podsolization and acid peat formation, becoming strongly acidic only during the late 19th and 20th centuries.

The land-use hypothesis owes most to the Norwegian forester, Rosenquist (1977), and on a local basis he has been able to provide instances which give his views strong credibility. Certainly, abandonment of marginal land has led to colonization by spruce and by other conifer species which tend to reinforce any natural soil acidity, once this tendency is no longer countered by active management. However, in a wide range of sites where pollen-analytical and documentary studies show there has been no significant land-use change over the last few centuries, fossil diatoms confirm the steep recent acidification trend that has become the hallmark of almost every palaeoecological study of the problem (e.g., Battarbee et al. 1985a; Jones et al. 1986).

Several of the tree species commonly used in commercial forestry plantations in upland Britain also tend to produce leaf litter which is poor in base status and slow to decompose. In consequence, they tend to promote increases in soil acidity. Some of the lakes first recorded as strongly acidified at the beginning of the debate had been surrounded by conifer plantations, hence the plausibility of the afforestation hypothesis. Battarbee and Flower (1983) addressed this question and were able to show that in afforested sites the trend to acidification predates preparation of the land for tree planting, and that unafforested lakes in the same area have experienced at least as strong a trend to acidification as have the planted ones.

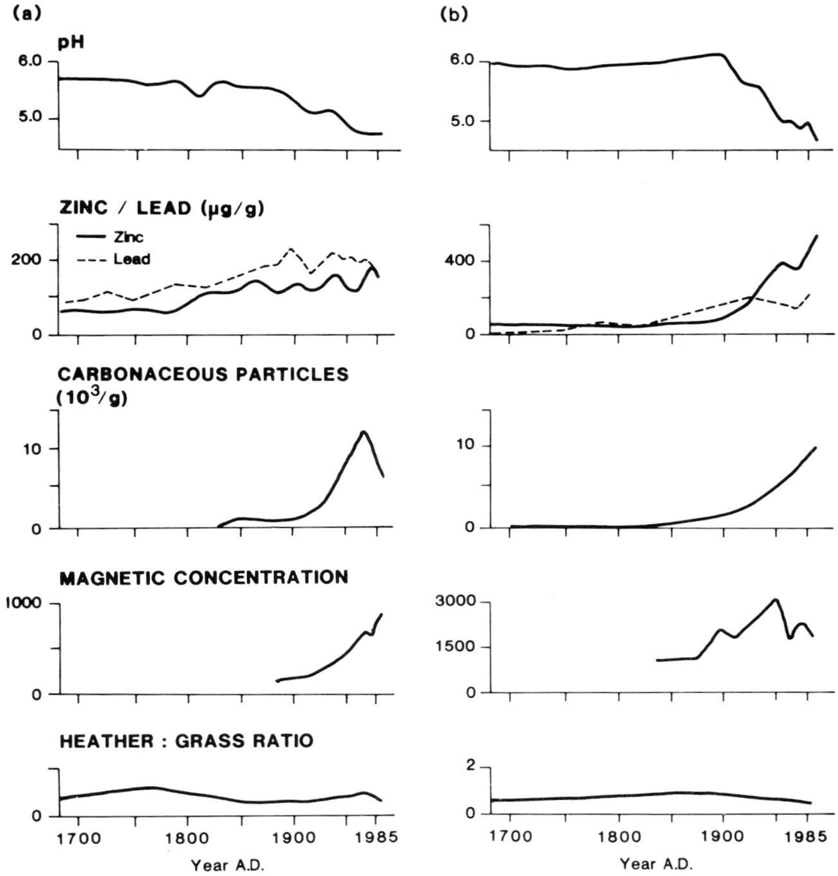

Fig. 1a,b. Acidification of two lakes as reconstructed from the sediment evidence. Fossil diatoms were used to trace the changes in pH. Zinc, lead, carbonaceous particles and magnetic remanence all show increases this century and provide reinforcing direct and indirect evidence for fallout from industrial emissions. The heather: grass ratio shows that there were no land use changes at either site that might explain the fall in pH. **a** Lochnagar, Scotland: **b** Llyn y Bi Wales. (After Battarbee et al. 1988)

Figure 1 summarizes just a fraction of the research that confirms a close parallel between the trend to acidification recorded in recent lake sediments and indicators of the deposition of industrially generated particles and chemical elements closely associated with fossil fuel combustion for power generation (Battarbee et al. 1988). Such parallels, encountered throughout the regions of lake water acidification in both Europe and North America (e.g., Davis 1987), make it virtually impossible to avoid proposing a direct link between recent industrial processes and acidification.

It is worth stressing several points at this stage. The degree of justifiable confidence in the final hypothesis could not have been achieved using documentary sources alone, for the crucial evidence regarding trends and timing exists only in the sedimentary record. Extensive and detailed use of documentary evidence for land-use history was necessary for testing the second hypothesis (Patrick 1987). More detailed exploration of the links between industrial emissions and acidification will need greater understanding of the implications of technological changes, such as chimney heights and combustion temperatures, for which documentary evidence will be required.

3 Requirements for Palaeoecology

The main requirements for palaeoecological studies of recent environmental history are summarized in Table 2. All these were met in the work outlined above, the results of which have led to the resolution of a central question in the acidification debate. In order to make the most of the environmental evidence, it is necessary to understand something of the methodology of palaeoecology. This calls for some elaboration of the points made in Table 2.

Lakes and their catchments form one of the most important contexts for study. Not only do lake sediments usually contain a sequential record of ecosystem changes, but this record can often be resolved in useful temporal detail. It can also be used to reconstruct the history both of the lake and its aquatic biota and of the catchment with its terrestrial plant communities and associated soils. In many areas of heavy human impact and resource management, reservoirs provide the best sediment record available for this type of study (Clark and Wasson 1986). Peat bogs, on the other hand, provide a record much more directly linked to atmospheric deposition. As such, they have an important role to play in the reconstruction of pollution history.

Given the availability of lakes, reservoirs, and peat bogs, the raw material for palaeoecological studies is provided by the sediments and peats still accumulating. Before full use can be made of them, methods of dating are needed which are applicable to peats and sediments over the recent time scales of greatest interest. In some situations, seasonally deposited laminae can be identified in the sediments and used to provide a chronological framework of quite exceptional

Table 2. Methodological requirements for palaeoecological studies

Requirement	Examples
1. Environmental contexts where evidence survives	Lakes, reservoirs, peat bogs, estuaries
2. Dating techniques	Laminae, pollen, radioisotopes (e.g. ^{14}C, ^{210}Pb, ^{137}Cs, ^{241}Am)
3. Extraction and analytical techniques	Coring, chemistry, magnetic properties, pollen, diatoms, soot particles
4. Theoretical framework for interpretation, explanation and reconstruction of the past	Spatial: lake-watershed ecosystems Temporal: successional or cyclic change; steady state shifts

detail (O'Sullivan 1983). In peat bogs the closest parallel to laminae are the seasonal growth increments in moss species such as *Polytrichum strictum* and *Sphagnum fuscum* which are important hummock-forming species in strongly seasonal, continental bogs (Pakarinen and Tolonen 1977).

In all too many sites, such opportunities for direct dating are absent. In these cases it is necessary to develop and test other dating techniques, most of which involve the measurement of radioactivity. The naturally occurring radioisotope of lead (lead-210) has been used successfully to date both sediments and peats deposited over the last 150 years or so (Appleby et al. 1988). For the period since 1954, radioisotopes such as caesium-137 and americium-241, both resulting from weapons testing, have also been used. Other chronological methods are dependent on knowledge gained from documentary sources. For example, pollen analysis clearly reveals a sequence of land use changes associated with the late 18th and early 19th century in northwest England and so can be used as an indirect dating method (Fig. 2). Figure 3 shows how similar historical and pollen analytical evidence can be used to verify the dating based on the radiometric measurements. Joostens (1985) has published results from infilled peasant peat pits in the Peel District of the Netherlands. Here he is able to establish a chronology based on pollen records of the period of mid-19th century buckwheat "fire" cultivation, the outbreak of dutch elm disease in 1919 and the widespread cultivation of maize, as well as on the presence at one site of a sand lens dating from an attempt in 1939 to exclude German tanks from the area by digging a canal. Sites such as these, with independently established dates, provide ideal situations for testing the chronologies derived from radiometric measurements.

The methods of extraction and analysis now available for use in studies of recent environmental history are numerous and varied. Some, like the counting of fossil diatom remains, relate to the history of biota within the lake. Diatoms have provided vital evidence for the history of acidification, as noted above, and of eutrophication, or pollution by nutrient overenrichment, as illustrated subsequently. Pollen analysis can provide a record of vegetation and land use change on the soils around a lake or peat bog and, used in conjunction with studies of stratigraphy and macrofossil content (e.g., seeds, fruits, wood, and leaves), can also add to our knowledge of the history of aquatic or mire vegetation within the lake or bog (Oldfield 1969a). Charcoal is often well preserved in peats and sediments and can be used to reconstruct recent fire history and impact (Tolonen 1978). Where charcoal bands can be linked to documented fire events, the charcoal record can contribute to dating (Clark 1986). Physical and chemical properties of sediments and peats can be used as a basis for reconstructing erosion (Mackereth 1966; Thompson et al. 1975) or trace-metal pollution histories (Oldfield et al. 1980a; Tolonen and Oldfield 1986). In the last few years, techniques of magnetic analysis have been developed that can contribute to studies of these same questions, since rocks, soils, and industrial particulates include magnetic minerals with distinctive signatures (Thompson and Oldfield 1986; Oldfield and Thompson 1986). This is far from an exhaustive list, but it illustrates the range of concerns that can be addressed by palaeoecological methods. The examples considered later illustrate and amplify this point.

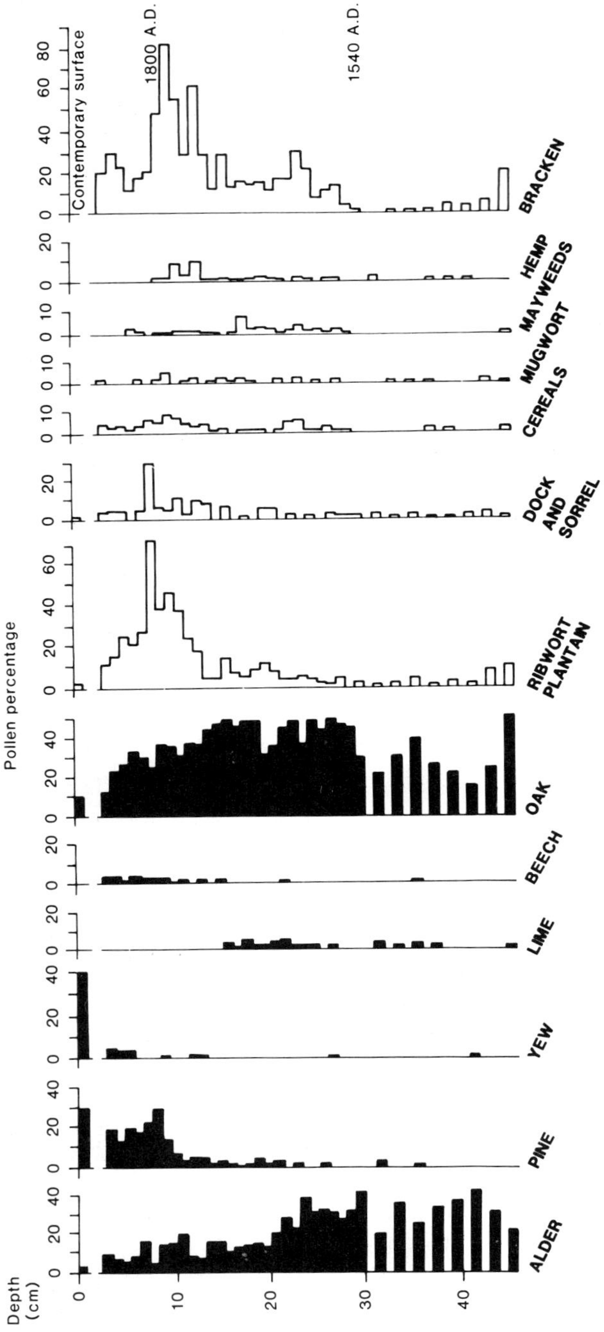
Contemporary surface
1800 A.D.
1540 A.D.
BRACKEN
HEMP
MAYWEEDS
MUGWORT
CEREALS
DOCK AND SORREL
RIBWORT PLANTAIN
OAK
BEECH
LIME
YEW
PINE
ALDER
Pollen percentage
Depth (cm)

Fig. 2.

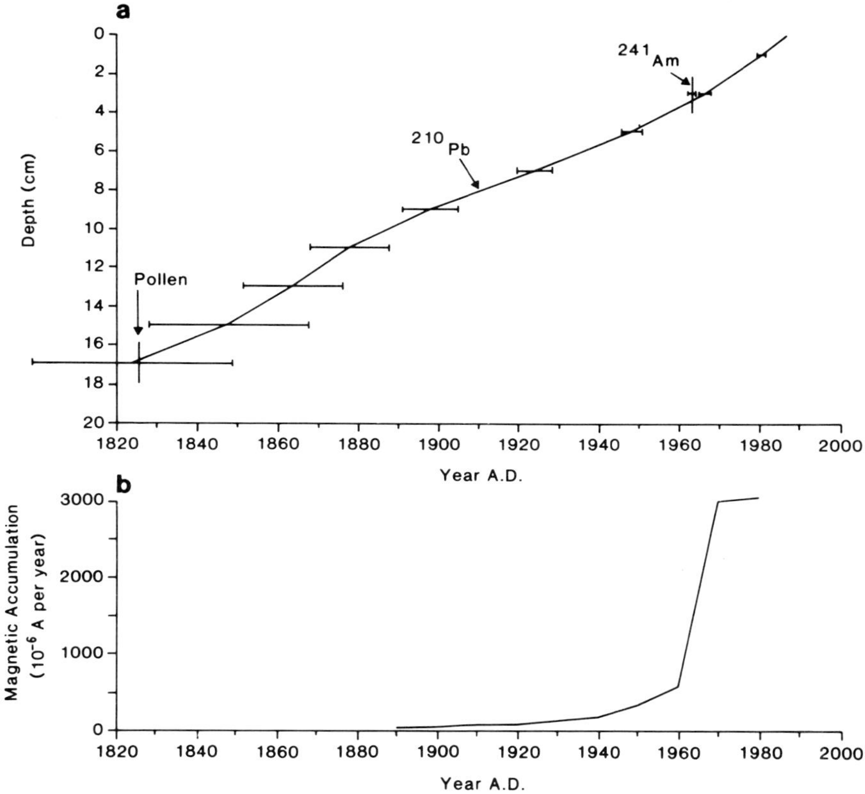

Fig. 3. **a** The chronology of peat accumulation and **b** the record of atmospheric deposition of magnetic minerals at Ringinglow Bog, Sheffield, England. The depth/age curve derived from *lead-210* measurements is confirmed by the position in the peat of the 1963 peak in *Americium-241* deposition from weapons testing and by the early 19th century changes in pollen at the base of the profile (cf. Fig. 2). Magnetic deposition is derived from measurements of magnetic remanence after magnetization in a field of 1 Tesla and is expressed in units of 10^{-6}A a^{-1}. The record indicates a gentle increase in deposition from the late 19th century followed by a steep rise after 1960. Although this trace may be modified by some solution of magnetic minerals in peat, it follows closely the pattern of deposition recorded in lakes in Britain and reflects increasing accumulation of the magnetic component of power station fly ash in the peat. (After Appleby et al. 1988 and Oldfield et al. 1987)

Fig. 2. Fossil pollen records from the upper levels of Deer Dyke Moss, Cumbria, England. Dates have been inferred from a comparison between documented land use history and changes in relative pollen frequencies (Oldfield 1969b). The decline of hemp cultivation in northwest England between 1750 and 1800, the virtual elimination of mugwort and mayweeds by deep ploughing soon after this, and the increase in planted trees such as pine and beech during the 18th and 19th centuries fix the 1800 A.D. date. The increase in broadly based mixed subsistence farming in the area after the dissolution of the monasteries is well recorded in the cereal and weed pollen frequencies above 30 cm and points to a mid-16th century date for this level

Finally, conceptual models are needed to form coherent frameworks both for problem definition and for the interpretation of results. One of the most useful spatial frameworks is the unit comprising a lake or reservoir and its drainage basin (Oldfield 1977), a unit for which Bormann and Likens (1969) have coined the term"lake-watershed ecosystem". This forms a partially "closed" system by virtue of the great extent to which material flows are bounded by the limits of the catchment. So far, the lake-watershed ecosystem approach has been underused in integrated studies of environmental history. In many cases it could form a workable and convenient unit for document, cartographic, and even archaeological study (Vaughan et al. 1985). This would allow directly parallel use of palaeoecological and historical evidence through the adoption of spatial definitions operational on either side of the methodological divide (Dearing et al. 1987).

Conceptual models of ecosystem change through time are equally important. Ecologists have often used notions of succession, or harmonic or cyclic change, as a basis for explaining both spatial (Watt 1947) and temporal (Iversen 1964; MacLulich 1937) patterns. Both models have important limitations and, increasingly, we have found it useful to deploy an alternative model in which major ecosystem changes are seen as shifts between steady states. Laszlo (1972) provides a stimulating introduction to this way of uniting concepts across the physical, biological, and social sciences and Oldfield (1983) illustrates and discusses this approach to ecosystems change. Figure 4 shows a generalized steady-state-shift model and examples of dramatic ecological changes that can be interpreted within the framework of such a model. The ecosystem changes from tidal to freshwater wetland on the Magela floodplain have been reconstructed largely from pollen analytical evidence (Clark and Guppy 1988). By contrast, the schematic reconstruction of ecosystem change over the last 200 years of human impact in Australian sheep grazing lands (Adamson and Fox 1982) rests largely on documentary evidence.

4 Palaeoecological Answers to Historical Questions

The use of environmental evidence to answer historical questions, either alone or in conjunction with other sources, can be illustrated by a series of examples that highlight the methodology and may stimulate ideas for other applications of palaeoecology to history.

4.1 The Eutrophication of Lough Neagh

During the summer of 1967, the waters of Lough Neagh in Northern Ireland became overenriched to an alarming degree. Blooms of blue-green algae formed, were rafted to the shore, and decomposed, with consequent serious oxygen depletion and threat to aquatic life, as well as loss of amenity. The algae also clogged water intake filters and reduced the value of the lake to industrial and

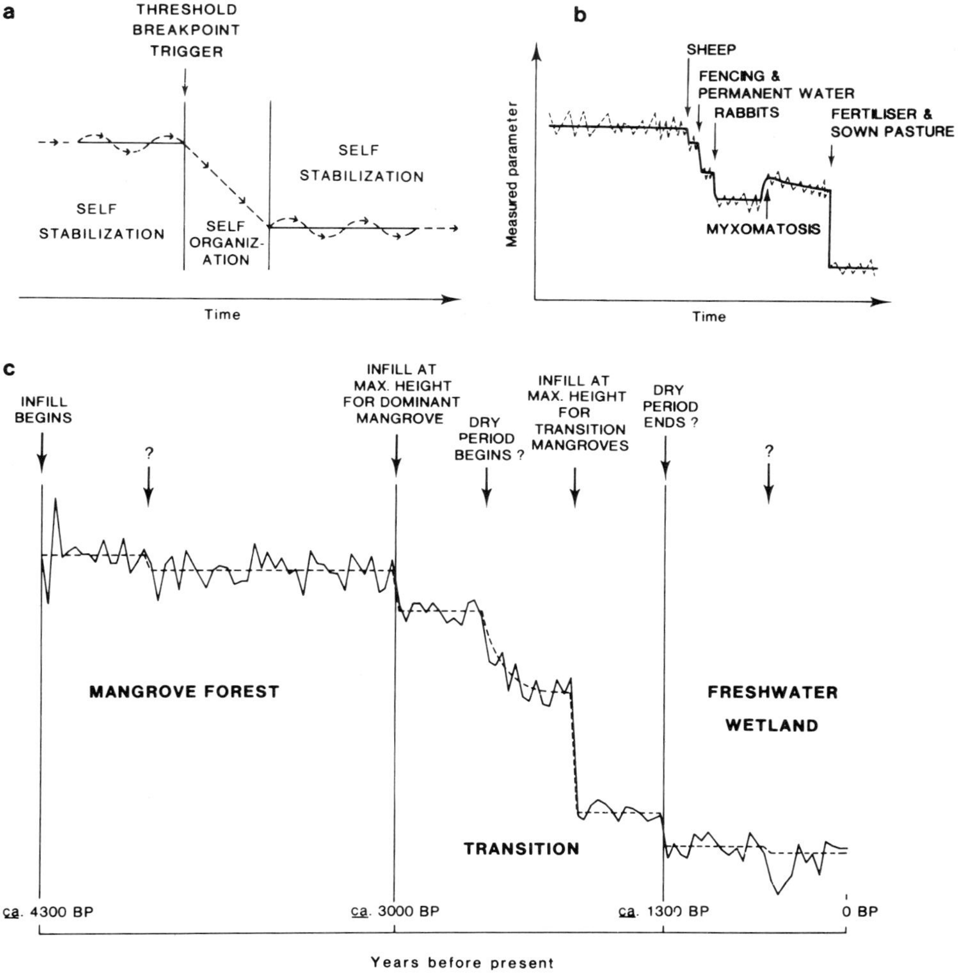

Fig. 4a-c. Applications of Laszlo's (1972) model of systems change to the environment. **a** The basic model showing two steady-state ecosystems, each varying around a mean over time. A transition phase between them is triggered by some external or internal factor, or combinations of factors. **b** Theoretical application of this model to ecosystem changes over the last 200 years in Australia's sheep grazing lands. The measured parameter in this case might be biomass or number of native plant species. (Adamson and Fox 1982). **c** An example using sedimentary pollen from a coastal freshwater wetland in the monsoon tropics of Australia. Two steady-state ecosystems, a mangrove forest and a freshwater wetland, have a complex transition between them, triggered by changes in the physical environment. Note that the time scale is not proportional. Dates have been determined by or extrapolated from carbon-14 ages. (After Clark and Guppy 1988)

domestic consumers of water. Coming at a time of heightened public awareness of the problems posed by parallel developments elsewhere, for example in Lake Erie in North America, these sudden and unexpected manifestations caused great public concern. Experimental studies and detailed observations quickly confirmed that phosphate availability was crucial to the present day expression of the problem, but evidence about its origins and development were completely lacking. Thus it was impossible to say whether the problem was recent, related to human activities and potentially treatable or reversible, or whether it was latent in the long-term nature of the lake and thus ready to appear at any time, given the right combination of summer weather conditions.

Eventually the question was answered through detailed studies of fossil diatom and insect remains in the sediment (Battarbee 1978; Carter 1977; Fig. 5). These studies showed that productivity increased gradually in the late 19th and early 20th centuries after a long period of very limited change and apparent stability. The trend steepened dramatically after World War II. Through comparison with the historical evidence it was possible to ascribe the first increase to the development of integrated sewerage systems in the towns around the lake, and the second, steeper increase largely to the use of phosphate-rich detergents. Demonstration that the problem was recent and potentially curable led to the

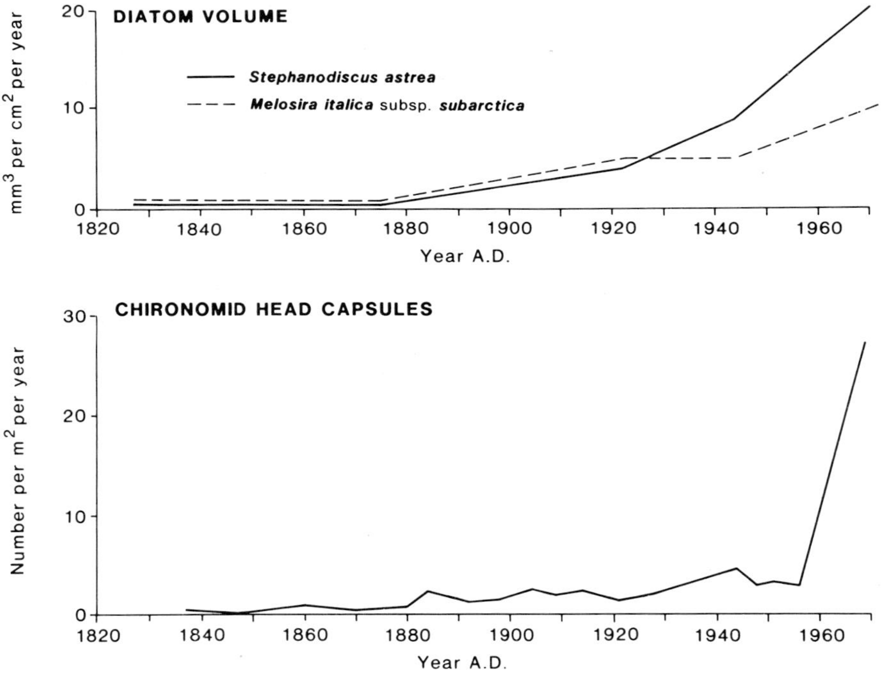

Fig. 5. Annual influx of two *diatom* species and of *chironomid* (midge) head capsules to the sediments of Lough Neagh, Northern Ireland. Both show small increases from about 1880 and very large increases in recent times, responding to slow and then rapid nutrient enrichment of the lake waters

installation of tertiary sewage treatment plants in surrounding towns so that the polluting phosphates could be removed from the waste water draining into the lake. Interestingly, an exhaustive search for records of the changing diatom flora in the lake during earlier decades revealed several accounts and actual microscope slides. These records, though by no means sufficiently reliable, comprehensive or continuous to establish the historical reconstructions made from the sediment record, were compatible with it (Battarbee 1979). Subsequent research identified similar trends in neighbouring lakes such as Lough Erne (Battarbee 1986) and also documented in much greater detail the links between the development of integrated sewerage systems and the early stage of eutrophication (Patrick 1986).

4.2 Changing Erosion Rates and Sediment Sources

Figure 6 summarizes data on sediment accumulation rates in six lakes in the Galloway region of southwest Scotland (Battarbee et al. 1985b). In the three cases where sedimentation rates have remained relatively low and constant, the drainage basins of the lakes have not been afforested. Instead, they are surrounded by moorland and rough pasture of long standing, and can therefore be used as controls for the other three cases. In each of these latter, a period of recent afforestation by conifers was preceded by ploughing designed to improve soil quality and drainage status prior to planting. The sedimentation rate increased at the time of preafforestation ploughing at each site.

In this study only one core per lake was used. Other research using multiple cores has shown that the quantitative expression of erosive input to the sediments of a lake may vary, depending on the location of the coring site with respect to the lake margin or to the area of catchment disturbance. This point is just as true for sites affected by changes in the pattern and intensity of subsistence agriculture (Oldfield et al. 1980b) as for those subject to plantation forestry (Appleby et al. 1985). Ideally, enough cores are needed from a site to allow whole-lake to whole-catchment comparisons, and the translation of total sediment input into total erosional loss. Several recent studies have succeeded in doing this (Bloemendal et al. 1979; Dearing et al. 1981, 1987; Bloemendal 1982; Oldfield et al. 1985; Wasson and Galloway 1986).

Even when a picture of total lake or reservoir sedimentation has been obtained, it is not always possible to translate this directly into catchment erosion because, especially in large and complex catchments, eroded material will often be stored for long periods at sites between the eroding areas where catchment derived material originates and the eventual sites of deposition in the lake. In the case of the Egari record (Fig. 7) from Papua New Guinea, the nature of the catchment, within the rim of a small volcanic crater, precludes this. The evidence provides an insight into processes for which no documentary records exist, since prehistory at the site ends, in effect, during World War II. The response of erosion rates to land clearance and intensive sweet potato cultivation are especially noteworthy. Alongside the site is another similarly situated lake, Pipiak, which at

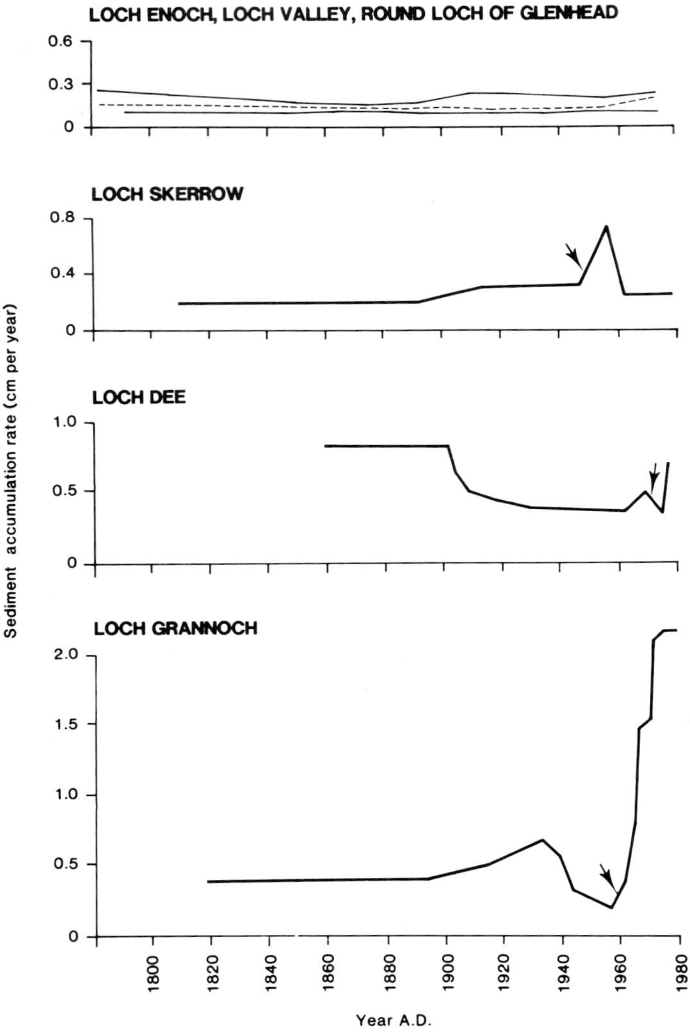

Fig. 6. Sediment accumulation rate in six Scottish lakes. Lochs Skerrow, Dee and Grannoch show increased erosion from their catchments from the time of preafforestation ploughing (*arrowed*). The other three lakes – Lochs Enoch and Valley, and the Round Loch of Glenhead – remain unplanted and show no change in erosion rates. (After Battarbee et al. 1985b)

the time of sampling in 1973 was still largely rimmed by forest. Lack of any acceleration in sedimentation rates at Pipiak strengthens the link between gardening and erosion at Egari. The lakes have, in effect, conducted a long-term controlled experiment.

Studies of artificial lakes in the English Midlands are noteworthy for their documentation of the relationship between sedimentation rates and human land

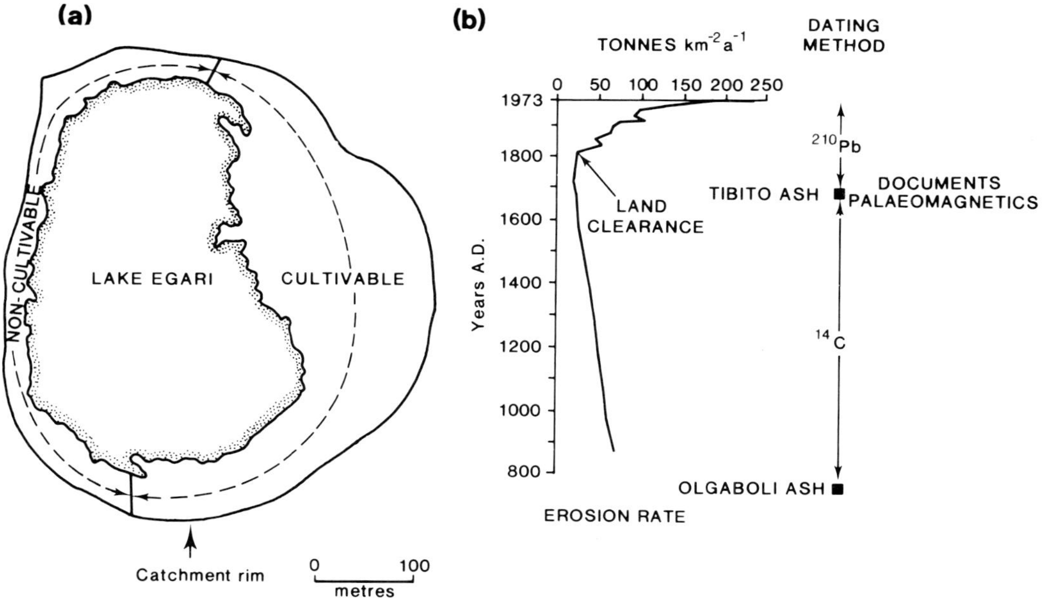

Fig. 7a,b. The history of soil erosion from the catchment recorded in the sediments of Lake Egari, a small volcanic crater lake in the Southern Highlands Province of Papua New Guinea. Several cores from the lake (**a**) were correlated by means of magnetic measurements and dated using a range of techniques (**b**). The two volcanic ash layers have been encountered and dated over a wide area of Papua New Guinea. Total sediment input for each dated time interval was calculated as mass loss from the catchment (**a**) and plotted in tonnes $km^{-2} a^{-1}$ against a time scale for the last 1200 years (**b**). (After Oldfield et al. 1980b)

use as these have varied both spatially and temporally (Fig. 8). In addition, these historical sediment studies have been linked in quantitative terms to contemporary monitoring of stream flow and sediment input (Foster et al. 1985). The latest stage of the research includes identification of dominant sediment sources for each sedimentation phase at each lake by using magnetic properties (Dearing and Foster 1987).

Figure 9 shows how a variety of techniques has been used to date sediments accumulating in the Burrinjuck Reservoir in Australia (Clark 1986). The first appearance of caesium-137, a radioisotope produced by testing nuclear weapons, indicates sediment deposited in 1954. Floods and drought-breaking rain, recorded instrumentally, were responsible for thick sediment layers in 1945 and 1925 (the latter not shown in this section). Visible charcoal bands in the sediment were matched with recorded fire years. Finally, grass pollen peaks could be used to distinguish annual layers in some parts of the sediment. Without such detailed dating from documentary, instrumental and environmental evidence, it would have been impossible to construct a record of sedimentation rate in the reservoir (Fig. 10). Since 1925, 56% of the sediment has been deposited in only 5% of the

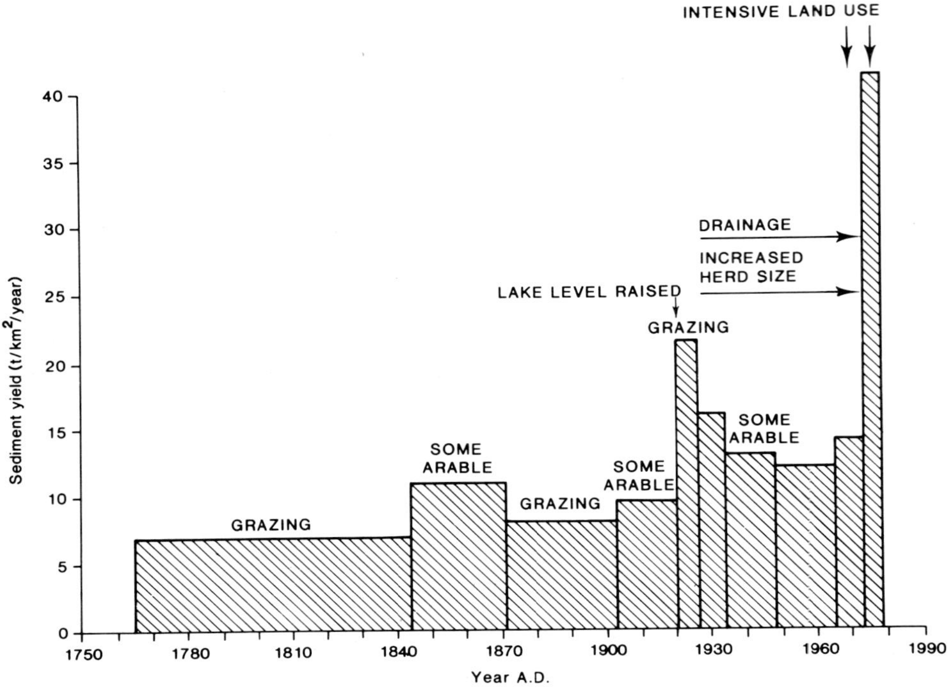

Fig. 8. Sediment yield from the catchment of Seeswood Pool in the English Midlands. Changes over time in the rate of sediment yield, as determined from the sedimentary record, are attributable to land-use changes in the catchment as reconstructed from documentary evidence. (After Dearing and Foster 1987)

years. Most of this rapidly deposited sediment was brought in by a major flood and intense local rainfall after periods of drought. A well-dated sediment record such as this, resolved in the kind of detail that rapidly accumulating sediments permit, provides a basis for reconstructing the impact and antecedent conditions of major events in the past, for comparing their effects through time, and for studying questions of magnitude and frequency as well as hazard. Clearly, once more, the sediment record is complementary to that available from documentary evidence. In funding research on Burrinjuck Reservoir, the New South Wales Water Resources Commission recognized this by appointing a historian as one of the research team. Documentary and sedimentary studies are interdependent, though the sediment record, once calibrated by documentary evidence for recent times, is then extendable further into history and prehistory, beyond the reach of written records.

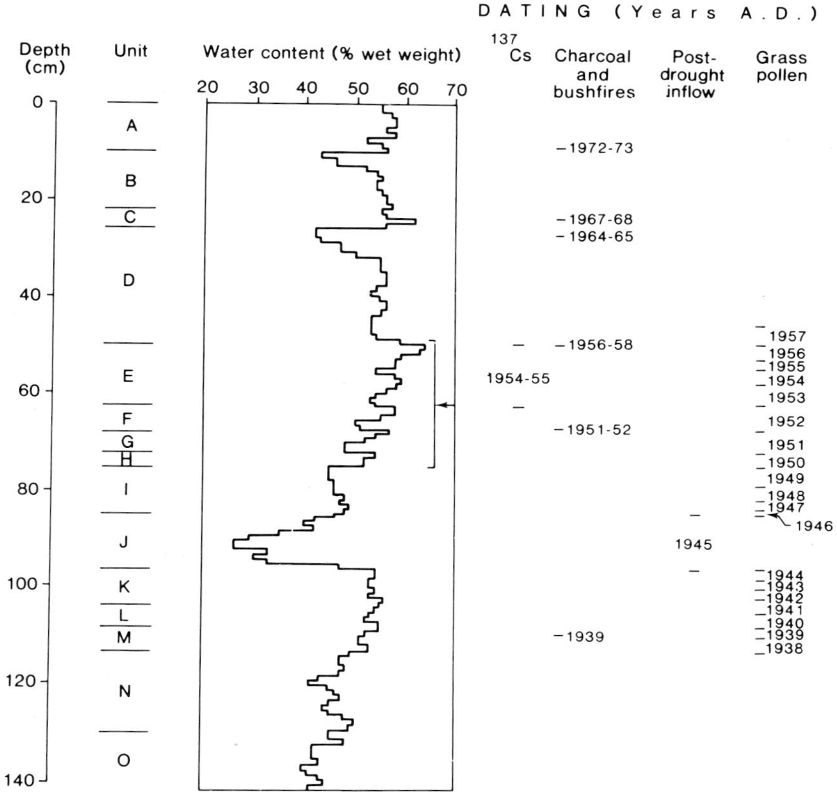

Fig. 9. Dating sediment from Burrinjuck Reservoir, Australia. This particular sediment core was taken in 1982, so that is the date at the top. The base of the core is above a thick layer of sediment deposited in the record flood of 1925. Visible stratigraphic units, or layers, are shown by the *letters A* to *O*. In parts of the core, these units match cycles in the water content of the sediment and, in the section indicated by an *arrow*, grass pollen peaks occurred at the same point on each cycle. Dating was determined by: the first appearance of the nuclear weapons testing fallout isotope caresium-137; matching visible charcoal bands with recorded bushfire years, and floods and intense rainfall with thick layers of larger particle size; and grass pollen peaks that indicate the late-spring to early-summer flowering

4.3 Integrated Catchment Studies

At Havgardsjon (Fig. 11), research results have provided a basis for elegant modelling of phosphorus flows in the catchment at different times and for linking the sedimentary record of changing erosion rates to the documentary evidence of land use and stocking density (Dearing et al. 1987). The study as a whole provides a pioneering example of what can be achieved within the lake-watershed ecosystem framework by collaborating palaeoecologists and historians.

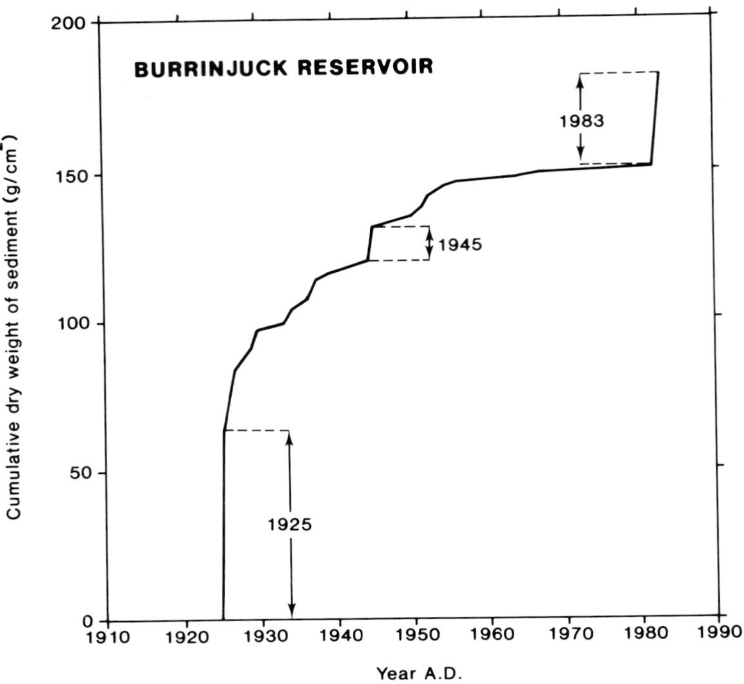

Fig. 10. Using the dated sediment record from Burrinjuck Reservoir (Fig. 9), annual sediment input can be calculated and expressed as cumulative sedimentation. This shows that the sedimentation rate has slowed since 1945, and that most sediment was deposited in three major events in 1925, 1945 and 1983

Dearing and his Swedish colleagues have pointed the way towards exciting developments in human ecology by basing several lines of complementary research on the lake and its catchment. Not only does the study unite rather divergent methodologies, its success implies that the lake-catchment ecosystem approach may also serve as a framework for uniting our perspectives on a range of issues. These include questions of carrying capacity, of long term demographic and environmental trends, and of the relationships in the dynamics of rural ecosystems between energy capture and conversion, limiting nutrient availability, aperiodic Malthusian controls and cultural mechanisms.

Multidisciplinary research in the region of Lac d'Annecy, eastern France, is addressing this type of issue through combining archival studies with soil and sediment analyses. The lake and its large drainage basin provide a beguiling mixture of opportunities and challenges. The quality of the documentary evidence for survey and census years from the 16th century onwards provides a superb archival basis for that aspect of the work (Jones and Siddle 1982). The sediments have accumulated rapidly in the form of seasonal calcareous laminations, which should eventually permit dating to a level of accuracy and precision comparable to that reflected in the archival sources. As yet, deriving a time scale

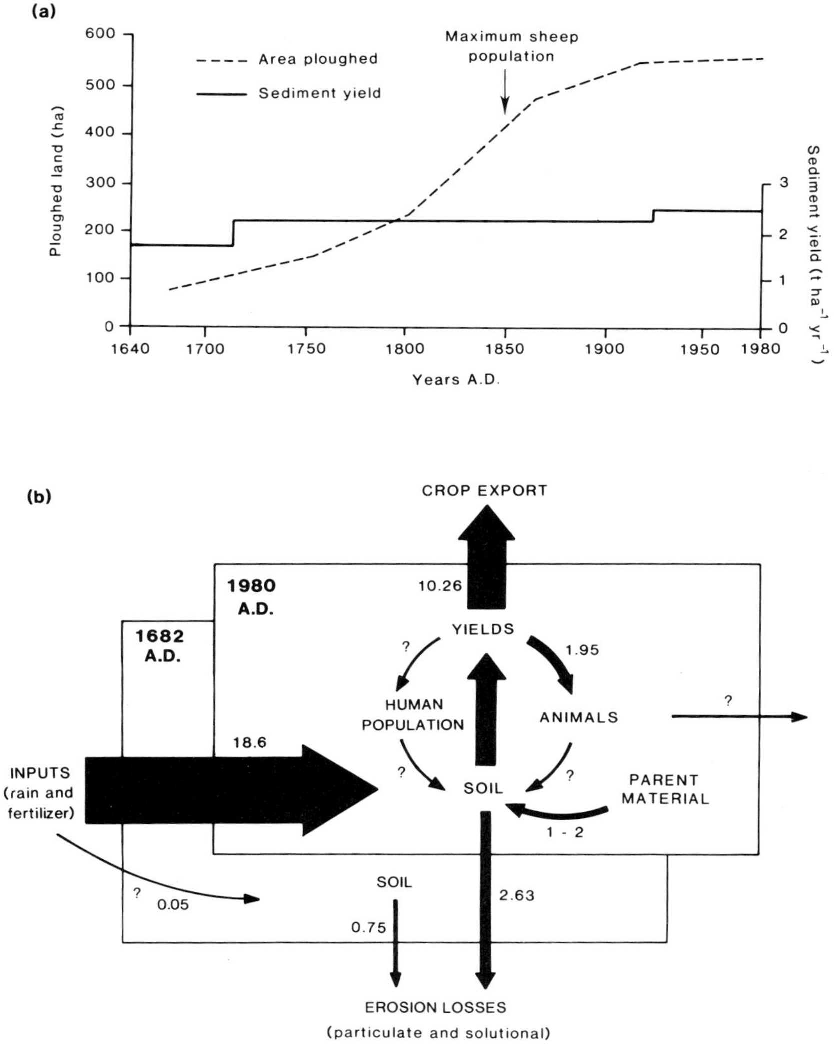

Fig. 11a,b. The sedimentary record of soil erosion in relation to land use since the seventeenth century at Havgardsjon in southern Sweden. **a** Sediment yield and total cultivated area in the catchment, showing the time of maximum sheep population. **b** Phosphorus inputs and outputs in 1682 and 1980; units are kg ha^{-1} yr^{-1}. (After Dearing et al. 1987)

from the sediment laminae has not proved possible, for technical reasons (Higgitt 1985), and differences in the spatial scale to which the sedimentary and the archival evidence apply constitute a serious impediment to a fully unified perspective. Eventually, it should be possible to reconstruct a finely resolved sedimentary record capable of addressing crucial problems of climatic effects and human impacts and responses.

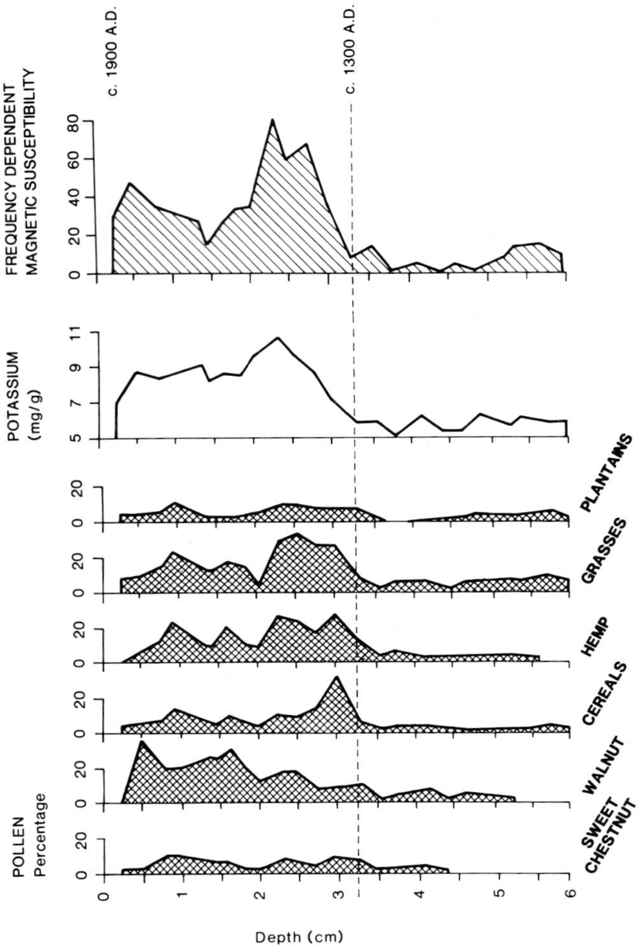

Fig. 12. The sedimentary record of the effects of land-use changed at Lac d'Annecy, France. Pollen shows increases in cultivated plants, grasses and weeds at the expense of forest species (not shown) after the establishment of monastic estates in the area in the late 13th century. Increased soil erosion as a result of settlement is indicated by potassium concentration and frequency dependent magnetic susceptibility (X_{fd}, here in units of $10^{-9} m^2 kg^{-1}$); the latter shows that topsoils, rather than subsoils or channel banks were eroding

For the present, the possibilities can best be illustrated by reference to Fig. 12 which shows in a highly selective way some of the parallel evidence furnished by the sediments. Sweet chestnut and walnut are among the orchard crops established by religious houses in the region from the 13th century onwards. Pollen from cereals and other cultivated crops such as hemp, and from pasture grasses and weeds such as plantains, reflect the surge and eventual decline of upland subsistence farming in the catchment. In response to this, there were sharp increases in the inwash to the sediments of potassium, an indicator of accelerated erosional input. The trace of frequency dependent susceptibility, X_{fd}, a magnetic property indicating the relative importance of very fine grained and almost exclusively soil derived iron oxides such as maghemite or magnetite, shows that erosion triggered by subsistence activities took the form of surface soil loss. The next stage in the study of the sediments must be to resolve the laminae into a secure chronology and focus on time intervals spanning the ages of the most useful archival material. Perhaps the ideal site is one that combines the more modest spatial scale of the Swedish study with the chronological resolution available from laminated sediments, such as those at Annecy and many other sites (O'Sullivan 1983).

5 Parallels and Complementarity

Any attempt to bring together the palaeoecological and historical approaches to environmental history must begin by dismissing the still all too prevalent stereotype of the inductive scientist working with levels of proof and certainty well beyond the reach of the professional historian. In reality, although the sources, evidence, and activities differ for the two groups, the basic structure of the two types of enquiry is remarkably similar. Figure 13 compares the chain of events leading to the survival, retrieval, and ultimate presentation of evidence in both types of study. There are clear parallels and rather comparable "filters" at all stages. In the present context, it will suffice to follow through the palaeoecology line, with some explanations and annotations at each stage.

It is first necessary to consider the way in which the sediment of a lake, for example, will actually retain evidence of some environmental event, whether it be a flood, a shift in mean summer temperature, forest clearance and farming, or industrial pollution. First, there must be some significant ecosystem response which leads to a change in output from the affected area. By output is meant, for example, the yield of pollen types from altered vegetation communities, the erosion of soil and soil-derived minerals and the transport of charcoal from a burnt area. These changes in output generate quantitative and qualitative changes in the nature of the material deposited on the lake bed. At each stage there will be no perfect "match" or transfer of information. Already, by the time the material for subsequent analyses reaches the bed of the lake, the effects of transport and survival will have been selective and perhaps a little arbitrary. During the period of storage between deposition and retrieval, much will happen to the deposited material. Some will be dissolved and recycled through the water, plants, and animals of the lake. Much will survive, but in a chemically altered form. Even that which remains preserved and recognizable will often be subject to processes of vertical mixing or lateral redeposition so that the quality of the sediment record will be significantly degraded.

Sample retrieval and analysis involve selectivity and, because of the risk of creating artefactual patterns and relationships, they rely heavily on the skill, experience, and integrity of the investigator. The final stages of the study can be

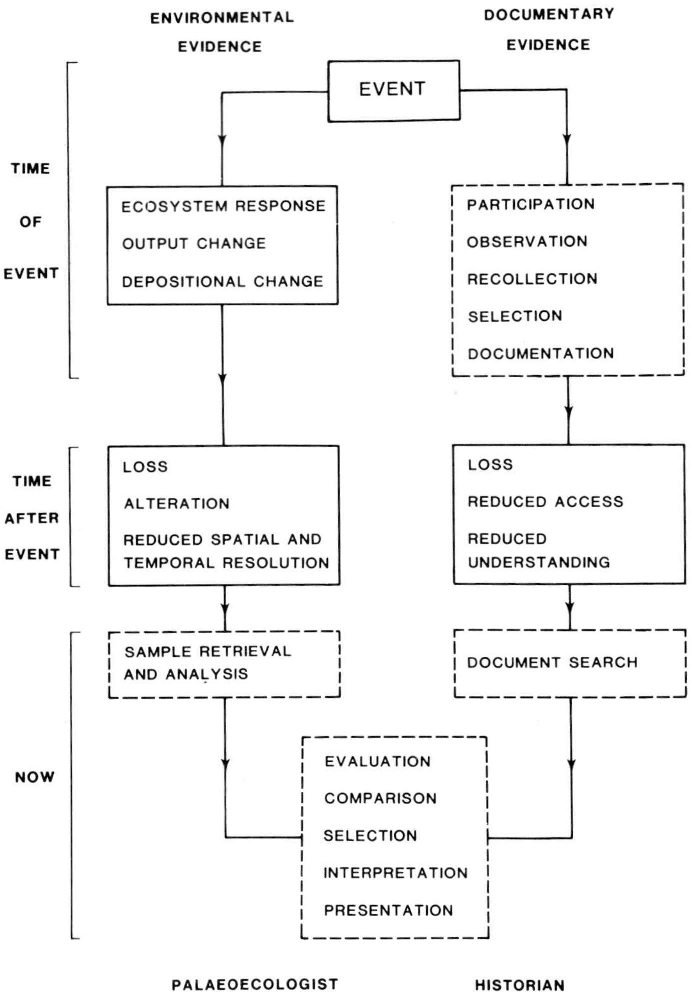

Fig. 13. A comparison of the "filters" environmental and documentary evidence go through between the event they record and interpretation by palaeoecologists or historians. The diagram is divided into three sections that show processes at work at the time of creation and preservation of the evidence, through the period of alteration or destruction afterwards, to present-day methodological limitations. *Boxes* bounded by *dashed lines* indicate processes that are solely dependent on humans; those bounded by *solid lines* are affected by environmental processes or human activities

described in similar terms for both history and palaeoecology. Throughout the palaeoecological sequence, parallels with the raw material and practice of environmental history *sensu stricto* are close and very important for they point to structural commonalities which belie the popular stereotypes.

Turning now to the ways in which palaeoecological and historical lines of study can complement each other (Table 3), we can begin with some simple,

Table 3. Comparison of some characteristics of environmental and documentary evidence

Environmental evidence	Documentary evidence
Hard to date or ascribe spatially: more spatially generalized	Specific for time and location: more arbitrarily particularized
Usually continuous	Usually discontinuous
Quantitative or semi-quantitative	Qualitative or descriptive
Interpretable in "scientific" terms	Poorly related to "scientific" constructs
Non-anthropocentric residues	Human artefacts
Includes human effects and consequences in an undifferentiated way	Directly records human effects and consequences

perhaps even oversimplified, generalizations. Peats capture mainly atmospheric material derived from a large area that can never be precisely defined. Lake sediments mostly record catchment changes without giving more than rather general guidance about source materials and locations. Despite the advances in chronology already noted, lake sediments and peats can still be difficult to date with the degree of reliability and precision ideally required. In all these respects the resolution afforded by archival material is often so much better, although the existence, survival and recovery of material may reflect many biassing or arbitrary processes, and the effects of thematic or spatial particularization may be biassed or arbitrary too.

The distinction on the basis of continuity versus discontinuity is a reasonable generalization, though it is important to remember that, especially under disturbed environmental conditions, the sediment record may contain gaps that can be hard to detect, let alone fill (O'Sullivan et al. 1972). On the other hand, long sequences of comparably compiled archival material are not unknown (Manley 1974) and can provide, in effect, a continuum of source material.

It is usually, but not always, possible to express palaeoecological results in quantitative terms; though the entities being so expressed will have passed through the various distorting filters already noted and the quantification may be misleading. The historian also strives for quantitative insights in many research areas and, although much of the evidence has a predominantly qualitative tenor, quantitative statements can often be made, if not always without controversy.

The distinction proposed in relation to "scientific" constructs is not intended, and should not be seen, as a simplistic, loaded value judgement. It implies neither more nor less than it says. Many "scientific" inferences about chronology or the nature of past environments require testing against well founded, though never in any conventional sense "scientific", historical evidence. Although the latter will often fail to provide a fully adequate basis for past environmental reconstruction, its careful use can provide vital tests of credibility. The quality of empirical insights into the past will never be simply commensurate with cost, scientific elegance or technological sophistication, but rather with the degree of care, rigour, and discrimination employed.

Inevitably, as the last two antitheses in Table 3 suggest, there is a contrast in the degree to which the two types of evidence are in a primary sense anthropocentric. This has often been at the root of the divergences in purpose, motivation and outlook of the two types of research worker, and hence, the misunderstanding which all too easily flows from these divergences. Clearly, as sources of evidence, documents are anthropocentric, sediments not, but the type of recent environmental history that is the concern of palaeoecologists is anthropocentric in a purposive sense, for the main objectives include a better understanding of present-day human problems and future prospects by setting them in historical perspective. Thus the divergence apparently implicit in the contrast noted here is not inevitable. Instead, we stress the complementarity of the two different kinds of evidence, for together they can join to form insights greater than the sum of their parts.

Out of the above, some parallels of outcome and purpose can be drawn between history and recent palaeoecology. Issues of hypothesis testing and the dependence of results and interpretations on current paradigms (Kuhn 1970) are especially important. Inevitably, both kinds of research have to create their own coherence from limited evidence. Every judgement is made within a zone of tension between the desire to contribute to, and significantly reinforce, emerging explanations and reconstructions, and the need to seek and recognize those crucial data which may serve to falsify some or all of the accepted views. Although opportunities to carry out crucial tests of falsification (Popper 1963) are rarer in retrospective studies than in the world of contemporary observation and experiment, they are by no means absent. Both the historical and environmental records have provided more tools for the disinterested student and the devil's advocate than are immediately obvious. The needs and opportunities for taking an "experimental" approach to palaeoecological studies are delightfully outlined in Ed Deevey's (1969) note on *Coaxing history to conduct experiments*.

Historians are often, by virtue of their training and the nature of their subject, more aware than scientists of the degree to which their priorities, approach and eventual interpretations are coloured by personal values and judgements. Palaeoecological work, especially in so far as it focuses on present day problems, does not avoid this limitation. Moreover, since so much of the work of palaeoecologists depends on integrating studies drawn from a range of quite disparate scientific fields, they need to be sensitive to the ways in which fashions and perspectives are changing around them.

6 Conclusions

The applicability of much recent environmental history to an understanding of present day problems has already been illustrated, as has the way in which the two different approaches are necessary and mutually interdependent in order to provide the best possible insights for decision makers at the present day and in the future. It is useful to recognize here Passmore's (1974) distinction between problems in ecology, that is, the intradisciplinary concerns of the professional ecologists, and ecological problems, those which affect society and are therefore by definition social, technological, economic, and political, too. Environmental history contributes in the former area by increasing our understanding of processes and systems interactions that operate on time scales beyond the reach of direct observation and experiment. It contributes to the latter by virtue of its focus on the most recent time interval, its frequent orientation to specific problems, and the extent to which any appraisal of contemporary problems requires an exploration of their immediate antecedents.

The time interval crucial to such applied studies is the last 200 years, the period of the most severe and dramatic environmental transformations through population expansion and industrial development. This is the very time interval that falls between the traditional concerns of the experimental and observation-oriented scientists on the one hand and the environmental reconstruction specialists on the other (Oldfield 1983), but it is a time interval familiar to the historian. Although only a relatively small number of palaeoecologists have begun to learn how to share in its exploration, their contribution should be recognized as a vital complement to, and extension of, the more familiar "historical" approach. Environmental history is naturally multi-disciplinary and now of far too great importance to be limited by quibbles about internal methodological boundaries.

References

Adamson DA, Fox MD (1982) Change in Australasian vegetation since European settlement. In: Smith JMB (ed) A history of Australasian vegetation. McGraw-Hill, Sydney, pp 109–146

Appleby PG, Dearing JA, Oldfield F (1985) Magnetic studies of erosion in a Scottish lake catchment I: core chronology and correlation. Limnol Oceanogr 30:67–75

Appleby PG, Nolan PJ, Oldfield F, Richardson N, Higgitt SR (1988) ^{210}Pb dating of lake sediments and ombrotrophic peats by gamma assay. Sci Total Environ 69:157–177

Battarbee RW (1978) Observations on the recent history of Lough Neagh and its drainage basin. Philos Trans R Soc London Sev B 281:303–345

Battarbee RW (1979) Early algological records – help or hindrance to palaeolimnology? Nova Hedw 64:379–394

Battarbee RW (1986) The eutrophication of Lough Erne inferred from changes in the diatom asemblages of ^{210}Pb and ^{137}Cs dated sediment cores. Proc R Ir Acad 86B:141–168

Battarbee RW, Flower RJ (1983) Diatom evidence for recent acidification of two Scottish lochs. Nature (London) 305:130–133

Battarbee RW, Flower RJ, Stevenson AC, Rippey B (1985a) Lake acidification in Galloway: a palaeoecological test of competing hypotheses. Nature (London) 314:350–352

Battarbee RW, Appleby PG, Odell K, Flower RJ (1985b) ^{210}Pb dating of Scottish lake sediments, afforestation and accelerated soil erosion. Earth Surface Process Landforms 10:137–142

Battarbee RW, Anderson MJ, Appleby PG, Flower RJ, Fritz SC, Haworth EY, Higgitt SR, Jones VJ, Kreiser A, Munro MAR, Natkanski J, Oldfield F, Patrick ST, Richardson NG, Rippey B, Stevenson AC (1988) Lake acidification in the United Kingdom 1800–1986. Evidence from analysis of lake sediments. Rep Dep Environ, Palaeoecol Res Unit, Univ College, London

Bloemendal J (1982) The quantification of rates of total sediment influx to Llyn Goddionduon, Gwynedd, North Wales. PhD Thesis Univ Liverpool

Bloemendal J, Oldfield F, Thompson R (1979) Magnetic measurements used to assess sediment influx at Llyn Goddionduon. Nature (London) 280:50–53

Bormann FH, Likens GE (1969) The watershed-ecosystem concept and studies of nutrient cycles. In: Van Dyne GM (ed) The ecosystem concept in natural resource management. Academic Press, New York London, pp 49–76
Carter CE (1977) The recent history of the chironomid fauna of Lough Neagh from the analysis of remains in sediment cores. Freshwater Biol 7:415–423
Clark RL (1986) Pollen as a chronometer and sediment tracer, Burrinjuck Reservoir, Australia. Hydrobiologia 143:63–69
Clark RL (1989) Ecological history for environmental mangement. In Saunders DA, Hopkins AJM, How R (eds) Australian Ecosystem: 200 years of utilization, degration, and reconstruction. Proc Ecol Soc Aust 16
Clark RL, Guppy JC (1988) A transition from mangrove forest to freshwater wetland in the monsoon tropics of Australia. J Biogeogr 15:665–684
Clark RL, Wasson RJ (1986) Reservoir sediments. In: De Deckker P, Williams WD (eds) Limnology in Australia. CSIRO, Melbourne; and Junk, Dordrecht, pp 497–507
Davis RB (1987) Palaeolimnological diatom studies of acidification of lakes by acid rain: an application of quaternary science. Quat Sci Rev 6:147–163
Dearing JA, Foster IDL (1987) Limnic sediments used to reconstruct sediment yields and sources in the English Midlands since 1765. In: Gardiner V (ed) International geomorphology. John Wiley & Sons, New York, pp 853–868
Dearing JA, Elner JK, Happey-Wood CM (1981) Recent sediment flux and erosional processes in a Welsh upland lake-catchment based on magnetic susceptibility measurements. Quat Res 16:356–372
Dearing JA, Håkansson H, Liedberg-Jönsson R, Perrson A, Skansjö S, Widholm D, El-Daoushy F (1987) Lake sediments used to quantify the erosional response to land use change in southern Sweden. Oikos 50:60–78
Deevey ES (1969) Coaxing history to conduct experiments. Bioscience 19:40–43
Foster IDL, Dearing JA, Simpson A, Carter AD, Appleby PG (1985) Lake catchment based studies of erosion and denudation in the Merevale Catchment, Warwickshire, UK. Earth Surface Process 10:45–68
Higgitt SR (1985) The palaeoecology of the Lac d'Annecy and its drainage basin. PhD Thesis, Univ Liverpool
Hutton J (1788) The theory of the earth. Trans R Soc Edinburgh 1:209–304
Iversen J (1964) Plant indicators of climate, soil and other factors during the Quaternary. Rep 5th Int Congr Quaternary, Warsaw, 1961, vol 2: Palaeobotanical section, pp 421–428
Jones AM, Siddle DJ (1982) Sources for the reconstruction of a man-environment ecosystem. Liverpool Pap Human Geogr 3, Univ Liverpool
Jones VJ (1987) A palaeoecological study of the post-glacial acidification of the Round Loch of Glenhead and its catchment. PhD Thesis, Univ London
Jones VJ, Stevenson AC, Battarbee RW (1986) Lake acidification and the land-use hypothesis: a mid-postglacial analogue. Nature (London) 322:157–158
Joostens JHJ (1985) A 130 year micro- and macro-fossil record from regeneration peat in former peasant peat pits in the Peel, the Netherlands: a palaeoecological study with agricultural and climatological implications. Palaeogeogr Palaeoclimatol Palaeoecol 49:277–312
Kuhn TS (1970) The structure of scientific revolutions, 2nd edn. Univ Press, Chicago
Laszlo E (1972) Introduction to systems philosophy. Harper & Row, New York
MacLulich DA (1937) Fluctuations in the numbers of the varying hare (*Lapus americanus*). Univ Toronto Stud Biol Ser 43, 136 p
Mackereth FJH (1966) Some chemical observations on post-glacial lake sediments. Philos Trans R Soc London Ser B 250:165–213
Manley G (1974) Central England temperatures: monthly means, 1659–1973. Q J R Meteorol Soc 100:389–405
Oldfield F (1969a) The ecological history of Blelham Bog National Nature Reserve. In: Walker D, West RG (eds) Studies in the vegetation history of the British Isles. Univ Press, Cambridge, pp 141–157
Oldfield F (1969b) Pollen analysis and the history of land use. Adv Sci 69:293–311

Oldfield F (1977) Lakes and their drainage basins as units of sediment-based ecological study. Prog Phys Geogr 1:460–504
Oldfield F (1983) Man's impact on the environment: some recent perspectives. Geography 68:245–256
Oldfield F (1987) The future of the past – a perspective on palaeoenvironmental study. In: Clark MJ, Gregory KJ, Gurnell AM (eds) Horizons in physical geography. Macmillan, Basingstoke, pp 10–26
Oldfield F, Thompson R (eds) (1986) Mineral magnetic studies. Phys Earth Planet Interiors 42:1–128
Oldfield F, Appleby PG, Petit D (1980a) A re-evaluation of lead-210 chronology and the history of total lead influx in a small south Belgian pond. Ambio 9:97–99
Oldfield F, Appleby PG, Thompson R (1980b) Palaeoecological studies of lakes in the Highlands of Papua New Guinea. J Ecol 68:457–477
Oldfield F, Appleby PG, Worsley AT (1985) Evidence from lake sediments for recent erosion rates in the Highlands of Papua New Guinea. In: Douglas I, Spencer E (eds) Environmental change and tropical geomorphology. Allen & Unwin, London, pp 185–195
Oldfield F, Appleby PG, Richardson N, Higgitt SR, Clymo RS, Hunt A, Jones JM, Pearson GW (1987) Magnetic and radiometric studies of sediments and peats. In: Surface water acidification programme. Proc Mid-Term Rev Conf, Bergen, Norw, pp 400–408
O'Sullivan PE (1983) Annually-laminated lake sediments and the study of Quaternary environmental changes – a review. Quat Sci Rev 1:245–313
O'Sullivan PE, Oldfield F, Battarbee RW (1972) Preliminary studies of Lough Neagh sediments I. Stratigraphy, chronology and pollen analysis. In: Birks HJB, West RG (eds) Quarternary plant ecology. Blackwell, Oxford, pp 267–278
Pakarinen P, Tolonen K (1977) On the growth rate and dating of surface peat. Suo 28:19–24
Passmore J (1974) Man's responsibility for nature. Duckworth, London
Patrick ST (1986) The influence of sanitary and other social changes on the eutrophication of Lough Erne, Ireland since c1850. PhD Thesis, Univ London
Patrick ST (1987) Palaeoecological evaluation of the recent acidification of Welsh lakes V. The significance of land use and land management. Palaeoecol Res Unit, Univ Coll London, Res Pap 21
Popper KR (1963) Conjectures and refutations: the growth of scientific knowledge. Routledge & Kegan Paul, London
Rosenquist I (1977) Acid soil–acid water. Ingenir forlaget, Oslo
Thompson R, Oldfield F (1986) Environmental magnetism. Allen & Unwin, London
Thompson R, Battarbee RW, O'Sullivan PE, Oldfield F (1975) Magnetic susceptibility of lake sediments. Limnol Oceanogr 20:687–698
Tolonen K, Oldfield F (1986) The record of magnetic mineral and heavy metal deposition at Regent Street Bog, Fredricton, New Brunswick, Canada. Phys Earth Planet Interiors 42:57–66
Tolonen M (1978) Palaeoecology of annually laminated sediments in Lake Ahvenainen, S Finland. I. Pollen and charcoal analyses and their relation to human impact. Ann Bot Fenn 15:177–208
Vaughan HH, Deevey ES, Garrett-Jones SE (1985) Pollen stratigraphy of two cores from the Peten Lake District. In: Pohl M (ed) Prehistoric lowland Maya environment and subsistence economy. Univ Press, Harvard, pp 73–89
Wasson RJ, Clark RL (1985) Environmental history for explanation and prediction. Search 16:258–263
Wasson RJ, Galloway RW (1986) Sediment yield in the Barrier Range before and after European settlement. Aust Rangelands J 8:79–90
Watt AS (1947) Pattern and process in the plant community. J Ecol 35:1–22

"Policey"* and Environment as a Form of "Social Discipline" in Early Modern Hamburg

N. LANGE[1]

The 16th century marked the beginning of Hamburg's heyday both as a commercial centre and as a bastion of political and religious tolerance. Many people were attracted by the relative prosperity and the stable political climate of the German town. Among these were numerous foreign traders and craftsmen. They had suffered religious or political persecution in their native countries following the 30-Year War and thus sought refuge in Hamburg. The hanseatic town's reputation as a trade centre was enhanced both by its newly gained trading connections and by the introduction of the migrants' modern commercial practices. Additionally, Hamburg also proved attractive to the less wealthy and powerful. Persistent political and economic crises had compelled many of these people to abandon their livelihoods in the country and to seek a new life in the town.

Simultaneously, the resultant growth of the population – at about 1560 some 25 000 people lived in Hamburg, by 1600 there were around 40 000 – led to a high population density. Since Hamburg had no real hinterland of its own and a city-centre that was traversed by waterways, the potential for expansion was severely restricted.

The resulting town-planning problems increased with the transition from the late Middle Ages to the Renaissance when, with the incipient breakdown of the feudal agrarian society and the beginnings of early capitalism, urban commercial production overhauled agricultural output. This led to radical changes in the status quo and, in particular, the creation of a more dynamic social infrastructure. Previously powerful institutions, e.g., the various craft and industrial corporations, suffered a considerable loss of influence.

Furthermore, towns, as the social form furthest removed from nature, became arenas for pollution and environmental damage. Thus the above social, economic, and demographic changes led directly to an aggravation of Hamburg's ecological problems.

In particular the level of hygiene deteriorated proportionally to the increases in urban development and population density. Every additional household within the city walls increased the demand for water and the number of pigs

*The concept of "Policey" refers to a state of, or the creation of, 'good or harmonious order' in society. It is a reflection of the Germanic concern for pattern and regularity.

[1] Heinrich-Barth-Str. 7–9, 2000 Hamburg 13, FRG

Brimblecombe, Pfister (Eds.)
The Silent Countdown

roaming free in the town, and thus the amount of dung and refuse lying about in the streets. Furthermore, private buildings often both infringed upon public space and constricted the thoroughfares, resulting in an increasing obstruction of traffic and the enhanced risk of fire spreading throughout the town. There was also intensive industrial exploitation of the Alster (a large, artificial lake in Hamburg) and its canals, notably by numerous mills, tanneries and dye-works. These contributed substantially to the pollution of Hamburg's water.

These problems were identified at the time. Both the first attempts at early modern town-planning and the utopian perspectives of the 16th and 17th centuries show consistent developments in the fields of both preventative and remedial urban hygiene. The concern of town-planners for hygiene, for example, in respect of water, air and refuse, also offers us insights into their general thinking about matters like water-supply and waste-disposal. Against a background of advancing urbanization, the awareness of the link between health on the one hand and the cleanliness of the water and air on the other led the Hamburg district council to pass a large number of conservational decrees. Private exploitation and influence on the public environment was thus curtailed and regulated.

This was achieved through the issue of decrees which were called *Burspraken* and later *Mandate*.

Through legislation, the authorities used the only means available to impose their political will on the everyday lives of the citizens of Hamburg. For this reason the legislation was generally easy to understand, detailed, of an explanatory nature, and directly relevant to specific problems. At least twice a year, at the beginning and end of the winter break in ship-traffic both new and existing decrees were proclaimed in public.

It is this frequent repetition of *Burspraken* and *Mandate* which enables the modern reader to trace the development of Hamburg's environmental politics. The decrees, which were issued over a period of centuries, show that certain problems like pollution of the streets and water could not be dealt with effectively. But it is not just the content of these regulations but also their language and quasi-clamouring tone which bear witness to the nature of the situation. Often they consist of set phrases which do little more than emphasize the seriousness of a problem through the medium of warnings, demands or punishments. Furthermore, the importance attached to environmental problems can be inferred from the severity of the penalties that were threatened. In some cases rewards were even offered for reports of especially severe cases of pollution.

Here are two illustrative decrees of the period. In a *Bursprake* of 1457 we read: "As it has had to do every year, the Council is compelled to proclaim and decree that nobody shall throw, or cause to throw, refuse into the Elbe or the Fleete (Hamburg's canals). The Council will make sure that this decree is observed. Whoever fails to comply will be fined three silver marks. Should he not be able to pay this, he will have to live on bread and water in prison for four weeks."

And in 1567 the Council proclaimed the following edict: "Everyone shall sweep their rubbish and when the cart comes shall help the driver to load it on."

We can draw the following conclusions from our analysis of the *Burspraken* and *Mandate*, the first such analysis to deal with their importance for environmental history. The decrees were not widely observed, with the result that their impact on environmental matters was limited. It speaks for itself that the Council struggled to deal with the same issues over a period of centuries and was effectively compelled to reissue the same decrees time and time again. The legislation regularly contained complaints about the disregard of decrees.

Without modern administrative machinery, the Council had no effective means of implementing its policies. Only a small number of public employees were entrusted with the task of enforcing the legislation. Furthermore, the threat of punishment had no real effect. Awareness of the need for effective waste-management was so limited that the respective public officials didn't even bother paying their fees. (Note: some Council duties were customarily leased out to members of the public. These paid for the privilege of carrying them out and thus for the prestige that the position entailed. This very mechanism also served to diminish the Council's real power.)

The Council recognized the inefficiency of its measures and gradually developed new strategies. With its growth as an administrative body, it thus extended its portfolio to directly cover environmental matters. Comprehensive regulations, like the "Street Regulations" of 1710 or the "Building Regulations" of 1730, targeted problem areas and reveal a new and broader awareness of ecological issues on the Council's part.

The environmental policies of the Hamburg Council, as seen in the various *Burspraken* and *Mandate* do not just reflect attempts at crisis-management; they are also the wider expression of the "Policey" thinking of the early modern period. This mode of thought emerged as a reaction against the political, economic and social turbulence of the time, and the resulting structural problems in society.

It was in towns that the modernization of production methods and the change in people's lifestyles most quickly took hold. It was for this reason that Town Councillors were the first to translate the idea of "Policey" into practical and political action. The political will to produce a state of "harmonious or good order" in the social sphere – and "Policey" had no other meaning at this stage – resulted from the failure of the traditional guarantors of order. These were no longer able to cope with the increasing complexity and variety of urban life; the urban authorities were thus able to step in and fill the power vacuum. As in the Germanic states, demands for an absolutist power structure (which themselves had their roots in the doctrine of sovereignty) ended with the formation of a centralized authoritarian body in towns.

The "Policey" regulations of the towns, as for example the Nuremburg orders of the 13th – 15th centuries, confirm emphatically that town authorities used this new concept of "Policey" to comprehensively regulate and control the social sphere. In order to create a state of "good order" in communal life they applied their regulations to every area of public life. There were regulations for clothing, for commerce and for trade, among many others. With the loss of its secular power, the Church also soon lost its responsibility for public morality. Matters like adultery, christenings, burials, weddings, begging, the distribution of alms, alcohol, seduction, and gambling thus soon came under the jurisdiction of the urban authorities. Regulations relating to street-cleaning, the prevention of fire, building, and traffic, as well as the vigilance towards hygiene, were the clearest evidence of the need for ecological action. The above should suffice to illustrate both the authorities' demands for complete jurisdiction and the failure of the citizens to distinguish between public and private spheres.

For the Chancellor of Saxony, Melchior von Osse, who wrote down his political legacy in 1555, a good "Policey" contained four elements: a ruler, a good and wise Council, an impartial and good judiciary and a pious and obedient people.

In Hamburg there was no such "ruler". Urban administration was performed collectively, i.e., the Council and City Parliament – which was organized by the districts and parishes – passed laws as *Rats und Bürgerbeschlüsse* (Council and Citizen's resolutions).

Nevertheless the notion of absolute power (*summa potestas*) also led to constitutional conflicts in Hamburg. In the light of the new perception of political rule, the old idea of divine authority became suspect and the Council and City Parliament accused each other of seeking absolute power. The dispute over the *summa potestas* became a permanent ingredient in the town's politics. It was not until the 18th century that a compromise was found to resolve the conflict. This was achieved not by invoking the doctrine of sovereignty but by applying it, contrary to its original meaning, to two bodies, the Council and the City Parliament. In essence they reverted to a relationship of reciprocal power and to a composite constitution.

But the Hamburg authorities also demanded an "obedient people". For whatever the formal composition of the ruling body, "good order" could not be established without public obedience and compliance. And even though the Hamburg Council did not present its laws specifically as "Policey" decrees, their language and aim were comparable to the "Policey" decrees of other towns and regions.

Ecological measures consequently became a permanent fixture of a policy whose aim was to guarantee a state of "good order" (in other words good "Policey") in the social sphere.

The Council did not achieve its aim. In the Hamburg of the 16th, 17th, and 18th centuries the streets and waterways were still being polluted.

But the idea of "Policey" did not remain completely without impact. Gerhard Oestreich has described the *zeitgeist* which is characterized by the political will to create good social conditions as "social disciplining". The following exposition is based on Oestreich's (1969, 1976) definition and the further formulations of Winfried Schulze (1987) (which he developed with reference to Oestreich's posthumous notes).

"Social disciplining" refers to a phenomenon which was recognized in the early modern period, but which retained its significance beyond this period.

The desire to regulate every sphere of human existence became decisive for the political behaviour of the time. The absolutist development of monopolies in political, economic, and military spheres was not adequate to replace that medieval power structure whose disintegration was complete by the early modern period. In particular, the numerous different problems which the above entailed forced changes in human behaviour. The individual had to undergo a process of socialization so that he could adapt to the different circumstances and be integrated into the social framework. Basing their ideas on the Stoic moral doctrine, people placed their faith in the *ratio* and above all in the *disciplina*. This process of disciplining was to have the effect of converting the individual into a social being so that he subordinated his own needs to those of the public at large, without concerning himself overly with the distribution of power which this entailed. Even though social divisions remained intact, the necessity to behave in a socially disciplined manner affected all social strata.

Owing to the urban transformation described above urbanization became per se a process of "self-disciplining". The shaping and structuring force of "Policey" influenced the political, and in this case the ecological, thinking of the general public.

The individual's perception of both himself and his social environment changed. He became aware of both those problems brought about by the increase in population density and greater complexity of urban society and of the regulating and moralistic nature of policies whose aim was good "Policey". This tension between obvious deficiencies in urban administration on the one hand and pressure exerted by the authorities on the other created a dilemma whereby he had to weigh his own needs, e.g., how to dispose of his refuse in the cheapest and most convenient manner, against the interests of the majority, who wanted the streets clear of obstruction.

The initial success of social-disciplining lay in the fact that the individual developed a new awareness of the need to adjust his behaviour.

References

Oestreich G (ed) (1969) Strukturprobleme des europäischen Absolutismus. In: Geist und Gestalt des frühmodernen Staates. Duncker und Humblot, Berlin, pp 179–197

Oestreich G (1976) Policey and Prudentia civiles in der barocken Gesellschaft. In: Schöne A (ed) Stadt, Schulz, Universität, Buchwesen und die deutsche Literatur im 17. Jahrhundert. Beck, München, pp 10–20.e

Schulze W (1987) Gerhard Oestreichs Begriff "Sozialdisziplinierung in der Frühen Neuzeit". Z Hist Forsch 14:265–302

Environmental Policy in 19th-Century Leyden

H. Diederiks and C. Jeurgens[1]

1 Introduction

Men have always been changing and reconstructing their physical environment. How far that implied an improvement or worsening is an important question more easily brought forward than answered. The Dutch are supposed to have created their own country after God had created the world. The struggle with the water not only resulted in a physical, but also in a moral geography (Schama 1987). The countryside and the urban landscape, especially in the western parts of the country, was gained by dispelling the water, the sea and the rivers which originated in Switzerland and France and found their way to the sea through the Netherlands. Dealing with the physical environment is for that reason largely a question of dealing with water. In the process of industrialization and the speeding up of the urbanization, changes occurred in other sectors of this man-made environment. The industry itself influenced the environment by its buildings and by the industrial processes. Water pollution was joined by air and ground pollution.

In a characterization of the global process, it has been brought forth that before the industrial revolution, industrial pollution was generally minimal, with the exception of the area in and around some cities where a dense population was combined with some industrial activities. During the preindustrial period, pollution was mainly caused by microbes. In the 19th century, the shaky balance that existed between the number of people and microbes was disturbed in a rather dramatic way. Because of the accumulation of many people with their cattle and domestic animals in cities, the water management was overburdened (de Swaan 1988). Drinking water and water needed for food processing industries, like beer brewing, became seriously polluted by waste-water. Territorial or urban governments were not able to cope with these problems. In a second period, starting roughly around 1830, the pollution by microbes diminished, but then industrial activities began to create considerable problems by polluting the air, water, and ground. People in the cities began to be psychologically burdened by pollution. A third and last period started after around 1950; a time of hyper-pollution started (Goubert 1974). Because of the retardation of industrialization in the Netherlands, problems related to large-scale industrial pollution occurred at the end of the 19th century.

[1] University of Leyden, Doelensteeg 16, Leyden, Netherlands

Brimblecombe, Pfister (Eds.)
The Silent Countdown

In this contribution, we will deal with the northern Netherlands during the 19th century. We mention explicitly the northern Netherlands because the southern Netherlands (known since 1830 as Belgium) were during the years 1815–1830 part of the newly established Kingdom of the Netherlands. We will (1) introduce some aspects of approaches of the physical urban environment in the existing literature; (2) we will make some remarks on the development of the legal framework; (3) some of the contemporary opinions on pollution and the environment will be dealt with; and (4) some results of recent research on urban ecology for the 19th century will be presented.

In dutch historiography the environment has not received a great deal of attention up to now, although problems of water pollution in relation to urban industrial development for the period before 1800 has been touched upon in a conference contribution as early as 1973 (Faber et al. 1973). Nevertheless, there are some indirect approaches to the problem which in modern terminology would have been called "environmental". These approaches can be characterized as medical, demographic, economic, geographical, and from the point of view of the consumer.

Medical doctors of the 19th century were already aware of negative environmental influences on the health situation. A number of medical "topographies" of towns, as they were called, deal with causes of death and of diseases and the authors point out the urban environment as a factor (Nieuwenhuijs 1816–1820; Coronel 1859). A number of medical doctors also published, mostly in the second half of the 19th century, studies on causes of death (Saltet 1868). In these studies, environmental factors, e.g., the circulation of surface water, are given a heavy weight. In more recent studies on the causes of death by social historians, the most important factor for the decline of death rates is the availability of food, and is only to a lesser degree the purity of surface water (Jansen and de Meere 1982).

Another study, dealing with Amsterdam in the 19th century from a medical-demographic and sociological point of view, mentions the environmental factor but nevertheless explains the lowered death rate after around 1880 by the breakthrough of the "modern, dynamic pattern of culture" (Verdoorn 1965; Leonhard 1986). In demographic studies, the ecological problems are given more weight. The situation in the western part of the Netherlands, consisting of the two provinces of Holland, which contain many moors and much brackish water, provided favourable breeding grounds for the malaria mosquito, and subsequently, for malaria. An increase of death in those areas during the first half of the 19th century was observed (Hofstee 1978).

From the point of view of the national, regional, or local economy, many studies also contain elements of importance leading to further knowledge of the physical environment. The construction of canals, the improvement of rivers, and the plans for land reclamation certainly contain elements related to the environment (Pool 1838). Some of these problems are dealt with by the author of a study generally categorized as a geographical study. This book, written in 1987, deals with the problem of spatial planning, or rather, the absence of that planning, in the Netherlands between 1798 and 1848. Based on secondary literature, the author tries to give a description of the rural and urban landscape during the first half of the 19th century. One of the questions he started with was the existence of deliberate urban planning. On the basis of his reading he concluded, a conclusion that certainly can be discussed, that there was no planning in that period (van der Woud 1987).

As a last category of "marginal" approaches to the problem of the physical environment, we mentioned studies analyzing consumption, especially the consumption of water. Because of the lack of wells and of clean surface water, the lack of sources of drinking water was a great problem, especially in the urbanized, western part of the Netherlands (Vogelsang 1956).

The only full-blooded historical study of ecology in the Netherlands deals only with nonindustrial pollution for the period 1850–1920 (van Zon 1986). Recently a project has been started studying the environmental problems in the industrial town of Leyden from 1850 until the present time. Some results of that project will be presented in the part of this contribution dealing with urban ecology (for a first review see Diederiks and Jeurgens 1988).

2 Legislation in the 19th Century

In the beginning of the 19th century, the first national legislation was passed to reduce the negative effects and damage caused by human activities. Before this century, many towns had enacted local rules, but there were no regulations on a regional or national scale. The first general law reducing pollution was enacted on October 15, 1810, by the emperor Napoleon, who incorporated the Netherlands into his empire. After the introduction of this general law, one needed permission to establish a factory or an institution emitting offensive odors. In this Napoleonic decree, three categories of factories were distinguished:

1. Factories which were situated on a great distance of dwellings. The entrepreneur had to inform the *prefect* (the governor of the department), who announced the request on placards in all municipalities within a radius of 5 km. In order to build this type of factories, a decree of the State Council was needed. If there were objections, the State Council was advised by the *conseil de préfecture* (the Country Council). If there were no objections made, the State Council was informed by the governor and by the Home Secretary. Factories and institutions which needed such a permission were, for instance, slaughter houses, zoological gardens, glue-works, pig farms, soda factories, and factories of bone-oil.

2. Factories and institutions which were allowed to be situated near dwellings, on the condition that the occupants would not experience any trouble. The entrepreneur had to ask permission of the "sous-préfect" of the district. This official handed over the request to the mayor of the municipality where the factory had to be built. The mayor was obliged to make inquiries *de commodo et incommodo* (about convenience and incovenience). The *sous-préfect* made an official report

on behalf of the governor of the department, who had to give permission. Opponents had the right to appeal to the State Council. Factories belonging into this cluster were, for instance, white-lead works, gin distilleries, bleacheries which made use of chlorine gas, and dye-works.

3. Factories which could be situated near dwellings without objections. They were inspected by the local police authorities. The *sous-préfect* had to give permission to build this kind of factories, after being advised by the mayor of the municipality where the factory would be built (MacLean 1982).

In 1824, this Napoleonic decree was replaced by the Order in Council of January 31, *rakende de vergunning ter oprigting van sommige fabrijken en trafijken*. It was almost a duplicate of the decree of 1810. The classification in three clusters was maintained. The authorities granting permission were the King, Country Council, and municipal authorities. Permission to raise a factory was only given after inquiries *de commodo et incommodo*.

The legality of this Order in Council has often been questioned, because the opponents were of the opinion that an imperial decree could only be replaced by a law and not by an Order. Although legality of this Order in Council was confirmed by the Supreme Court in 1846, the continuous criticism on this Order stimulated the achievement of the so-called *Hinderwet* (nuisance act) of 1875 (Aalders 1984). This law prohibited the raising of factories without permission of the municipality. Permission could be refused if a factory or an institution was a danger, or if it was regarded as harmful to people's health, or when it caused damage to property (especially to dwellings), e.g., by diffusion of smoke or noise. Appeal to the State Council was possible (MacLean 1982; Aalders 1984). This law remained in operation (with minor changes introduced in 1896) until 1952. The opponents of the *Hinderwet* of 1875 feared a curtailing of the possibilities to establish industry. Manufacturers were of the opinion that people had to accept the few negative consequences of industrialization. In 1875, the Maastricht manufacturer L. Regout said in the Provincial States of Limburg, referring to the pollution of the river Geul, that he regretted "to hear, that industry has to be second to the demands of a beautiful nature and to the preservation of trout and crawfish. That would be unfavourable for the whole region. What would have become of the city Luik (Liége in Belgium, quite near Maastricht) if people there had argued in this way. Because of the nearness of the coalmines, there have been subsidences in some places. On asking the inhabitants of Luik, if they would be liberated of this, they answer: yes, but we are brooking it because our prosperity is depending on it" (cited in Philips 1981).

In reality the *Hinderwet* did not cause much trouble for industry. Although the intention of the law was the prevention of harmful practices, the reality was different. Only when there was evidence of danger or damage, did the authorities take action against factories. A lot of industries even started production without permission of the local authorities, and many factories sought permission only after complaints from the neighbours (Aalders 1984). Common to all these laws

(1810, 1824, and 1875) was a specification stipulating that factories had to ask permission. Some kinds of pollution could be reduced, but other sorts could not be tackled by these laws. Households belonged to the category of the greatest polluters in the 19th century, and they remained outside the reach of these laws. In the absence of sewers or a well-organized system of garbage collection, people just threw their waste in the streets, gardens or canals. These circumstances were disastrous for people's health. Legislation to reduce this kind of pollution was very ineffective. By a decree of March 20, 1804, committees of medical inspection were established. The committees were on a local level, appointed by the municipalities to spot epidemic diseases and to inform the departmental committees, which in their turn had to keep the central government posted. In 1818, this decree was replaced by a very similar law; local committees were still established by the municipalities, the provincial committees by the king. One of their main tasks was to be alert on the arising of contagious diseases. As we will see in the next paragraph, contagious diseases were, in perception of the 19th century, caused by polluted air, soil, and water. So the inspectors paid attention to the quality of these three elements. However, the effectivity of the law depended on the willingness of the local authorities to take measures when abuses were noticed by the committees. The committees themselves had no power to take measures. The situation was ameliorated by a law of 1851, because all municipalities were obliged to make regulations in this field. The committees were then given a criterion, and offenders of the local regulations could be prosecuted. In the reports of the local committees, the sanity inspectors constantly point out the pollution of water, nuisances caused by factories, and the bad quality of dwellings (Querido 1965). The local authorities were either unwilling or unable to tackle the abuses. That was the main reason why a new law was enacted in 1865, which remained into operation until 1901. The committees were abolished and replaced by seven inspectors and seven assistant inspectors. These inspectors had to advise the local authorities and give information to the minister. Since the new law gave them no power, the main result of their work was a long list of noticed abuses. When local authorities were unwilling (despite the act of 1851) to change the abuses, the sanitary inspectors were helpless. Only when they noticed trespasses of the regulations, could they take action, as shown by the following example: in the official report of the inspection of Limburg and the eastern part of North Brabant of 1894, the inspector described an enormous pollution of the river Voer in Eysden, near Maastricht. This pollution was caused by two slaughter houses which threw their waste, offal and sheepheads, in a large pit just on the banks of the river Voer. Because the abbatoirs did not possess a permission in virtue of the *Hinderwet*, they could be closed. However the inspectors were mainly powerless. The same inspector wondered how it was possible that the abuses in a factor of zinc in Herkenbosch were still tolerated. In his report he noticed that the air in and in the surrounding of the factory was laden with sulphur fumes to a degree that people could not stand it (Querido 1965). The descriptions are numerous where the inspectors report an enormous amount of pollution caused by the absence of a sewerage system.

In 1897 a governmental committee had to report on the environmental problems as a preparation for rules concerning the public surface water. The committee made an inventory of all the regulations in existence. It distinguished:

1. Rules forbidding to let blood, garbage, dead animals etc. into the public water;
2. Rules concerning the letting of blood in sewerage systems;
3. Rules related to the letting of faeces into public water;
4. Rules forbidding to weaken skins in public water;
5. Rules concerning flax in public waters.

The committee searched all municipalities – 1121 in total – and found that 25% had no rule at all; 70% had regulations for the letting of blood into the public water; 21% had rules concerning faeces. There were differences in the occurrence of rules. In the provinces of North- and South Holland, only 15% of the towns had no rules at all.[1]

3 Opinions About Environmental Pollution in the 19th Century

Pollution of water, soil, and air are considered to be a great threat for our contemporary society. How did people in the Netherlands look upon pollution in the 19th century? Did people see it as a problem?

Undoubtedly there was pollution in the 19th century. Numerous descriptions tell us about the dreadful smell produced by the canals in the cities. Water played a crucial role in the western part of the Netherlands, where most cities had a system of canals. The water circulated in these canals and was mostly carried off to a river. The cheapest way to get rid of waste was to dump it in these canals, which became public sewers.

Especially in summer, when there was little water in the canals, these public sewers became a problem for the inhabitants. In many cities it was quite normal to drink the water of the canals. But in summer people regarded this as dangerous. Then fresh water became scarce, and when rainwater, which was collected in numerous casks (this water was often plumbiferous, because the casks were made of lead), was nearly exhausted, the city government had to take measures. In some cities, water pollution was a problem during the whole year. In Amsterdam, the water of the canals had not been potable since the 17th century. The water was not only polluted by faeces, but also by industries, such as tanneries, sugar refineries and dye-works, draining their waste in the canals (Verdoorn 1965, p. 221). In 1860, the "Directeur der Maatschappij tot Redding van Drenkelingen" (the managing director of the society to save-drowning persons) noticed that drowning persons died who were saved out of the Oosterdok, even when they had

[1] Verslag van de staatscommissie ingesteld bij K.B. van 18 october 1897, no. 32, tot voorbereiding van maatregelen tegen verontreininging van openbare wateren, Den Haag 1901.

Report of the governmental commission established by a Royal Decree of 18th of October 1897, no. 32, for the preparation of bylaws against the pollution of public waters, The Hague 1901.

been only for a short time in the water. He assumed that the water was extremely polluted and as a probable cause he mentioned the numerous ships. He could not establish the nature of the pollution (Kool 1961). In Amsterdam drinking water had to be supplied from other places, like the river Vecht near Weesp. In 1844 the "Versch Water Sociëteit" (Fresh Water Society) had 23 ships to carry fresh water from the river Vecht to Amsterdam (Verdoorn 1965, p. 208).

After 1853 the situation for Amsterdam improved because drinking water was supplied by conduit pipes from the dunes near Haarlem. The situation in this town was exceptional. In most towns, the water had to be taken out of the dirty canals. As long as water was flowing, it was not regarded as harmful to people's health (van der Woud 1987, p. 399). This opinion has to be seen in the framework of the medical knowledge of that time. The occurrence of many diseases like cholera, typhoid fever, and malaria, were explained by the so-called miasmatic theory. This theory assumed that diseases were carried by air, i.e. polluted by bad vapours (Morris 1976, Diederiks and Jeurgens 1988). Stagnant water was regarded to be very conducive to the origination of these vapours, but polluted soil could also result in the origination of miasmas (bad vapours). This miasmatic theory was generally accepted until the 1880's (Kooij 1982).

People tried to protect themselves against these dangerous miasmas in many ways, for example with the cupboard-bed, which was regarded as an effective protection against air currents containing bad vapours. In 1770, it was considered as useful when a town had a wall in between the town and the canal around it, assumingly providing good protection against the foggy vapours coming out of that canal (van der Woud 1987, p. 391).

Because people were aware of the danger of stagnant water, many towns endeavoured to improve their hydraulic works. The most famous plan came from the architect of the city of Rotterdam, W.N. Rose. He presented in 1842 his so-called waterproject, an attempt to reduce the enormous pollution of the water in the city canals. He designed a system of canals that could be flushed with water from the river Maas. In this plan, the many slaughter-houses were banned from the centre of the town. Because of technical and political problems, this project was only partly realized (de Graaf 1954).

In 1837, the city of Leyden learned about the intention of the government to reclaim the Haarlemmermeer (Lake of Haarlem), a lake of over 44 000 acres, situated between the cites of Amsterdam, Haarlem, and Leyden. The city government of Leyden strongly protested against these plans. This great lake caused the much-needed flushing of the dirty city-canals. If this lake were turned into a polder, the flushing would disappear and the water in and round the town would become a stagnant stinking pool, dangerous for people's health. Because the reclamation of the Haarlemmermeer meant a reduction of the system of reservoirs for superfluous water (a reduction of 80%!), the government of Leyden feared they would be confronted with an enormous pollution problem in summertime, because the water level would be lower in summer. The city government was afraid that the consequences would be disastrous. In a letter to the government, the mayor of Leyden predicted an economic disaster if the reclamation became a reality. In summer, the water would be polluted so much, that it could not be used anymore by the numerous factories in the town needing fresh water (textile-industry, breweries, etc.) This would lead to unemployment, and the labourers would have to apply for their maintenance to the poor-relief system. Inhabitants would fall ill because of the miasmas rising out of the canals (Gemeente-Archief Leiden 1837).

The many reclamations in the western part of the Netherlands, implied a special sort of risks for health. People were very apprehensive of pools with a boggy soil. In the final phase of reclamation of such pools, the bog would get in touch with open air. Especially during hot weather, people feared the occurrence of the so-called polderdiseases, a general term for all sorts of fevers. Indeed, these diseases occurred often in these "wet" areas, but people were not aware that they were caused by the malaria mosquito, bred in brackish water. The reclamation of the enormous Haarlemmermeer (1840–1852) certainly had the attention of the authorities, because this lake was situated in the most densely populated area of the Netherlands. The politicians concluded that there was no danger if the final draining took place in a season with low temperatures (Algemeen Rijksarchief 1837). The physicians J. van Geuns and G.J. Pool were not convinced that this was a sufficient guarantee to prevent the occurrence of the so-called polder-diseases. They recommended the government to take extra precautions to insure the health of the laborers. Between the cottages of these workers and the lake, the government had to plant trees to protect the labourers against the dangerous miasmas coming out of the lake. Fires had to be kindled to drive away the bad vapours (Pool 1838; von Geuns 1839). The arguments of both physicians are summarized in Gevers van Endegeest (1861). Industrial wastes were not regarded as highly dangerous in the miasmatic theory, although in 1827 the chemist M.J. Mulder argued that industrial activities were harmful. In his opinion, every form of industry damaged to the atmosphere and therefore also people's health. He condemned the burning of coal, wood and, peat because of the bad vapours, which were not only harmful to breathe, but could also pollute the drinking water. Mulder insisted on scientific research in order to give the government data to develop a responsible industrial policy (Mulder 1827).

After all this, one cannot deny that the 19th century populace took care in relation to pollution problems. We can confirm the quotation of the Dutch historian Van Zon, who wrote that, in previous centuries "complaints about bad smell and other nuisances or dangers are not less serious than in our time. The motives were maybe of a less dangerous kind than present-day. For the experience of the 'victims' this implied probably no difference" (van Zon 1986, p. 9).

However, it was not only the bad vapours which worried the people in the 19th century. The government also paid attention to forms of pollution which are still regarded as very dangerous. A good example is the concern of the Dutch government with regard to the German waste-drainings in the last quarter of the 19th century. The opening of the Nieuwe Waterweg in 1872 gave Rotterdam a direct connection to the North Sea and meant a faster connection with the sea for the developing German Ruhrgebiet. One of the prospering branches of industry in Germany produced chemical paint. This analine production replaced the traditional paint industry. A disadvantage of the chemical paint production was the often dangerous poisonous waste-products. The flourishing German aniline industry (like the Badische Analin und Soda Fabrik BASF) used as raw material coal, tar and arsenic. The waste-products subsequently contained much arsenic. This was thickened with lime, put in barrels and brought by ship to Hamburg to be dumped in the Baltic Sea. After April 1873, the poisonous waste was transported to Rotterdam, to be loaded in sea-going vessels which dumped the load in the North Sea (de Graaff 1982). When this became known in the Netherlands, the authorities tried to prevent these transports. The German authorities reacted furiously to the Dutch intention to prevent the import and transit of this dangerous material, and they threatened to throw the barrels into the Rhine, just on the German side of the border. In 1877, after a lot of political haggling, an Order in Council provided strict regulations in relation to the packing of the poison and how the barrels were loaded in the ships in view of possible accidents on Dutch territory. Although fishermen warned the authorities against the danger of the poison, the Dutch government could not prevent the North Sea from becoming the garbage heap for this chemical waste. Some fishermen said that they had had a heavy barrel several times in their nets, and the fish they caught on that occasion were all red and already dead (de Graaff 1982). The Dutch government acknowledged the danger of this poison, but it was for mainly political reasons that the enactment of a total prohibition of the import from Germany and transit of this dangerous waste through the Netherlands was prevented.

It is important to notice that people in the 19th century were not worried about pollution of the environment because they considered that the environment as such was being threatened. Their anxiety was almost always related to their own health. Only when there was a direct link with people's health, did governments take measures. The first movements founded for the preservation of nature were born only at the turn of the century. The appreciation of nature, however, can be traced back much earlier. In the middle of the 19th century we can find an increasing interest of people in cities to visit unspoilt nature: the woods, the dunes and the heather (Beckers 1983). This may have been the beginning of an appreciation of nature, resulting in the acknowledgment that it is important to preserve nature because it contains many known and unknown values for (human) life.

Although not only water was polluted, other forms of pollution did not draw much attention from people in the 19th century. Only when epidemic diseases, such as cholera, were feared, did the authorities conduct campaigns against pollution. Streets were cleaned, garbage heaps were cleared away and dirty slum dwellings demolished. These periods of cleaning-up, were over as soon as the epidemic diseases were over (de Swaan 1988). In the last part of the century we can recognize an increasing interest in cleaning the towns. The first municipal public cleaning services were established in the last quarter of the 19th century. It was not only because of the increasing pollution problems, caused by a growing population, that authorities tried to find solutions, but the interest in cleaning also increased because of the economic value of refuse. Faeces, ashes, etc. could be used as dung to improve the quality of agricultural soil.

Since 1848, the Amsterdam municipality leased the right to collect faeces, ashes, and garbage to the "Maatschappij tot bevordering van den landbouw en landontginning" (Society of the promotion of agriculture and reclamation). The Society, however, did not perform her task of waste collection satisfactory. Since this service functioned badly, people continued to throw their waste into the canals. That was a thorn in the flesh of the local authorities, and the municipality decided to establish a municipal public cleaning service in 1877 (Verdoorn 1965; van Zon 1986).

4 The Urban Landscape

To get some insight into the problems in an urban environment we will take the case of Leyden in the 19th century. First we will sketch in a very short outline of Leyden during the second half of the 19th century. The period of great prosperity of the town lasted during the second half of the 17th century, while the decline started in the 1680's (Diederiks 1985). During that period, about 70% of the employed population was involved in industry, mainly textiles. In the mid- 19th century, the textile industry was still the greatest employer in Leyden. In 1859, and also at the beginning of the 20th century, more than half of the employed population worked in the industrial sector. Of this half, about a quarter was involved in the textile branch, 17% worked in the building trades, 15% in the sector of ready-made clothing and cleaning, 15% in the food processing industry, and about 10% in metallurgy (Branderhorst 1962). From 1829 till 1912, the plants increased the number of workers per plant. For instance, in the textile industry we find in 1859 an average of 27 workers in a plant, and in 1912, 166. This increase in the work force per unit of production was of course also reflected in the amount of mechanization and the introduction of steam engines.[2]

Demographically, Leyden had a small increase of the population, but it never regained during the 19th century the number of inhabitants of the 17th century. The increase of the population during the 19th century was less than in nearby Haarlem or Amsterdam. Birth and death rates were higher in Leyden than in Haarlem and Amsterdam, and they were also higher than the national average. In the second half of the 19th century, more people left Leyden during most years than entered it, while the total turnover was very high compared to Haarlem and Amsterdam, where more people established themselves than left the towns. One may conclude from these rough demographic data a very unfavourable picture of the social and economic situation in Leyden (Diederiks and Jeurgens 1988).

So, generally speaking we may describe the town of Leyden during the second half of the 19th century as an industrial town which was not very attractive for immigrants. Nevertheless the town kept, maybe because of its industrial past,

[2] De fabrieken en werkplaatsen vallende onder de veiligheidswet, samengesteld uit het centrale kaartregister der Arbeidsinspectie. Den Haag 1912

Listing of the factories and workshops liable to the Factory Act, compiled from the central registration of the Board of Inspection, The Hague 1912.

industrial functions and even developed those functions and broadened its industrial base. Modern industrial methods, mechanization, and large-scale production were introduced. Traditionally, the industry was located in the northern and eastern part of the town (the last expansion of seventeenth century) and there we also find the newly established industries of the 19th century. Research concerning the first 30 years since the mid-century has revealed more than 2000 requests for permission to introduce a steam engine, or to change an already existing one, to construct a forge or a depository for paraffin, or to establish a factory as mentioned in the laws of 1824 or of 1875. A quarter of the requests concerned a steam engine or the change or expansion of an existing one in several sectors of industry, mostly in the textile branche; 28% of all requests came from the textile industry, 21% from the rather new food processing industry. The metallurgy followed with 16% of the requests handed in between 1850 and 1880 (Jeurgens et al. 1988). The other branches concerned depositories of paraffin and enterprises of cleaning tapistries. In most cases, one may guess about the nature of the polluting materials, either in case of storage, or in case of production. One may wonder what happened with the cinders of the coal used in the steam engines, or with printing ink lost from the many printshops. Paraffin and sulphur are two products, mostly mentioned in requests for permission of storage, that may have caused pollution of the air or soil. About 150 requests, that is about 7.5%, concern those two products. What about the procedures of the requests and the ways to handle in case of a refusal by the local authorities? In 1824, the royal decree provided regulations for very dangerous enterprises, which had to apply at the central level, the crown; for the establishment of less dangerous ones, one had to apply with the provincial authorities, and the more simple cases were dealt with by the local, i.e., urban authorities. In Leyden there was a "committee of manufacturing" which had to provide advice for the mayor and the board of eldermen. The committee organized hearings, *de commodo et incommodo*, inviting the neighbours to bring forward objections or complaints. If the request was refused, the applicant could appeal to the Council of State. Next to the national laws there were local regulations, laid down in the "general police regulations" (APV). For instance, in 1873 it was forbidden in Leyden by a new article in these regulations to pollute the public space. The regulation mentions in this respect a great number of products: bones, manure, soot, refuse, etc., and it states that it is forbidden to deposit those in the streets or in public waters.

For a steam engine, the request had to be brought at the provincial level. The provincial committee asked advice from a local, urban committee. The role of the mayor and the board of eldermen as responsible to the municipal council was very important; although, as a last resort, an appeal at the Council of State, was possible. So, who do we find in the municipal council, controlling the mayor and his eldermen? Can we expect a conservative or a more liberal policy? A policy favourable for the environment or for the industry? During the 30 years under consideration (1850–1880) we know of about 75 persons who made up the municipal council in varying combinations; 20% were industrial entrepreneurs, 34% were lawyers, 14% were university professors, 12% were merchants, and 12% officials. The industrial entrepreneurs were mostly engaged in textile industry, two were saltmakers, two were distillers of liquor and one was a printer. Three of the merchants were active in the wool trade and may have had common interests with the textile entrepreneurs. Of most of the university professors we know that they were liberals, in the sense of the word that they supported a laissez-faire policy.

We may conclude that the composition of the town council indicates an attitude that, generally speaking, would support a policy favourable to industry. In the national parliament, we find as representatives for the district of Leyden (at that time the system of districts was still in use in the Netherlands) two conservative persons. In general, the conservatives supported measures protecting the physical environment. In the parliamentary discussions on the new nuisance act of 1875, these conservatives maintained that the water of the polders surrounding the towns should be kept free of all kinds of pollution, and they combated the view that the local mayors should decide on the establishment of industries such as gasworks. These gasworks were polluting the public water; not only the supply of drinking water, but also the cheese- and butter industry was harmed. In parliament, the discussions on the new nuisance act was focussed on the aspects of repression or prevention. The conservatives were in favour of prevention, while the liberals thought that the effects of the law had to be repressive. The last point of view implied only a reaction of the authorities in case of complaints. In the end, the repressive approach became the prevailing one, although the preventive effects of the law must also have been there, however, these preventive affects are not easily measured. On the basis of some of the appeals at the Council of State in the Hague from the decisions taken by the mayor and eldermen of Leyden, we can explore the nature of the requests as well as the policy of the local government and that of the national government.

We can make a distinction into four categories of complaints: (1) those concerning the introduction of steam engines; (2) those concerning stench and dust; (3) the danger of fire or explosions; and (4) the pollution of drinking water. Of the appeals at the Council of State in regard to steam engines, we find two examples, both concerning a textile mill. In 1869, the mayor and his board of eldermen had given permission to an entrepreneur to build a steam engine. During the hearings, using the procedure of *de commodo et incommodo*, the party appellant had not appeared; however, he approached the provincial committee after permission was already given to the entrepreneur. The party appellant mentioned in his report that people living next to the factory already had abandoned their houses and that other neighbours had also many complaints. A neighbouring house had many cracks and people were afraid that it would fall down. The Council of State decided that the owner of the textile mill had to construct a wall to prevent noise. In the argument of the Council we read that "although the interests of the industry implies that the establishment of factories within the city area could not being prevented, the owner of a factory had the duty, as far as possible, to make arrangements minimalizing the nuisance for the neighbours." In 1878 a neighbour of the same mill complained about the introduction of a new steam engine because of the damage to his health by the noise of the engine. This appeal was refused by the Council of State.

Complaints about stench are almost all related to the storage of bones originating from butchers and slaughter houses. The mayor and eldermen of Leyden refused a request for the establishment of such a storehouse in a rather well-to-do residential area of the town. The entrepreneurs appealed to the Council of State, arguing that an industrial town like Leyden was full of annoying institutions, and that for that reason it was unjust to destroy one section of the urban economy.

Another problem was caused by the annual cleaning of dwellings. Carpet cleaning produced a lot of dust, and because of the dust there was always a danger of pollution of the gutters and subsequently of the rainwater, which still made up an important share of the drinking water supply. The owner of the carpet cleaning company answered that the rainwater was already polluted by a steam engine in a nearby factory. A medical report also discusses this point and brings forward that the dust from the carpets was much worse than the dust from the chimney of the steam engine. The complaints were accepted and the carpet cleaning enterprise was not allowed to be established at the requested location.

Leyden was traditionally a garrison town, and in 1913 there were plans to build a storage for cartridges. The future neighbours protested because of the danger of explosion. These complaints were dropped because the cartridges would never explode all at once.

Problems in regard to the drinking-water supply have been found among other reports on pollution in 1864. The municipal gasworks were held responsible for a leaking gaspipe which polluted the well of an inhabitant of Leyden. The gasworks investigated the case and concluded that frost had damaged the gas pipe. The well was cleaned and the gasworks paid for that, and also for the sand used for the cleaning.

The storage of paraffin was limited in 1873 by the General Police Regulations to 350 liters, in exceptional cases, to 450 liters. The danger of fire was the reason for this rule, but during some hearings, *de commodo et incommodo* neighbours brought forward the danger of leaking tanks and the pollution of the drinking water (Diederius and Jeurgens 1988).

5 Conclusion

This article contains merely preliminary observations in the field of environmental history of the Netherlands, mainly in the 19th century. The history of spatial planning and of environmental problems is still in its infancy. The dutch were always confronted in their history with water as an environmental problem and that resulted in water management at an early stage, implying some forms of spatial planning. The high degree of urbanization in the western parts of the country before the 19th century also created urban ecological problems. The absence of a centralized government before the end of the 18th century prevented the establishment of rules of a national level in respect to the physical environment. The Napoleonic laws (1810) and those of the Kingdom of the Netherlands (1824, 1875, 1896) brought stricter regulations on a national level. In practice, local authorities had a very great influence in the application of those rules. How far the magistrates of the 19th century were aware of "objective" dangers is another matter. Certainly the interests of the industry were not greatly hampered by a strict and rigid use of the rules. To what degree the rules had a preventive effect, is difficult to measure. The population at large was sometimes aware of threats from the deterioration of its environment, e.g., during epidemics. Nineteenth-century people were certainly alarmed by some developments related to the physical environment. Whether this concern was based on a real or a subjective argument, may be open for discussion; in some cases, one may doubt the "objective" basis. However, the problem of objectivity in this matter is not limited to the 19th century.

References

Aalders MVC (1984) Industrie, milieu en wetgeving. De Hinderwet tussen symboliek en effectiviteit. Kobra Amsterdam, pp 131 and 134–143

ARA–Algemeen Rijksarchief (ed) (1837) Binnenlandse Zaken, Nijverheid, 2.02.23/239 Rapp Min Binnenlandse Zaken, Buitenlandse Zaken en Financiën aan de koning, dd Dec 1837

Beckers T (1983) Planning voor vrijheid; een historisch sociologische studie van de overheidsinterventie in rekreatie en vrije tijd. Dept of Sociology of the Agrarian Univ, Wageningen, pp 89–90

Branderhorst HAC (1962) Het economische leven. Leyden 1860–1960. JI Groen, Leyden, p 386

Coronel S, Sr (1859) Middelburg voorheen en thans. Bijdrage tot de kennis van den voormaligen en tetenwoordigen toestand van het armwezen aldaar. van Benthem en Jutting, Middelburg

Diederiks HA (1985) Beroepsstructuur en sociale stratificatie in Leyden in het midden van de achttiende eeuw. In: Diederiks HA et al. Nordaam DJ, Tjalsma (eds) Armoede en sociale spanning. Sociaal historische studies over Leyden in de achttiende eeuw. Hollandse studien, vol 17. Hilversum, pp 45–67

Diederiks HA, Jeurgens Ch (1988) Milieu, bedrijf en overheid; Leyden in de negentiende eeuw. Jaarb Geschiedenis Bedrijf Tech Ver Loren 5:427–445

Faber JA, Diederiks HA, Hart S (1973) Urbanisering, industrialisering en milieuaantasting in Nederland in de periode van 1500–1800. AAG Bijdragen 58:251–271

GAL–Gemeente-Archief Leyden (ed) (1837) Notulen B&W II, bijlage 240, Secretarie-archief II. Nota houdende bezwaren waaraan de stad Leyden bij het droogmaken van de Meer kan blootgesteld zijn, dd 10 Sept 1837

van Geuns J (1839) Natuur- en geneeskundige beschouwingen van moerassen en moerasziekten in verband met de vraag of de droogmaking van het Haarlemmermeer nadeelige gevolgen hebben zal voor de gezondheid der bewoners van nabijgelegene plaatsen en der arbeiders aan het meer. Müller Amsterdam

Gevers van Endegeest DT (1861) Over de droogmaking van het Haarlemmermeer, pt 3. Müller Amsterdam, pp 159–161

Goubert JP (1974) Environment et urbanisation. 6e Congr Int Histoire économique, Copenhague 19–23 août 1974, Rapp Gen, pp 59–69

de Graaff CAA (1954) Rose's waterproject. Rotterd Jaarb 2:177–208

de Graaff MGHA (1982) Milieuvervuiling; een oud probleem. Spiegel Hist 17:90 and 95–96

Hofstee EW (1978) De demografische ontwikkeling van Nederland in de eertse helft van de negentiende eeuw, historisch-demografische en sociologische studie. NIDI:115–168

Jansen PC, de Meere JMM (1982) Het sterfte patroon in Amsterdam 1774–1930. Een analyse van de doodsoorzaken'. Tijdschr Soc Geschiedenis 26:180–223

Jeurgens Ch, Diederiks HA, Doorn PK (1988) Historische bedrijfslokaties en bodemvervuiling in de gemeente Leyden. Onderzoeksverslag over de periode 1850–1880. Univ Leyden, Leyden

Kooij P (1982) Gezondheidszorg in Groningen 1870–1914. Tijdschr Soc Geschiedenis 8:112–155

Kool JA (1961) Bijdragen tot de geschiedenis van het drinkwater te Amsterdam. Schat Gezondh 4:253–254

Leonard J (1986) Archives du corps. La santé au XIXe siécle. Ouest-France

MacLean J (1982) Hinderwetten in de 19e eeuw. Chem Weekbl Mag 473–475

Morris RJ (1976) Cholera 1832. The social response to an epidemic. Croom Helm London, pp 170–184

Mulder MJ (1827) Verhandeling over de wateren en de lucht der stad Amsterdam en aangrenzende deelen van ons vaderland. Amsterdam

Nieuwenhuijs Cs Js (1816–1820) Proeve eener geneeskundige plaatsbeschrijving (topographie) der stad Amsterdam, 4 vols. Amsterdam

Philips R (1981) Het ontstaan van het moderne milieubewustzijn. Spiegel Hist 16:184

Pool GJ (1838) De droogmaking van de Haarlemmermeer, mits met de noodige voorzorgen in het werk gesteld, voor de gezondheid der naburige bewoners en arbeiders niet schadelijk. Sulpke Amsterdam

Querido A (1965) Een eeuw staatstoezicht op de volksgezondheid. Staatsuitgeverÿ Den Haag, pp 23 and 129–130

Saltet RH (1868) De sterfte aan besmettelijke ziekten in Nederland en in Amsterdam van 1866–1905. Tijdschr Geneeskd:399 ff

Schama S (1987) The embarrassment of riches, an interpretation of Dutch culture in the Golden Age. Collins London, pp 15–50

de Swaan A (1988) Aantekeningen uit het ondergrondse. Over de stedelijke waterhuishouding in de negentiende eeuw. Tijdschr Geschiedenis 101:341–342

Verdoorn JA (1965) Volksgezondheid en sociale ontwikkeling, Beschouwingen over het gezondheidswezen te Amsterdam in de 19e eeuw. Spectrum Utrecht Antwerpen

Vogelsang I (1956) De drinkwatervoorziening van Nederland voor de aanleg van de drinkwaterleidingen. Gouda

van der Woud A (1987) Het lege Land, de ruimtelijke orde van Nederland 1798–1848. Meulenhoff Amsterdam

van Zon H (1986) Een zeer onfrisse geschiedenis, studies over niet-industriële vervuiling in Nederland, 1850–1920. Ministerie van VROM Groningen

Air Pollution in York 1850–1900

P. Brimblecombe and C. Bowler[1]

1 Introduction

There was considerable concern about urban sanitation in Victorian England (classically Chadwick 1842). This interest was driven by observations of increases in mortality as the population shifted its locus into cities. Brenner has argued that the statutory regulation of pollution during the last century formed part of a general movement to improve the sanitary conditions of urban life (Brenner 1974). Like other sanitary concerns, agitation about city smoke lead to the inclusion of pollution control clauses in 19th-century sanitary legislation. Despite the passage of such well-meant laws, they seem to have been largely unsuccessful in bringing about any marked reduction of the smoke in British cities.

This chapter is concerned with a particularly interesting period of air pollution history in the United Kingdom. The broad framework of the development of air pollution in London is treated in a book (Brimblecombe 1987) while the development of 19th-century law, with much emphasis on the metropolis has also been the subject of a book (Ashby and Anderson 1981). Here we look at a more provincial situation, examining the way in which the civic administration of York approached the issue of smoke abatement in the second half of the 19th century. The problems encountered in this city no doubt serve as a model for those of many smaller cities in the UK last century (Bowler and Brimblecombe 1990a).

2 Air Pollution in Victorian York

York is usually regarded as a nonindustrial city, away from the polluted heart of Yorkshire and towns such as Leeds or Bradford. Last-century visitors tended to perceive it as "smoke free" (e.g. Kohl 1844), although this was only true by comparison with the major industrial cities. There had been a few isolated incidents concerning smoke pollution dating back to the 16th century, but there was no systematic legislative action until the 1850's.

Some cities, such as Manchester, adopted smoke control regulations of their own in the first part of the 19th century (Malcolm 1977), but a formalized

[1] School of Environmental Sciences, University of East Anglia, Norwich NR4 7TJ, Great Britain

Brimblecombe, Pfister (Eds.)
The Silent Countdown

approach in York awaited the passage of the Health of Towns Act 1853, which allowed the appointment of a Sanitary Inspector. The smoke clauses of this Act did not pass into law unnoticed by the public. This might be gauged by the fact that Mary Gaskell, in her novel *North and South* of 1855, talks of "unparliamentary smoke".

It is often argued that the Victorian society was anxious not to burden industry with pollution abatement regulations (e.g., Brenner 1974). However it is clear that equating smoke with wealth was by no means universally accepted. In York, for instance, a general public interest in the mid-19th century was shown by the publications of articles in the *Yorkshire Gazette* (29/10/1842) such as the one on smoke burning by W. Beckett, who argued that while anxious for the commercial prosperity of York "we cannot but regret the great increase in steam engine chimnies. . ." Even broader concern about the matter is evident from the submission of "A Smoke Nuisance Petition" by hundreds of residents in the 1850s. The petition was addressed to the Lord Mayor, Aldermen, and Councillors and stated that this nuisance was not the "necessary result of the extension of commerce" and as such the residents were not prepared to put up with it. They asked that the Council as guardians of the interest and comfort of their constituents "obtain such adequate legislative provision from Parliament, as may give them a due supervision and control in this matter, such as in other towns and cities has been found sufficient to remedy the smoke nuisance (Smoke–York City Archives)". Clearly the public were in tune, at an early date, with the increasing feeling that air pollution should not be seen as the inevitable outcome of industrialization.

A deputation from York City Council went to Manchester on November 1, 1852 to see how the police dealt with the procedures for abating nuisance and to inspect factories where patent smoke removal devices had been installed. The deputation concluded that patent devices often did not work. They formally reported their findings in a document of the same year: "the Smoke Nuisance with evidence as to the Causes of the Nuisance and the Entire Practicability and Economy of Prevention".

The deputation concluded that it was not necessary to give examples of the economy of smoke prevention; "On the contrary they feel strongly that even if the prevention of smoke entailed an outlay upon the proprietors of furnaces without any equivalent advantage, their neighbours and fellow citizens may reasonably expect that every effort should be made to abate the nuisance, in consideration of the incalculable injury it inflicts upon their health, comfort and property. Families in every grade of society, and of every calling, but especially large numbers of the industrious poor, are under an imperative necessity to reside in the city. The soot injures their clothing, their furniture, their goods, prevents the due ventilation of their houses, by compelling them to exclude the soot, and in various ways inflicts domestic and personal discomfort. Such of the wealthier inhabitants as are able, leave the city to avoid the nuisance, and thus valuable house property is depreciated in value, and the poor and other rates proportionally increased in the localities infected with the nuisance. It must be admitted that several proprietors of furnaces in the city have felt the force of these facts, and have at some expense and trouble endeavoured to remedy the evil, but it is also certain that isolated efforts are comparatively useless, and that ALL should follow the same line of conduct to ensure a real and permanent good."

The momentum from this activity was apparently not lost because in 1853 there was pressure on the council to introduce a "smoke consumption" clause into the York Drainage Sanitary Improvement and Foss Navigation Bill. After much discussion between Lord Redesdale, local manufacturers and smoke abatement enthusiasts, a clause was agreed upon:

"That every fire-place or furnace constructed after the passing of this Act in order to be used within the limits of this Act in the working of Engines by Steam or in any Mill, Factory, Dyehouse, Brewery, Gas Work -or in any manufactory whatsoever (although a Steam Engine be not used or employed thereon) shall in every case where the same shall be practicable be so constructed as to consume the smoke arising from the fuel used in such fire-place or furnace, and every such fire-place or furnace existing within the said limits at the date of the passing of this Act, used for the purposes aforesaid and not so constructed as to consume the smoke arising from such fire-place or furnace shall within two years after the passing of this Act in every case where the same shall be practicable be so altered in its construction as to consume smoke, and if, after such period, any person use for any of the purposes aforesaid, any fire-place or furnace not so constructed as aforesaid, or if at any time, any person use any such fire-place or furnace constructed after the passing of this act and not so constructed as aforesaid or so negligently use any such fire-place or furnace as not to consume the smoke arising from the fuel used therein, every person so offending shall be liable to a penalty of not exceeding 40 shillings for every week during any part of which such furnace or fire-place shall be so used and continued after one months notice in writing shall have been given to the owner or occupier of such furnace or fire-place by the Inspector of Nuisances within the City to remedy or discontinue the use of the same Provided always that in every case before such Notice be given a complaint shall have been first made to the Local Board of Health of the said City by two Inhabitant Householders who are aggrieved thereby, and that the Committee of such Board shall have directed such Notice to be given."

By 1857 two men had complained to the council that smoke from the furnace of Messrs Baines and Duffield "is not consumed" and that the Town Clerk was to give notice in accordance with the Act of Parliament.

The Local Board of Health (LBH) minutes for 1858 contained a "Return of Chimneys within the District of the Local Board which emitted smoke on the day and hour reported so as to be a Nuisance to the Public". There are 19 chimneys on the list and each received a request to take early measures to abate the nuisance (LBH 1858). This rather weak response appears typical of the reaction to complaints about the smoke nuisance in the 1850's. The ineffectiveness of council actions in no way hampered public pressure for more stringent controls. In April 1859, for example, a memorial was presented by a deputation complaining of the nuisance from the railway works, William Elmsleys mill on the Mount and the works of both Mr Steward and of Baines and Duffield (*City Minutes* 1859).

These early attempts at battling with the smoke problem were merely sporadic confrontations. A complaint from members of the public was lodged before the council. Following limited discussion the matter ended with an admonitory (and probably ineffectual) letter to the owners of the offending chimney. One can hardly recognize a great deal of concern from the council over these breaches of the smoke abatement legislation.

It is much more illuminating to look at a lengthy campaigns against a factory that later became a continued source of complaint. In these circumstances a wider group of official bodies were involved such that the matter could not simply be dismissed and forgotten about. Thus we are able to see a genuine official concern about smoke abatement for the first time. A good opportunity to examine a broader confrontation arises with reactions to the emissions from Messrs Close Ayre and Nicholson, Iron Founders, York.

3 Close Ayre and Nicholson

The firm of Close Ayre and Nicholson bought the business of Edwin Thompson, Iron Founder, Fishergate in 1859 (*Yorkshire Gazette* 1859). Thompson's foundry had already been the subject of complaint in 1855 (LBH 1855–1861). Close Ayre and Nicholson removed the name of Nicholson from its title in the early 1870s following the death of Nicholson in Ventor on the Isle of Wight in 1866 (*Yorkshire Gazette* 1866), but continued to be an important aspect of metal working in York until the mid-1880s.

Close Ayre and Nicholson had no less problem with smoke than the earlier owner. Soon after they took over the firm it was said that the new owners' foundry chimney at Fishergate "emitted Smoke which is a great nuisance" (LBH 1855–1861; *City Minutes* 1859–1862). By the early 1870s, the firms new foundry was disturbing local residents who complained of the noise, smoke and untidiness. The council's actions were as usual ineffective. This time the citizens did not intend to let matters rest. They placed the issue before the Local Government Office in Whitehall, petitioning the Rt. Hon. H.A. Bruce. The Local Government Office wrote to the York City Council about this matter in late May 1870. The council deferred making any decisions in August, so by December there was a further letter from Whitehall about foundry smoke. The letter from Whitehall was written in response to a further complaint from local residents who claimed that "the Nuisance from Smoke was more intolerable than ever and with the low Chimneys that were in use must continue so."

In December 1870, the Town Clerk reported that he had called the attention of Ayre Close and Nicholson to the matter and that the Inspector of Nuisances had visited the foundry. A letter from the Phoenix Iron Works of the firm of Close, Ayre and Nicholson to Mr Wilkinson the Town Clerk reads:

"...we are working two puddling and two Ball furnaces for which Chimnies are made as high as practicable and we believe fully as high as any other Factory in any other part of the Kingdom. We have three ordinary Steam Boilers and the furnaces connected with them consume the Smoke as far as possible. We have also several Smith's fires for which. . . we have erected a large Chimney Stack and taken flues into it from Fires nearer the dwellings of the people who could have any reasonable complaint." This has cost the firm several hundreds of pounds and "in every way used our best endeavours to prevent any nuisance."

Samuel Healey, the Inspector of Nuisances, reported to the Board of Health: "[I] found 4 low chimneys belonging to different furnaces which are left open and the Smoke unconsumed – two of these are puddling furnaces and the other two are ball furnaces. The proprietors assured me they were all constructed in the only way in which furnaces for such purposes could be constructed and that the Chimneys were as high as the nature of the operation would admit, and so far as I am able to judge I have no reason to doubt the accuracy of this statement."

He described the flues taking smoke to two large chimneys and "in my opinion the Firm has done all they possibly can to prevent the issuing of black smoke so as to be a nuisance." Healey drew attention to the 45th Section of the 1858 Local Govt. Act which shows that the powers contained in it shall not extend to the puddling of iron or to the melting and casting of iron.

Shortly after, Whitehall replied to the council's deliberations in a lengthy letter indicating that the council:

". . .are probably right in their contention that the Nuisance arising from the Steam Hammer is not one within the jurisdiction of the Local Board and the Memorialists shall be so informed. With regard to the nuisance from the Chimnies attached to the furnaces I am to point out that though the Inspector of Nuisances states in his report they 'were all constructed in the only way in which furnaces for such purposes could be constructed and that the Chimnies were as high as the nature of the operations would admit' this assurance of the proprietor of the Works and this belief of the Inspector do not seem sufficient grounds for the Board's acquiesing in the continuance of the nuisance without any further effort for its abatement."

The letter calls attention to the Inspector of Nuisances's claim regarding the 45th Section of the Logcal Government Act:

"The Local Board will find that the 45th Section only provides that the Smoke section (108) of the Town's Improvement Clauses Act shall not extend to compel the consumption of all Smoke, in the case of certain processes including the puddling and casting of iron in any District where the provisions of the said act for the prevention of smoke are not now in force, in which the Local Board shall resolve that any one or more of such processes shall be exempted from penalty for not consuming all smoke for any time specified in such resolution not exceeding ten years which may be annually renewed for a similar or any shorter term if the Board shall think fit. I am to ask if the Local Board have passed any resolution exempting processes of puddling or casting iron from penalties under the powers of this Section."

But the Act goes on to indicate the conditions on which the Justice may remit the penalty on the parties causing such Nuisance – "and any Justice or Justices before whom any person shall be summoned may remit the penalty in any case within such District in which he or they shall be of opinion that such person has adopted the best known means for preventing any nuisance from Smoke and has carefully attended to the same so as to consume as far as possible the Smoke arising from the process so exempted during such time as any such resolution shall extend to unless an Order shall be issued by one of H.M. Principal Secretaries of State directing that such exemption shall no longer be continued in such District to such processes or any of them after a time specified in such Order.

It would no doubt be a satisfaction to the Local Board as to those who suffer from this nuisance, to have proved to the satisfaction of the Justices that of which the proprietors of the Works have assured the Inspector of Nuisances- (furnaces all constructed in only way in which furnaces for such purposes could be and that chimneys were as high as the nature of the operation would allow)... I am therefore to point out to the Local Board that it is their duty to put the parties causing the Nuisance to proof before the Justices that they are within the protection of the 45th Section of the Local Govt. Act 1858".

Whitehall accepted that steam hammers were not within the jurisdiction of the Act, and although it agreed on the point of law concerning exemptions, it hardly seemed convinced that the council had acted particularly diligently. In January, the York Sanitary Committee resolved that it had carefully considered the matters and that Ayre Close and Nicholson had constructed their furnaces so as to consume "as far as practicable" the smoke and the Committee "does not consider it expedient to commence any legal proceedings". It is interesting to note at this point that the council's inertia contravenes not only the Local Govt. Act of 1858. They might also be regarded as failing to meet the requirement of the Sanitary Act (1866) which made it mandatory for local authorities to proceed against those who discharged "Black Smoke in such Quantity as to be a Nuisance". The council could only escape from the requirements of this 1866 Act by insisting that the best possible means had been used to abate the smoke.

On January 9th, Whitehall wrote repeating the question of the earlier letter, asking if the York Board of Health had passed any resolution exempting processes of puddling or casting iron from penalties under the powers of the 45th Section and if not, "they should do so without loss of time." (York City Archives-YD1 1871). In February a further letter from Whitehall described the receipt of another complaint from the residents of George St. and the neighbourhood concerning the smoke from the Phoenix Foundry. The letter continued that until the Local Board of Health passed the necessary resolution "it cannot legally avail itself of the resolution communicated to me... as a reason for not compelling the abatement of the nuisance" (York City Archives-YD1 1871).

The letter also included a copy of the memorial:

"Memorial from the inhabitants of the parish of St George to the Local Govt Act Office, Whitehall – We the undersigned Inhabitants and Ratepayers of the Parish of St George in the City of York having appealed to you for your Honorable assistance in the matter of a Smoke nuisance... caused by the firm of Close, Ayre and Nicholson of the Phoenix Foundry that we could not get the Board of Health Committee to look to, you no doubt have reported to them our complaint a few weeks ago and we have been informed from very good authority, not from the Board of Health, that they have sent Mr Healey the Inspector of Nuisances to the place and he reports he sees nothing wrong and the Board of Health can do nothing more in the matter which they will no doubt have reported to that effect Sir to you.

Now Mr Healey being a Servant of the Board of Health he knowing of these complaints and their views of the matter he is wishful to please them all he can no doubt as we referred to before to fact of the intimate connection of members of the Board with the firm of Ayre Close and Nicholson for one member if not a silent partner is owner of part of the property and Mr Close was a member of the Council. In a case like this it is a very easy matter for the Board of Health to arrange for the Inspector to go down to view the place and then for the interested member to acquaint the firm of the same and for them to have the furnaces well burnt up and no Smoke issuing out and other things set right a little and him to find nothing amiss as he would not remain many minutes.

We think that an Inspector of the matters complained of which are a daily source of annoyance to the whole of the residents of the neighbourhood should be a stranger and act impartially so as justice might be done.

We should be truly sorry to make our application to you, Sir, if we had not just cause. If a stranger to all parties could come unknown to any one and remain for a few days near the Foundry he would have a chance of seeing things as they really are. . .

We humbly beg and request your Honorable assistance and pray that you will send down an Inspector to see the place and cause remedies to be applied to abate if not to remove the nuisances complained of, as the nuisances have been brought to the houses and not them to the nuisances for there has not been any house built for many years before the nuisances commenced but a number taken down to enable the firm to enlarge the Foundry."

It went on to outline other types of Council bias towards the firm in that the factory had been laying "dross" and rubbish on a piece of land adjoining the river which it had blocked, but although a resolution to take legal proceedings was passed and other firms were stopped from doing this, the Ayre Close and Nicholson continued to do so.

The residents of the Parish of St. George were clearly not willing to let this matter rest. They sensed the unwillingness of the council to act and sensed collusion between the council and local industry. The continued pressure they put on the council through the Local Government Office is novel, as is Whitehall's willingness to chastise the local administration over these issues. It would seem that smoke abatement was no longer regarded as unimportant.

Under this external pressure, the Sanitary Committee of York City Council resolved "That the attention of the Inspector of Nuisances be again specially directed to the premises of Messrs Ayre Close and Nicholson and that he do report to this Committee whether in his opinion there exists thereon such a Nuisance arising from the emission of Smoke as to justify the Local Board taking legal proceedings against Messrs Ayre Close and Nicholson in reference thereto" (*City Minutes* 1870–1872).

This time Mr. Healey was careful to take along the assistant inspector P.C. Johnathan Atkinson to the Ayre Close and Nicholson works. A manuscript copy of their observations is extant in the York Archives. Observations were made from

March 9–11 and detailed periods of dense black smoke, moderate black smoke and little or no smoke. The inspector concluded that "the quantity of Smoke emitted from Chimneys as above stated was a Nuisance."

The Sanitary Committee felt the facts stated in the above Report were sufficient to justify the commencement of legal proceedings against Messrs Ayre Close and Nicholson and that the Town Clerk should take such proceedings, accordingly.

The firm were aware that Healey and Atkinson were making observations and probably wrote to the Town Clerk while they were still underway (the date 11.3.71 is uncertain because the letter is poorly preserved). The letter reads something like:

"Having received your notice of the Board of Health to prosecute us for the nuisance that arose from our two Engine Chimneys and being desirous to do everything in our power to obviate the complaints which have been laid before the Board. We have consulted with eminent engineers. . . the best known methods of consuming the smoke generated in the two chimneys complained of and as the result we have determined to introduce the method (which after long consideration has been adopted and applied by the North Eastern Railway Company at their works at Leeds) and which we have every reason to believe from the authorities will efficiently accomplish the objects required. As the application of this method will require the suspension of the Works, the period required for the necessary alterations. We trust the Board will allow us ten (?) week for making such alterations as there is always broken time. . . that the workmen would suffer less to do themselves less inconvenience. In the meantime we will introduce Stiasa Jep (?) into the Chimneys and in all aspects endeavor as far as possible to prevent any complaints" (York City Archives-YD1 1871).

There was also some activity on the part of an enthusiast within the council. A letter of March 17, with lacunae, from W.A. Martin & Co. Fleet St. London, reads:

"My Dear Joe, Martin & Co. cannot cure the Smoke in a Puddling furnace for the reason that puddlers will not work a new invention and the masters cannot make them. What we did was most successful as far as it went, but the men said that it required more working and was likely to cause a decrease in production for which the masters would not pay. This was all fancy on their parts, but they are such a set of hounds that when on the scent they are the devil to lick . . . . Martin expresses [the belief that] smoke can be cured or at any rate much reduced in a puddling furnace, but its to be an open question, but we, Martin & Co. are quite ready to try it . . . works if they will pay the cost of . . . ten pounds. I think we should do it but the . . . missed work . . . was no more labor & them in fact. Much obliged to you old fellow and I hope some master will let us have a furnace. We are dead sure on steam boilers. I'll guarantee to cure any boiler fire ever put up for ten pounds cost of repairing and altering the flues."

Other letters and papers implied a reluctance to adopt smoke abatement procedures on the part of industrialists. Close Ayre and Nicholson did not appear to satisfy the council because they decided to start proceedings against them at the meeting of 30.4.71.

The growth of the firm does not appear to have been particularly hindered by environmental pressures, as there are plans for new buildings, which have very low chimneys, dating from 1873 (York Library 1873). The council may have been concerned about the proposed buildings, because the plans bear the handwritten annotation: ". . . will be carried out in accordance with the byelaws of the Urban Sanitary Authority." While the annotation stipulating conformity to council byelaws might be a common feature on submitted plans, it could also indicate a specific official concern over smoke from the foundry. Certainly their works continued to be the subject of occasional complaints (York Urban Sanitary Committee 1879).

Another possible council action resulting from the problems with Close Ayre and Nicholson, may have been the decision to "print and circulate a public notice calling attention to the provisions of the Law with reference to the prevention and consumption of Smoke and of the intention of the Board strictly to enforce such provisions after the expiration of three months of such notice (*City Minutes* 1872–3; *City Minutes* 1873–4).

The case described above shows how weak the Local Government and Health of Town Acts were. They could, by and large, be ignored by industry and local administrators. It was not until the Public Health Act of 1875 that the role of Local Government in terms of smoke abatement issues became defined and the appointment of an inspector of nuisances was required by law (BC 17/1, 19/8/1872).

4 Johnathan Atkinson

York was particularly fortunate in gaining a diligent and enthusiastic Inspector of Nuisances who was active in implementing the changes that arose in the final quarter of the 19th century. It was argued in nearby Beverley that the inspector was to be "an active, energetic and intelligent man, worthy of the confidence and support of the Board, capable of doing his duty with firmness yet careful to avoid giving unnecessary offence" (Beverley Local Board of Health 1853). These qualities were certainly to be found in Atkinson.

Samuel Healey, who had replaced Robert Chalk as Sanitary inspector in 1862 (*City Minutes* 1859–62), was Chief Constable of York. It was normal for sanitary inspectors to be drawn from the police force in mid-Victorian times. Johnathan Atkinson arrived in York in 1866 to take up a position as a part-time policeman and soon found himself replacing the Assistant Sanitary Inspector, Henry Kinson (BC 16/1, 20/10/1866). However, even at this stage there were doubts about the wisdom of appointing police officers as sanitary inspectors and in November of that year there was a move to obtain the "appointment of a competent person as Assistant Inspector of Nuisances in lieu of the Police Officer at present employed" (BC 16/1, 13/11/1866).

Under the early Public Health Act 1872, Urban Sanitary Authorities were required to appoint inspectors (BC 17/1, 18/8/1872). The administration in York wished to retain Healey as the Inspector, but the Local Government Board found the appointment of police officers as Inspectors of Nuisance objectionable. They were "not able to sanction the appointment of the Chief Constable [Healey] to the office . . .". This being so it was decided Johnathan Atkinson would be appointed as Inspector of Nuisances from March 1, 1873 (BC 17/1, 21/1/1874).

We have little idea of the daily workings of the Sanitary Inspector from the Council Minutes. Only one of the volumes of the "Inspector's Journal" survives in the York Archives (Acc. 157/2, May 1896–July 1897). Daily entries go into columns in this large account book that deals mostly with households rather than industries. The inspector often visits four sites a day. Letters are sent out after the visit and later the premises are checked for compliance. It is probable that Atkinson wrote fortnightly reports to the Sanitary Committee rather than sitting on it and one of his handwritten letters to the members exists in the archives (YD1 I 106/1). Although Atkinson was a busy man supervising an assistant inspector and a clerk, he still had time at the end of the 1880s to accompany a meat inspector into the field (York City Archives-YD1 I 106/1).

In 1904, a year after his retirement, Atkinson was presented with a silver tea service for his work. However, he was to be found aiding the Sanitary Committee for many years to come, since he remained a consulting inspector until the time of his death at the age of 81 in November 1916 (*Yorkshire Gazette* 1916). His appointment as the first Sanitary Inspector remained of sufficient significance to be alluded to at various times in this century (e.g., Birch 1928; *Yorkshire Gazette* 21/10/1937).

5 The Public Health Act 1875

Legislation of the 1870s clarified the position of local authorities in relation to air pollution (see upper portion of Fig. 1 for the evolution of smoke-control legislation). There were now well-defined powers that would, in theory, enable enforcement to go beyond written warnings to persistent offenders. The appointment of Atkinson as the Inspector came just before the Public Health Act (1875). Section 91 of this Act clarified the unlawful emission of smoke: "Any fireplace or furnace which does not as far as practicable consume the smoke arising from the combustible used . . . and any chimney (not being the chimney of a private dwelling-house) sending forth black smoke in such quantity as to be a nuisance." However courts could dismiss the complaint if it could be "satisfied that such fireplace or furnace is constructed in such a manner as to consume as far as practicable . . . all smoke arising therefrom, and that the fireplace or furnace has been carefully attended to by the person having charge thereof." (Ashby and Anderson 1981).

The battles between the council and industry in the last two decades of the 19th century relied heavily on interpretation of the 1875 Act. Following the appointment of Atkinson as Inspector, the Sanitary Committee minute books began increasingly to give detailed lists of all those industries observed by him, with particular attention paid to those considered to be emitting smoke in such quantities as to be deemed a nuisance within the meaning of the Act. The outcome

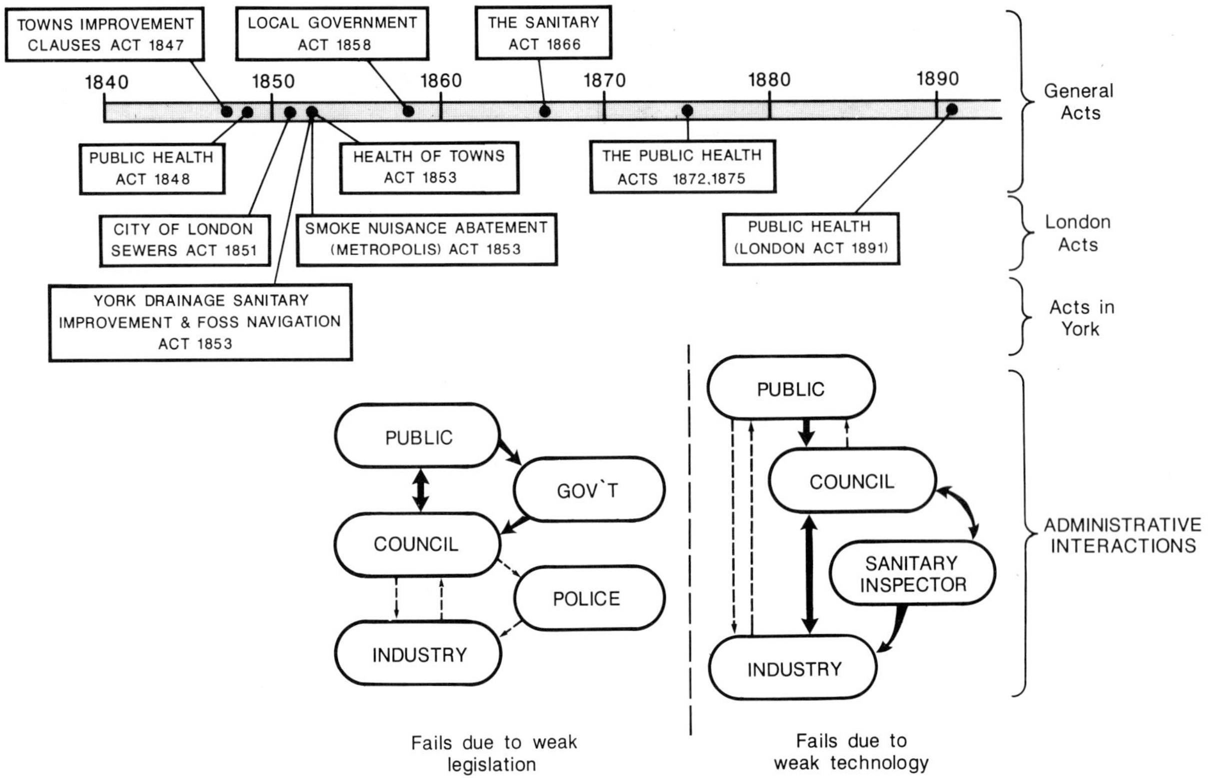

Fig. 1. Administrative mechanisms for abating smoke in late 19th century York. The *upper part* shows some of the more important acts of parliament. The strength of administrative interactions in York before and after the Public Health Acts are shown in the *lower part* of the diagrams with weaker interactions indicated by *dotted arrows*

of these offences was usually a letter from the Town Clerk to the firm in question, asking them to abate the nuisance. In return, the manufacturer concerned had to prove that all practicable means were being taken to consume the smoke.

The responses of manufacturers to the warnings varied, but most commonly they gave the reason for the smoke as (1) use of poor quality coal; (2) bad stoking or firing of the furnace; (3) insufficient ventilation of the furnace. The remedy for this was most commonly seen to be (1) change of fuel from coal to coke; (2) warning or sacking the fireman; (3) installation of a patent smoke apparatus. While the first two may well have been effective in reducing the smoke, the third remedy was no panacea, since smoke-consuming apparatuses, although readily available at this time, were not necessarily effective unless used with great care.

The years that immediately followed the passage of the Act seem to have been a relatively quiet period for the Sanitary Inspector, and each year saw only one or two references to smoke nuisance being investigated, although in 1878 a memorial was presented to the Urban Sanitary Committee by 79 inhabitants of Tanners Moat concerning the working of the Rowntree factory situated in Tanner Row (York Urban Sanitary Committee 1880).

In contrast to this the 1880s were a time of considerable activity. Despite the clarification made by the 1875 Act, York's Inspector of Nuisances and the City Council still found it difficult to obtain satisfactory reductions in smoke emissions from factories such as Rowntree's or Leetham's and most notably the North Eastern Railway Co (NER). Sufficient documentary evidence survives for a full treatment of the problems with the NER and these are described in Bowler and Brimblecombe (1990a). The way in which the council treated the NER contrasts markedly with their attitude to Close Ayre and Nicholson, who created so much of a smoke nuisance prior to the Act of 1875. In their arguements with the NER, the council made it very clear that they regarded the NER to be seriously in breach of the 1875 Act. They seemed to be determined to take a strong line. The actions of the council were not always entirely effective, but there is no feeling that they are colluding with industry. The genuiness of the concern is in line with the wider-scale concern over sanitary reform of the times; the health of towns was regarded as having a high administrative priority (Brenner 1974).

For their part the NER seemed willing to try to reduce their emissions. When this proved difficult, probably through lack of appropriate technology, they resorted to legal wrangling and fine words, no doubt sheltering behind their statutory authorization. However it must be admitted that a sense of frustration over the difficulties seems to characterise all the parties involved.

In reply to an accusation of smoke nuisance, in 1883 Rowntree agreed that there was "cause for great complaint". They promised to employ an experienced boiler setter from the West Riding to undertake work which would alleviate the problem (York Urban Sanitary Committee 1883). Observations of smoke "in such quantities as to be a nuisance" from their factory continued to be recorded by Atkinson and warning letters issued. In 1889, they replied to one such letter, assuring him that they were to install a large Green's Economizer on the chimney of their Wellington Row factory which would help abate the smoke (*City Minutes* 1889). Four years later, after continued observations of dark smoke by the

Inspector, Rowntree wrote to him again about the smoke from Wellington Row saying that it was "due to carelessness on the part of the stoker, and we are today giving him a week's notice to leave" (*City Minutes* 1893). Other conciliatory letters passed between Rowntree and the Sanitary Inspector, but in 1896 complaints had again been made of the chimneys in North Street and Tanner Row; the excuse given on this occasion was that at the time of the complaints, they were experimenting with one of their boilers a device to "reduce to a minimum the smoke nuisance" (*City Minutes* 1896).

Atkinson's other main sources of continued complaints were the various chimneys of the North Eastern Railway Company, York Confectionary Works, J.J. Hunt's Brewery, the Yorkshire Herald and Yorkshire Chronicle newspaper factories, and Leetham's Flour Mill. The latter was the subject of much wrangling during the 1880s–1900 (see Bowler and Brimblecombe 1990b, Chap. 5). In May 1899, notice was served on Leetham's to abate the smoke nuisance, and in June of the same year, a magistrate's order was issued against them "prohibiting the recurrence by the smoke emitted from their chimney in Hungate" (*City Minutes* 1899). This seems to have had a limited effect, and in May 1900 it was resolved by the Sanitary Committee to institute proceedings against them. A summons was taken out but adjourned until November in order to give them a chance to complete and test certain apparatus. Meanwhile Atkinson and his colleagues were becoming stricter and in May also "the Committee instructed the Town Clerk to intimate to the firms their intention to enforce the law in regard to smoke nuisance" (*City Minutes* 1900).

The role of Sanitary Inspector was clearly becoming more complex. There are hints that A.F. Drummond, Atkinson's successor, may have been more desk bound. Under Drummond (from 1904) the "Inspector's Journal" appears to be replaced by the "Assistant Inspector's Journal" which has a very similar form. Drummond merely initials the entries. It seems that the period of pioneering was over.

6 Conclusion

There were important changes in the way local administrators handled air pollution problems in the second half of the 19th century in England. The importance of the Public Health Act of 1875 is particularly clear in York. Prior to this act, there are indications that the Local Board of Health did not take a strong stand on air pollution issues and there were claims that the Sanitary Inspector colluded with industry.

With the implementation of the Public Health Acts the sanitary matters were the direct concern of the council's own sanitary inspector rather than the police. Johnathan Atkinson, with the support of the Town Clerk, made it clear that the emission of smoke was a serious matter. The failure to cure York's smoke problem appears to derive more from a lack of appropriate technology than a lack of resolve on the part of the administration at the close of the century.

References

Ashby E, Anderson M (1981) The politics of clean air. Univ Press, Oxford
Atkinson J (1890) Report of the inspector of nuisances upon the arrangements made by owners of furnaces in connection with manufacturing and other premises in the city of York, for the consumption of smoke, and other matters. City Counc Minutes, 1889/90 York
BC 16/1 Corporation of York local board of health minutes, York City Arch
BC 17/1 Corporation of York local board of health minutes, York City Arch
BC 18/1 Corporation of York local board of health minutes, York City Arch
Birch WH (1928) Public health and the prevention of disease. In: Morrell JB, Watson AG (eds) How York governs itself. Allen & Unwin, London, 87–111
Beverley Local Board of Health (1853) Beverley local board of health minutes, vol 1. Humberside Public Records Office (BC IV/11/1)
Bowler C, Brimblecombe P (1990a) Difficulties in abating smoke in late victorian York. Atmosph Environ Urban Atmosph 24A:49–55
Bowler C, Brimblecombe P (1990b) History of air pollution and building damage at York and Beverley. National Power Report (in press)
Brenner JL (1974) Nuisance law and the industrial revolution. J Legal Stud 3:403–433
Brimblecombe P (1987) The big smoke. Methuen, London
Chadwick E (1842) Report on the sanitary condition of the labouring population of Great Britain. Univ Press, Edinburgh, 1965
Kohl JG (1845) Travels in England and Wales. London, T. Roscoe, Bristol
Local Board of Health (LBH) (1858) A return of the chimneys within the district of the local board which emitted smoke on the day and hour reported so as to be a nuisance to the public. Local Board of Health Committee Minutes 1855–61, York
MacLeod RM (1965) The Alkali Acts Administration 1863–84: the emergence of the civil scientist. Victorian Stud 9:86–112
Malcolm CV (1977) Smokeless zones – the history of their development. Clean Air 6:1–14
York Urban Sanitary Committee (1876) York urban sanitary committee minutes. May 23, 1876. York City Arch
York Health of Towns Committee (1852) Report of the committee on the smoke nuisance with evidence as to the causes of the nuisance and the entire practicability and economy of prevention in York. York Libr
York Library (1873) Close and Ayre's iron foundry plans. November 1873. York Libr (Y942.7418 LEE C275)

Experts in the Smelter Smoke Debate

E. SCHRAMM[1]

1 Early Measures Against Industrial Emissions

In the late 19th century there were more scientific papers on industrial emissions damage published in German-speaking countries than elsewhere. This was due not only to the lead enjoyed by German science in several fields, in particular, plant physiology and agricultural chemistry; it was also due, more significantly, to the fact that the social and ecological problems caused by industrial emissions were, for various reasons, far worse in German-speaking Europe than elsewhere.

First of all, the process of industrialization had begun in most parts of Germany and Austria later than in western European countries. In Great Britain or Belgium, moreover, legal measures had lead to a reduction of emissions and thus of emissions damage [cf. Reuss (1907, p. 19); Ashby and Anderson (1981)]. Although these earlier measures containing the effects of industry on the environment were frequently taken note of by governments in Germany and Austria, these governments did not, because of their liberal industrial regulatory policies, adopt any of these measures themselves (Schramm 1984). This enabled the Belgian firm Vielle Montagne, for example, to transfer its production facilities located at Liege (Belgium), which had to be closed in 1856 for ecological reasons, to (Essen-)Borbeck, Germany. There this zinc-smelting plant continued to produce, despite protests, for over 100 years until 1968 [cf. Vohl 1863; Pappenheim 1865; Reg. Düsseldorf 10725 (Hauptstaatsarchiv Düsseldorf), Parisius 1984].

There is at least one more reason which led to the more intensive scientific consideration of smoke damage in German-speaking countries, i.e., the "particularism" characteristic of Germany before it was unified in 1871. Prior to the founding of the German Empire, each German state was, legally speaking, an area of autonomous jurisdiction. As a result, emissions laws developed separately in the individual states according to their particular socio-economic requirements, and were unified only in 1900 in the Civil Code (§ 906 BGB).

Saxony was in the avant-garde. Already in 1852, it was stated in the draft version of a Saxonian Civil Code that "... no one may bring onto his land devices that transmit damaging steam, fumes, smoke or soot to neighbouring property. Industrial facilities are subject to political laws".

[1]Institut für sozial-ökologische Forschung, Hamburger Allee 45, 6000 Frankfurt a.M., FRG

Brimblecombe, Pfister (Eds.)
The Silent Countdown

This passage is regarded as "the oldest provision in German law dealing with emissions" being "comparable to the later § 906 BGB" (Palmer 1979, p. 55). In 1861, this principle was incorporated into the Industrial Regulations of the Kingdom of Saxony.[1]

This fact suggests several things. To begin with, apparently such cases were already so common in Saxony that there was a need for such regulation. The implementation provision, with its reference to political laws, reveals moreover the problems of liberalism. As Geheimrat Held, author of the draft version, argued, in order to promote the public interest in a free and undisturbed development of industry, it is often necessary to impose severe restrictions on one's neighbours.

Frequently it was not possible to ignore industrial emissions. In the early industrial regions of Germany, the corresponding damage appeared mostly in very small areas around the metal-smelting works. For the local residents, the smoke rising from these works was not a sign of economic growth; rather it signified the ecological destruction that was undermining their own economy.

2 The German Smelter Smoke Debate

2.1 The Freiberg Experience: Identification of Sulfur Dioxide and Technical Solutions End of Stack

In Saxony's traditional smelting town, Freiberg, emissions damage had led to serious conflict, which already began in 1846. As the chemist Kohlschütter (1918, p. 26) recalled: "The smelting works lay deep in the burned out basin of the valley, with not a blade of greening grass around. The rooftops were covered with the sediment from the poisonous smoke".

Freiberg's silver- and lead-smelting works were owned not by private capital, but by the government of Saxony, which received a request from 14 citizens and the council of one of the neighbouring villages, Halsbrücke, to take some action. The residents, as "poor mine and smelting workers", were not able to compensate "in the long run" for the losses resulting from the emissions damage. Because the "smelter smoke" again and again destroyed their crops, they were "facing gradual impoverishment." (Staatsarchiv Dresden/Freiburg Gg No. 32, Vol. 1 after Andersen and Brüggemeyer 1987). At first, the community's petition was rejected. But after a short while more communities joined in, and soon Saxony's parliament was also brought into the picture. Such widespread protest could no longer be ignored. Adolph Stöckhardt, an agricultural chemist from the nearby

[1] The relevant sections of Saxony's Industrial Regulations were then adopted by some neighbouring lands: Saxony-Coburg-Gotha, in 1862; Saxony-Weimar, in 1863). The proposal for a set of industrial regulations drafted for the "Norddeutscher Bund" (North German League) lacked, however, this kind of emissions law. It was introduced only after the intervention of Members of Parliament, cf. Mieck (1981) and Palmer (1979, p 60ff).

Agricultural Academy of Tharandt, was appointed to carry out a study of the problem.

The processing of ores containing sulfur had increased in particular. The sulfur was forced out of these ores by a process of "roasting". Stöckhardt (1850) concentrated on this issue. He considered the acidic component of the smelter smoke (sulfur dioxide) to be the real pollutant and thus ignored the traditional pollutants, especially arsenic and the heavy-metal dust. Considerations of solubility suggested to him, moreover, that an indirect damage to vegetation due to a "poisoning of the soil" was neglect. The relevant issue was the "direct and acute poisoning of plants due to their coming into contact with gas or vaporized acid." In addition to damage to agricultural plants, he showed that some diseases of cattle and other domestic animals were caused by the smelter smoke, too (cf. Freytag 1873).

During the period from 1855–1865 the Freiberg smelter paid out 94 234 Taler in compensation for lost crops and 24 784 Taler for damage to the forests. But this considerable sum covered only a part of the claims. The chemists at the smelting works, drawing on Stöckhardt's study, proposed to detoxify the emissions from the works. Their suggestions were, however, only partly put into practice, mainly due to an ambivalent attitude of the technicians and to economic reasons. This was true in particular for the introduction of "end of stack" technologies. The exhaust fumes which arise during the roasting of ores containing sulfur were in part processed into sulfuric acid. In order to avoid overcapacities in the production of sulfuric acid, however, most of the sulfur dioxide emissions was simply blown out into the atmosphere through high smokestacks (cf. Andersen et al. 1986). Stöckhardt had already hinted at this method of "neutralization" through dilution in an ambivalent argument that referred to the high smokestacks used in the British soda factories.[2]

A 60-m-high smokestack did not, however, lessen the damage. Foresters discovered in an as yet undamaged wood 6 km away "a conspicuous sickening and dying off" of trees (Stöckhardt 1871). In 1871, the chemist Moritz Freytag (1881) noticed emissions damage "outside the immediate vicinity", and again in fields 3 km away. There "trees [were seen with] dried branches and tops and with corroded leaves and needles as well."

However, the strategy of relying on high smokestacks was perfected. In 1899 the management of the smelting works had a second smokestack built, this time 140 m tall. Technicians celebrated this as the successful completion of a model restructuring of the smelting works into a nonpolluting production process (cf. Ost 1907, Kohlschütter 1918, p. 26). In particular, the poisoning of cattle, which had led to an especially tense relation between small farmers and the management of the smelting works, no longer occurred. However, considerable crop and forest damage was now found at further distances; in 1900 as far as 10 km away (cf. Schröter 1908; Wobst 1925).

[2] On the other hand, soda manufacturers did not regard the dilution of acidic gases as a perfect solution. Cf. Hasenclever (1879).

2.2 *Expertise in the Harz Region: The Emergence of a Paradigma*

In the Harz region, with its old silver production, emission damages were already noticeable even before the height of industrialization. This was attributable above all to heavy-metal dust (and sand) pollution. Even the antlers of deer were regularly stunted or deformed as a result of damage due to heavy metals (cf. Wentzel 1984). There was also sporadic discussion of the damage to vegetation brought about since the beginning of the 19th century by the sulfur dioxide in the smelter smoke (cf. Meyer 1822).

The extent of the damage increased continually due to the expansions of ore-smelting operations after 1860. Since the smoke could only slowly escape from the narrow valleys surrounding the smelting works, the main problem was the high concentration of polluting factors. For example, around the Clausthal smelting works "the mountains lying closest in all directions . . . had become barren and sterile due to the effect of the smelt-smoke". In the valleys, damage to vegetation could be clearly seen; the leaves and needles of trees were damaged several miles away (Freytag 1881).

Moreover, the types of trees had to be altered. Earlier, where spruces could thrive without problem, now only pines could grow with great difficulty. The foresters worried that "in ten years not a trunk will still exist." In the direct vicinity of the smelting works nothing at all grew any longer and the vegetationless ground was partially eroded by rainfall. As a result of this situation, a conflict arose between two government agencies in the Prussian province of Hannover, the forestry administration and the smelting works administration.

Using a wide ranging expertise, based on scientific and practical knowledge, a crisis management strategy was worked out. The proceedings in Saxony were chosen as model; therefore, the chemist Julius von Schroeder from Tharandt was appointed to carry out the study. He was able to fall back on the same expert strategies that had pacified the Freiberg conflict. The head forester, Carl Reuss, who was not employed by the Prussian civil service but by the neighbouring town of Goslar, was to cooperate with him as a forest specialist.

Von Schroeder was satisfied with the state of affairs that had been achieved in the Freiberg smelting works. The techniques employed in the Harz region lagged, in comparison, far behind [cf. von Schroeder and Reuss 1883, p. 227; von Schroeder and Schertel 1884]. For Reuss (1896, p. 3), on the other hand, "the forestry industry's struggle against the smoke . . . [was] an act of self-preservation. It must be once and for all carried out in order to protect the forest from gradual destruction by the everywhere advancing industrial pollution. This struggle will end victoriously only when industry is forced to compensate for its damages".

In 1877, at Reuss's suggestion, the forestry service began cultivation experiments near the smelting works. They were carried out energetically, especially in the district of Grund near the Clausthal smelting works. Different plant species were planted in order to learn about their capacity to withstand the smoke. In these test areas "maple, ash, birch, beech and oak, the later mostly in the first and second year of cultivation," seemed at first to succeed. Willows and poplars also thrived. In expanded tests in this and other districts, it was possible to establish a range of resistance: deciduous trees were more smoke resistant than conifers.[3]

The foresters in the forests affected by smelter smoke had felt abandoned by the agency for some time. Therefore they welcomed the experts' support and adopted the views of the two experts. The head forester of Zellerfeld referred to the conclusions reached by the chemist von Schroeder thusly: both the smelting works administration and the forestry administration must contribute to the "halting and even reduction of the smelter smoke calamity". In addition to an emission-reducing modification of the production process ("by introducing the modified smelting process being sought at present by the Altenau smelting works") the composition of the forests must be adapted to environmental pollution "by cultivating deciduous trees with greater reproductive capacity and with more resistant leaf organs".

Behind this strategy of the foresters lay a particular assumption about the cause of emission damage, i.e., the pollutants were supposedly taken in by the trees directly out of the air and not through the detour of soil and roots. The foresters sought support for this thesis from the chemist von Schroeder, who, given his and Stöckhardt's experience in Freiberg, was concentrating on sulfur dioxide as the pollutant in question. "The damage does *not* result through the mediation of the *soil*; thiosulfate, which is formed in the soil, is indissoluble and thus incapable of being taken in by the roots of plants. Rather it results from infusion of pollutants by the leaf organs. The theory that the soil is poisoned, which was widely held before more exact research was undertaken, and which was very convincing to the laity in light of the areas of more or less devastation by smelter smoke, with their gradually receding vegetation [...], is therefore false".[4]

A cultivation experiment undertaken by Reuss in his garden also seemed to confirm that the theory of soil poisoning was false and that it was not necessary to take the heavy metals into consideration. In this emissionless garden, trees grew in soil from a vegetationless place. This seemed to confirm that only the smelt smoke caused damage by impairing the leaves which assimilated pollutants (von Schroeder and Reuss 1883, p. 56ff).[5]

The cooperation between the agricultural chemist and the forester achieved further successes. Up until this point, in addition to the barrenness caused by the smoke, only the corrosion of the leaves and needles was recognized unequivocally and indisputably as damage from smelter smoke. Such corrosion was in evidence in particular where the emissions were strongest, in the direct vicinity of the works. Somewhat further away from the emitting stacks, however, the growth rate of the forests also lagged behind expected levels. Reuss and von Schroeder considered this also to be emission damage, if the level of sulfur found in the plants was

[3] Letter of the Oberförsterei Grund (July 19, 1879) in Hauptstaatsarchiv Hannover 180 Hildesheim 17217; cf. von Schroeder and Reuß 1883, p. 306 ff.

[4] Letter of the Oberförsterei Zellerfeld (October 10, 1879) in Hauptstaatsarchiv Hannover 180 Hildesheim 17217.

[5] Reuss' original report to the government is documented in Hauptstaatsarchiv Hannover 180 Hildesheim/17215, 125 ff.

beyond the average. That is, without a consideration of the ecological peculiarities of the situation, the agricultural-chemical analysis alone was supposed to be adequate.

With financial support from the Prussian Ministry of Agriculture, von Schroeder and Reuss were able to publish their findings on "Smoke Caused Injury to Plants and Smelter smoke Damage in the Upper Harz" in book form. This comprehensive monograph left the scientific as well as the political public with the impression that the view of emission damage developed by von Schroeder and Reuss in Tharandt was unassailable. Almost all of the publications and expert opinions up until 1895 were based on the viewpoint presented in this monograph, which was further developed at the turn of the century by von Schroeder's successor, Hans Wislicenus. Wislicenus (1914) returned in particular to the question, already raised by Stöckhardt, of the exact determination of the phytopathological threshold values for sulfur dioxide.

2.3 Polycausality and the Controversy in Upper Silesia

About 1885, one point of view had come to dominate the technological discussion being developed mainly in Freiberg. Sulfuric acid plants were regarded as the best "end of the stack" technology used to avoid the smelter smoke damage. The idea that it was necessary "to capture and store the unused sulfur content of the mixing ores that [was otherwise lost] as worthless precipitation." This goal was pursued in particular by the chemist Robert Hasenclever, whose factory *Rhenania* "had erected . . . several lead chambers and Glover-towers" in order to process the sulfur dioxide to sulfuric acid (Kosmann 1893). As a result of these measures, the heavy emission damage such as corrosion of leaves and needles could be observed only on the directly adjoining plots of land (cf. Schramm 1986).

By 1891, 11% of the sulfuric acid in the roasting gases was already being recovered in Prussia. But this trend was not uniform: one must clearly distinguish Upper Silesia from the western Prussian provinces where the industry tried to develop scientific production processes and had a scientific interaction with Great Britain, France and Belgium. In Upper Silesia, with two exceptions, the technological "solution" to the problem, a combination of improved roasting furnaces and chemical "end of the stack" technology had not been put into practice.

The Upper Silesian smelting works attempted to continue their production with large-scale scientific innovation. One could also discern a reactionary attitude with respect to the protection of workers. But in addition to the hard-line attitude of the Upper Silesian businessmen, one must also discuss certain natural causes of this backwardness. This region "was able to enjoy, especially because of its lower production, a sufficient supply of Galmei [a low sulfur ore] . . . until the middle of the 1870's" (Kosmann 1893). So the smelter smoke problem became visible with a more than 20-year delay.

Thus in the Aachen and Ruhr districts, most of the roasting gases arising from ores containing sulfur were already being processed into sulfuric acid by the time the first court case appeared in Upper Silesia.[6] The Upper Silesian government had been licensing smelting works since 1878 only if they had rendered at least a part of their emissions chemically harmless. Here instead, the sulfur dioxide was precipitated as gypsum, which enjoyed only a low market price as fertilizer (cf. Kosmann 1888, p. 199). In addition, since 1887, tall smokestacks were required in order to dilute the remaining emissions.

However, beginning at the latest in 1893, vehement protests stirred against Upper Silesia's smelting industry. These did not remain confined to property owners who were directly affected, but led instead to a widespread public discussion. In the same year the members of the Silesian Forestry Association journeyed to the woodlands owned by Count Donnersmarck in Halemba. The foresters were provoked, because the woodland seemed "already [equivalent to] a wasteland" and was "increasingly [being] destroyed by the sulfur-containing smoke from the zinc and other smelting works" (after Kosmann 1893). These reports reached the general public, too.

The situation seemed similar in the forest directly adjoining to Count Donnersmarck's woodlands. In 1891, the owner of this forest had asked Reuss, as a recognized expert in this field, to come and assess the already existing emission damage. On this trip Reuss encountered a difficulty that had not appeared until then in Germany (though it was known in the British soda district): "A simple drive through the [Upper Silesian] industrial district . . with its remnants of destroyed forestry plots, both large and small, already had to make it clear to me that in this area it was impossible to speak, without further investigation, of the damage due to an individual factory or smelting works. That here, where smokestack stands next to smokestack, the sum total of all the smokestacks was involved in all of the damage; and that it would be unfair . . to attribute damage to any single smoke source without scientific foundation" (Reuss 1896, p. 3). Here it was not only, as in the Harz, the parts of the forests directly adjoining the factories that were damaged. In the direct vicinity, the damage was only the most visible; even invisible damage should occur. Reuss (1893) discovered a below-average growth rate for trees throughout the entire district, which he, using von Schroeder's chemical method, attributed to the emissions. Reuss' detailed report was published by the contractor, von Tiele-Winckler, who owned not only a large estate, but had shares in numerous mines. Tiele-Winckler acted on the basis of Reuss' opinion, who held that the entire 3500-ha forest was damaged due to "the chronic effects of smoke". Thus the landlord directed his lawsuit, claiming almost 200 000 Marks in compensation, at 19 individual and corporate owners of 33 smelting works, mines, chemical plants, and brickworks whose property bordered his forests.

The appropriateness of these lawsuits was also supported by some metalurgists, such as Kosmann. The "threatening" level of damage assessed in both districts "complement, as it were, one another, in that the one reveals the

[6] cf. Bericht über die vom Gewerberat Dr. Bernoulli ausgeführte Bereisung der Blei- und Zinkhütten des Harzes, Westfalens und des Rheinlandes. In: Hauptstaatsarchiv Düsseldorf, Reg. Aachen 7999.

destruction in the western part, the other that in the eastern part of the smelting districts, which together [add up] to a 12-kilometer-long by 6-kilometer-wide zone." For Kosmann (1893), this damage was "the bill" for having taken the Gypsum technological path in Upper Silesia.

The smelting industry got Bernard Borggreve to act as a counter-expert during the court proceedings. Borggreve had been trained as a forester and biologist, by Ratzeburg among others, and had led one of the two Prussian Forestry Academies from 1879 until his forced transfer in 1891. He had a good reputation among advocates of natural forestation as well as with the bird protection movement (cf. Ratzeburg 1872, p. 64ff.; Schwerdtfeger 1973; Barhelmeß 1981, p. 140).

Borggreve had as little trust in foresters who could not tell the difference between making observations and drawing conclusions, as in scientists who had had littled empirical experience. He wrote a comprehensive counter-report that recognized only a part of the emissions damage. Basically, Borggreve tore apart the report from the practically minded Reuss, showing where he had relied on scientifically questionable assumptions. But Borggreve's arguments were directed at the entire research program of the agricultural chemists in the field of emission damage, because these overly simplified the ecological complexity of the situation.

Moreover, Borggreve showed in particular that a part of the reported phytopathological appearances in the Kattowice forest could not be unequivocally understood as having been caused, in a monolinear way, by emissions.[7] Borggreve found phytopathological insects, especially *Nematus abietum* and *Tortrix hercyiana*, in parts of the forest. These could also have caused the damage.[8] Borggreve did not discuss whether it was possible, and if so, to what extent, these caterpillars had attacked the forests as a result of the emissions damage. There was no precise knowledge at that time supporting the existence of the latter kind of connection; therefore it did not seem impossible that the insect attack was the cause of the damage, for which the plaintiffs were suing. According to the German law, von Tiele-Winckler had to furnish unequivocal proof of the industrial cause of the damage. Thus the judges would be acting properly in accepting the counter-expert's arguments of multilinearity and complexity.

This kind of strategy has already been used in 1873 by the chemist Freytag in an expertise prepared for the Vielle Montagne in Essen-Borbeck (cf. Freytag 1881). Freytag's position was supported by prominent chemists, such as August Keckule. Borggreve had filed this strategy during a court case involving the Gute Hoffnungshütte, in 1875. The arguments in the latter case were then adopted by many industrialists to defend themselves against lawsuits brought by their neighbours.[9]

---

[7] cf. Borggreve 1895, especially p. 66ff., and the too short discussion by Reuss 1893, p. 25f.

[8] According to contemporary theory, this is an instance of secondary cause. cf. Rubner and Landa 1987.

[9] e.g., Grillo's *Actien-Gesellschaft für chemische Industrie* in Schalke 1883 or the *Maschinenfabrik Haniel und Lueg* in Düsseldorf 1896; cf. Haniel-Archiv, Duisburg 300112/10, ff. 145–158, 311–315, 339–344.

Borggreve seized on every old hypothesis that had not yet been refuted by field tests in his skilled defense of industry. Thus he pushed the corrosion hypothesis, introduced by the British Alkali Inspector and chemist Angus Smith and defended in Germany by Freytag. According to this conception, the acidic gases were supposed to directly etch the leaves' surface and to have a more destructive effect on the cells on wet days than on dry ones. Borggreve also stressed the role of the soil in pollutant intake, standing in opposition on this point to all of the experts in the field of agricultural chemistry.

The public viewed the court action against the Upper Silesian metal industry as a "monster process". But in the end the experts were not heard by the court. The landlord died and his heir withdrew the suit. "Only the owner of the Kunigunden-Hütte, who had been previously sued, had to pay compensation for woodland's damage [caused], as had been proven . . . by the smelting works" (Heß 1900, p 589).[10]

During the numerous court cases over crop compensation, the only experts, given the state of agricultural science at the time, dealing with the problem were gardeners, foresters, landowners, and agricultural chemists. The monolinear conception of causality held by the Tharandt chemists was confirmed in simplified experiments in the laboratory and jibed with the thinking of the judges. Joseph König and Emil Haselhoff in particular developed the results from Tharandt, without questioning any part of the theses, in a practical manner in the agricultural stations at Münster and Harleshausen (cf. Haselhoff and Lindau 1903; König 1896).

3 Wieler's Development of a Complex Model

The attention of the public motivated botanists to undertake specific experiments on the smelter smoke problem for the first time. The biological field workers worked at least with complex models of causality, but they were unable to confirm it by experiments. On the other hand, a polycausal theory without broad empirical proofs was not suitable to gain the judges' help in protecting a forest or other ecosystem. However, most biologists had not troubled themselves at all with questions of accuracy in their field studies.

By 1895 neither the insistence on polycausality nor the theory of soil poisoning was the stimulus for botanical investigations on smelter smoke. Borggreve's rejection of the chronical or invisible smoke damage though, did lead to the systematic involvement of botanists with the problem of smoke emission. The forest-botanist Robert Hartig (1896) had been made aware of the problem raised by Borggreve's work and believed he had found a microscope analysis for "invisible smoke damage", i.e., the reddening of the stomata in the leaves. This

[10] Recently environmentalists have argued that Borggreve's strategy resulted in this case being lost. But Borggreve accepted Reuss' proof of the Kunigunden-Hütte damage; cf. Wentzel (1985).

claim was disputed for almost a decade before being finally refuted after intensive anatomical studies by Sorauer (1911).[11]

In addition to the well-known phytopathologist, Paul Sorauer, only one other botanist, Arwed Wieler, professor at the Technical Institute of Aachen, worked systematically on emission damages. He was not involved in any of the legal or bureaucratic disputes on the industrial emissions; he noticed that the biological questions were answered by chemical experts neglecting phytophysiological knowledge. Thus Wieler received financial support from industrial lobbies as well as the Prussian government beginning in 1900 to work on the smelter smoke problem.[12]

Wieler's interest was not to investigate the "acute damage" in the nearest vicinity, but to concentrate his efforts on the so-called invisible damage caused by emissions arising from normal operations. In contrast to the Tharandt tradition of emissions research, for Wieler, laboratory tests played only a confirming role. Wieler, moreover, in contrast to previous practice, did not want to generalize from results achieved in one place. His line of thought was supported by the Prussian Ministries, because in the Harz region, despite the report from von Schroeder and Reuss and the restructuring of smelting operations, the emissions damage continued to grow.

In the summer of 1901 Wieler took his first measurements in the Schulenberg district of the Harz mountains.[13] In contrast to other authors of the time, Wieler was the first to have not limited himself in studying smoke damage to an agricultural-chemical, technical or anatomical point of view. He was interested in integrating physiological and ecological research in a scientifically adequate way.

Incredibly, Wieler's exactitude and his methodological breadth confirmed a part of the observations and hypotheses obtained by Freytag and Borggreve which had been attacked by the experts from the Tharandt school. He also found that the level of damage to leaves due to the direct intake of sulfur dioxide was partially affected by weather conditions. But another finding was to become more important: Wieler (1905, p. 289) regarded the extinction of the plant stratum under deciduous trees, the latter being particularly damaged, as "the most characteristic symptom of smoke damage".

Wieler (1905, p. 290) rejected Borggreve's view "that the acid from the air was supposed to enter the roots through the soil and from the roots then reach the leaves". In the face of this "extreme point of view it was a healthy reaction" to have left the soil completely out of consideration as a conduit of pollution in the research conducted in Tharandt and Münster. "Because one believed the amount of acid that rained on the ground to be very limited, one thought that one could ignore it and thereby fall into the other extreme of thinking that one could

[11] cf. Hartig (1896), Wieler (1897); summarized in Sorauer (1911).
[12] cf. Zentrales Staatsarchiv Merseburg, Rep. 120 BB IIa 2 No. 28 adh. 1, vol 7–8 (I'm indebted to Dr. Arne Andersen, Bremen, for his kind cooperation).
[13] cf. Wieler's Letter of July 17, 1901. In: Hauptstaatsarchiv Hannover 180/17216.

evaluate smoke damage without considering the soil at all". Wieler, on the other hand, believed that "even small amounts of acid could cause real changes as long as it continued to reach the soil for a long enough time, and that the effects of the acid could possibly be . . . indirect".

Wieler had been interested since 1903 in testing the idea that soil degradation and acidification could be caused indirectly by acidic smoke gases. It was, he suspected, the "large amount of acid" that "falls on the ground yearly in the smoke damage areas" that was probably one cause for the chronic emissions damage (Wieler 1905, p. 320).

Between 1905 and 1914, Wieler continued his research into aspects of this problem, again with governmental support.[14] A central part of this research was the attempt to compensate for the ecological damage by fertilizing the degraded soil with lime. In this test series, the Aachen industrial area was compared with that in the upper part of the Harz. The results then became the standard used in Wieler's further field and laboratory tests. Wieler (1912) also considered bacteriological aspects.

Different growth rates for trees in the limed and unlimed areas, and indications of water stress, confirmed Wieler's view that the "chronic" emissions damage was in part caused by the soil. In particular, in one Harz district, a renewed high growth rate for spruces was observed after the soil had been fertilized with lime. "The vegetation died off because of an insufficient water supply to the trees, which was caused in turn by the de-liming of the soil due to the acid-rain. The situation is made even more complex by the fact that, with the increasing de-liming of the soil, the trees' leafage removes even more lime from the soil which is not returned to it." (Wieler 1912, p. 120) This is so because the decomposition of the fallen leaves is slowed as a result of the change in (or even extinction of a part of) the soil's microflora due to the acid rain. The undecomposed leaves then draw less water into the soil, thereby increasing the water stress of the trees.

Wieler's model was ecologically more realistic than other models. However, it never achieved wide acceptance, being ignored both by the agricultural chemists and by those experts being friendly to industry. The traditional expert fractions continued to swear by the high smokestack policy. In Tharandt, Hans Wislicenus designed improved smokestacks which were also built into factories,[15] and Borggreve was still claiming in 1910, in an expertise for a Harz smelting works, that "sulfurous acid experiences evaporation over 1.5 km such that damage is no longer demonstrable". The noticed damage at a distance of 6–8 km should, therefore, be caused by other toxins.[16]

[14] cf. Letter and notice of the Prussian Minister für Landwirtschaft, Domänen und Forsten. In: Hauptstaatsarchiv Hannover 180 Hildesheim 17216, 9–11ff.

[15] cf. Archives of the *Metallgesellschaft AG*, Frankfurt a.M., W 13 Jur. Büro, Prof. Dr. H. Wislicenus.

[16] Remark of the Braunschweigisch-Lüneburgische Kammer (April 16, 1910) in Hauptstaatsarchiv Hannover 180 Hildesheim 17216.

4 Urban Smoke: Changed Focus of Attention

Wieler's theory did not permit faith in a self-purification of the atmosphere or in a dilution of the sulfur dioxide to a point where it caused no damage. But such an ecologically naive point of view was in fact necessary in order to enable the fight against the smoke where it provoked the greatest opposition. This was not in the countryside but in fact in the cities, where residents no longer wanted to live with soot and smoke.

In 1888, German architects and engineers had already regarded the question of smoke abatement in the big cities as pressing. As in Great Britain (cf. Flick 1980; Brimblecombe 1987) or in the United States (cf. Grinder 1980), an early environmental movement was developing. Its political imagination, however, only extended to technological solutions, especially smoke prevention measures, such as better fuel economy, use of improved smokestacks and ovens, electrification, etc.

In contrast to that urban movement, the foresters and biologists saw the limits of a technical solution of the smoke problems. "The conversion of sulfur dioxide into sulfuric acid .. is only possible in the largest works, and is only partially effective in that a considerable part of the sulfur dioxide escapes in any case. High chimneys, through which one leads the smoke into higher levels of the air in order to try to dilute them there to nonpolluting levels . . . have proven to be somewhat advantageous to the immediate environs but enlarge the circle in which noticeable damage is found." (Fürst 1912, p. 189) Hope in technological progress was replaced by hope in the judges. Before the outbreak of World-War I there was a "large increase of smoke-damage lawsuits" (Niggemeyer 1915, p. 4).

Those legal measures could not, however, lead to fundamental change, since there was no critical discussion on the limitations of the air pollution policy and of the applied technological solutions. Notably, the arguments of all chemical and technical experts pointed in the direction of intensifying the search for better "end of stack"-technologies.

Acknowledgement. This paper is based on a research project at the Institut für Philosophie, Technische Hochschule, Darmstadt, granted by Stiftung Volkswagenwerk, Hannover.

References

Andersen A, Brüggemeyer FJ (1987) Gase, Rauch und Saurer Regen. In: Brüggemeyer FJ, Rommelsbacher T (eds) Besiegte Natur. Beck, München, pp 64–85

Andersen A, Schramm E, Ott R (1986) Der Freiberger Hüttenrauch 1849–1865. Technikgeschichte 53:169–200

Ashby E, Anderson M (1981) The politics of clean air. Clarendon, Oxford

Barthelmeß A (1981) Vögel. Lebendige Umwelt. Alber, Freiburg München (Orbis academicus Sonderband vol 2.3)

Borggreve B (1895) Waldschäden im oberschlesischen Industriegebiet. Sauerländer, Frankfurt a.M.

Brimblecombe P (1987) The big smoke. Methuen, London

Flick C (1980) The movement for smoke abatement in 19th century Britain. Technol Cult 21:29–50

Freytag M (1873) Wissenschaftliches Gutachten über den Einfluß des Hüttenrauches. Jahrbuch für das Berg- und Hüttenwesen in Sachsen, pp Abh 3–86
Freytag M (1881) Die schädlichen Bestandtheile des Hüttenrauchs, der Kupfer-, Blei- und Zink-Hütten und ihre Beseitigung. Landwirtsch Jahrb 11:315–357
Fürst H (1912) Die Lehre vom Waldschutz. Parey, Berlin
Grinder R, Dale (1980) The battle for clean air. The smoke problem in post-civil war America. In: Melosi MV (ed) Pollution and reform in American cities. University of Texas Press, Austin, pp 82–103
Hartig R (1896) Einwirkungen des Hütten- und Steinkohlenrauchs auf die Gesundheit der Nadelwaldbäume. Forstl Naturwiss Z 5:245–289
Haselhoff E, Lindau G (1903) Die Beschädigung der Vegetation. Bornträger, Berlin
Hasenclever R (1879) Über die Beschädigung der Vegetation durch saure Gase. Chem Ind 2:225–231, 275–280
Heß R (1900) Forstschutz, vol 2. Teubner, Leipzig
Kohlschütter V (1918) Nebel, Rauch und Staub. Haupt, Bern
König J (1896) Die landwirthschaftliche Versuchs-Station in Münster i.W. in ihrer Entwicklung während der ersten 25 Jahre. Denkschr. Der Westfale AG, Münster
Kosmann B (1888) Oberschlesien, sein Land und seine Industrie. Verein Deutscher Ingenieure, Gleiwitz
Kosmann B (1893) Über die Beschädigungen der Waldungen durch Hüttenrauch. Glückauf 29:1579–1580
Meyer FGW (1822) Die Verheerungen der Innerste. Beiträge zur chorographischen Kenntnis des Flußgebiets der Innerste, vol 1, Vandenhoeck und Ruprecht Göttingen
Mieck I (1981) Luftverunreinigung und Immissionsschutz in Frankreich und Preußen zur Zeit der frühen Industrialisierung. Technikgeschichte 48:239–251
Niggemeyer H (1915) Die Beschädigung der Vegetation durch Rauch mit besonderer Berücksichtigung des rheinisch-westfälischen Industriegebiets. Thesis, Univ Münster
Ost H (1907) Der Kampf gegen schädliche Industriegase. Z Angew Chem 20:1689–1693
Palmer I (1979) Die Entwicklung des deutschen privatrechtlichen Immissionsrechts im 19. Jahrhundert verglichen mit dem französischen Recht. Thesis, Ruhr-Univ, Bochum
Pappenheim L (1865) Bericht über die Zinkhütten in Borbeck und Eppinghoven. Verh Ver Beförderung Gewerbefleißes 1865:65–75
Parisius B (1984) Lebenswege im Revier. Henselowsk, Essen
Ratzeburg JTC (1872) Forstwissenschaftliches Schriftsteller-Lexikon. Nicolai, Berlin
Reuss C (1893) Rauchbeschädigung in dem von Tiele-Winckler'schen Forstreviere Myslowitz-Kattowitz. Insbesondere Ermittlung, Bewerthung und Vertheilung des Rauchschadens. Jäger, Goslar
Reuss C (1896) Rauchbeschädigung in dem Gräflich v. Tiele-Winckler'schen Forstreviere Myslowitz-Kattowitz. Nachtrag Jäger, Goslar
Reuss C (1907) Maßnahmen gegen die Ausbreitung von Hüttenrauchschäden im Walde. 8th Int Landwirtsch Kongr, Wien. Referate, vol 4. Vernay, Wien
Rubner H, Landa M (1987) Die Anfänge der Rauchschadensforschung in Mitteleuropa. Allg Forstz 42:625–626
Schramm E (1984) Soda-Industrie und Umwelt im 19. Jahrhundert. Technikgeschichte 51:190–216
Schramm E (1986) Der Aufstieg der chemischen Industrie:Umweltschäden greifen aus. Bild der Wissenschaft 8/1986:86–90
von Schroeder J, Reuss C (1883) Die Beschädigung der Vegetation durch Rauch und die Oberharzer Hüttenrauchschäden. Parey, Berlin
von Schroeder J, Schertel A (1884) Die Rauchschäden in den Wäldern der Umgebung der fiscalischen Hüttenwerke bei Freiberg. Jahrbuch für das Berg- und Hüttenwesen in Sachsen 1884, pp Abh 93–120
Schröter E (1908) Die Rauchquellen im Königreiche Sachsen und ihr Einfluß auf die Forstwirtschaft. Sammlung von Abhandlungen über Abgase und Rauchschäden, vol 2. Parey, Berlin
Schwerdtfeger F (1973) Forest entomology. In: Smith RF, Mittler TE, Smith CN (eds) History of entomology. Ann Rev Entomol Suppl:361–386

Sorauer P (1911) Die mikroskopische Analyse rauchbeschädigter Pflanzen. Sammlung von Abhandlungen über Abgase und Rauchschäden, vol 7. Parey, Berlin

Stöckhardt A (1850) Über die Einwirkung des Rauches der Silberhütten auf die benachbarte Vegetation usw. Polytech Centralbl 16:257–278

Stöckhardt A (1871) Untersuchungen über die schädlichen Einwirkungen des Hütten- und Steinkohlenrauches auf das Wachsthum der Pflanzen, insbesondere der Fichte und Tanne. Tharandter Forstl Jahrb 21:218–254

Vohl H (1863) Über den schädlichen Einfluß der Ausdünstungen (Gase und Dämpfe), welche sich bei Zink- und Bleihütten entwickeln. Polytech J 169:204–207

Wentzel KF (1984) Wirkungen der Luftverunreinigung auf Wald und Wild in Geschichte und Gegenwart. Allg Forstz 39:517–522

Wentzel KF (1985) Oberförster Carl Reuß aus Goslar–Bleibende Erkenntnisse der "Rauchschadensforschung" im 19. Jahrhundert. VDI-Ber 560:9–19

Wieler A (1897) Über unsichtbare Rauchschäden bei Nadelbäumen. Z Forst Jagdw 29:513–529

Wieler A (1905) Untersuchungen über die Einwirkung schwefliger Säure auf Pflanzen. Bornträger, Berlin

Wieler A (1912) Pflanzenwachstum und Kalkmangel im Boden. Bornträger, Berlin

Wislicenus H, Schwarz O, Sertz H, Schröder F, Müller F, Bender F (1914) Experimentelle Rauchschäden. Sammlung von Abhandlungen über Abgase und Rauchschäden, vol 10. Parey, Berlin

Wobst A (1925) Die Hüttenrauchkrankheiten im Freiberger Bezirk. Thesis, Univ Leipzig

Unpublished Archivalia

Hauptstaatsarchiv Düsseldorf: Reg Aachen 7999, Reg Düsseldorf 10725
Hauptstaatsarchiv Hannover: 180 Hildesheim D 17215; 17216; 17217
Zentrales Staatsarchiv Merseburg: Rep. 120 BB IIa 2 No. 28 adh. 1, vol 7–8
Archives of the Metallgesellschaft Frankfurt/M.: W 13 Jur. Büro – Prof. Dr. H. Wislicenus
Haniel-Archiv, Duisburg, Dep Oberhausen: 300112/10

The Ruhr Basin 1850–1980: A Case of Large-Scale Environmental Pollution

F.-J. Brüggemeier[1]

1 Introduction

The industrial area along the Ruhr has had a dramatic and impressive history. For the last 100 years, it has been the biggest industrial region not just in Germany, but in the whole of Europe; its industrial strength was envied and it was admired, especially since the region became industrialized within a few decades. In fact, industrialization happened so rapidly that the area was called Prussia's Wild West: contemporaries were only able to describe what happened by referring to the American example.

Around 1850, the Ruhr was still dominated by agriculture. Apart from a few towns like Essen or Dortmund, which had approximately 10 000 inhabitants each, the area was only sparsely populated. Steel and iron works and mining had been going on for a long time, but they were carried out on a rather modest scale; in 1850, there were just 12 000 miners who produced 1.5 M tons of coal. The steel and iron industry, too, was at an early stage in its development.

Within a few decades, the situation changed dramatically. In 1910, the number of miners exceeded 400 000, while the production of coal had reached more than 110 million tons; a single modern mine produced more coal than the whole area 60 years before, and the Krupp steel works in Essen alone employed more than 30 000 workers. The population of Essen had risen to more than 300 000 inhabitants, and the sleepy countryside had turned almost overnight into an urban agglomerate (McCreary 1963; Fischer 1965; Wiel 1970; Brüggemeier 1984).

This process of expansion was not planned, let alone coordinated: each town, village, or company built houses, streets, and factories as they saw fit or could afford. The result was a chaotic mixture of railway lines, canals, streets, workers' colonies, buildings, factories, mines, etc. interrupted by fields and meadows (Crew 1979; Niethammer 1979; Vonde 1989). Farming did not stop suddenly, it was gradually replaced by industry, but still occupied large areas, especially in the northern part (Altkemper 1905; Avereck 1913).

The sheer momentum of these processes is impressive, and it goes without saying that they had far reaching consequences for the environment. So far, however, very little work has been done on this subject; the ecological history of

[1] Arbeitsbereich Neuere Geschichte, Fernuniversität Hagen, Feithsh. 140, 5800 Hagen, FRG

Brimblecombe, Pfister (Eds.)
The Silent Countdown

the Ruhr has only begun to be researched. In this essay, therefore, I will present first results and hypotheses; based upon the examples of water and air pollution, the essay will

1. describe the developments that took place between 1860 and 1980, in order to give an idea of the particularly marked ecological problems;
2. try to explain how these problems were perceived;
3. ask what was done to deal with the ecological consequences of the rapid industrialization.

In many ways, I will describe a very special case – mainly because of the enormity and the speed of the changes. At the same time, the development in the Ruhr region is instructive for the whole of Germany, since the basic institutions, laws, and procedures that applied in the Ruhr applied in the rest of Germany as well (Wey 1982). In the Ruhr, however, these elements combined to produce a special result: the Ruhr, so my thesis goes, emerged as an area where industry was protected, not nature, or the environment. While attempts to establish nature preserves were not too successful in Germany, the Ruhr region effectively developed into an industrial preserve (Brüggemeier and Rommelspacher 1990).

2 Water Pollution: The Case of the Emscher

The consequences of industrialization were first and most dramatically felt in the case of water. The new industries, mainly iron and steel works and coal mines, needed huge quantities of water, which they took from the two main rivers, the Ruhr and the Emscher. Having used it, they poured the waste water back into these rivers with hardly a process of clarification. The waste water from the mines was especially problematic. Here, water was largely used to wash the coal; in addition, enormous quantities of groundwater were pumped from underground, to prevent the mines from flooding, so that the water was very salty and extremely dirty, occasionally turning into a sticky sort of mud (Imhoff 1910; Heilmann 1914).

To make things worse, underground mining resulted in widespread dislocations on the surface with disastrous results in the Emscher area. On a length of approximately 70 miles, the Emscher had a fall of only 121.5 m, 40 m of which were on the first 6 miles. Even before industrialization began, the flow had been a problem; now it turned into a disaster. The waste water of the mines was so dirty that it hardly flowed, while the dislocations of the surface had more or less destroyed the river bed. As a consequence, a vast number of ponds, puddles, and lakes had formed, filled up with a dirty, stinky, watery sort of substance (Bach 1910; Bette 1928).

In some places, the flow had stopped almost completely, and the area had become so muddy "that solid ground was only reached 3 m deep. Here we find a black, thick, swampy, rotten and fermenting manure, which hardly moves; during summer, gas bubbles burst, poisoning the surrounding area. A unique situation – nowhere else to be seen in the whole wide world – arises from the fact that the fermentation and putrification of these enormous quantities of disgustingly dirty and muddy waste water is intensified to the highest degree by the hot water from innumerable steam engines; as a consequence, the waste water often reaches breeding temperature. The waste water to be found in Gelsenkirchen is the most disgustful liquid in the world" (Emmerich 1906, p. 145).

These conditions were described in the aftermath of a typhoid epidemic, which in 1901 resulted in more than 3000 sick and claimed 300 lives. The author, Prof. Emmerich, had been called in to investigate the causes of the epidemic; in his report he continued: "I was about to leave this dreadful scene, but my guides asked me to follow them, and after a few steps, hardly a minute later, I was standing – struck by horror – in front of an even bigger, stinky manure lake of about 500 m^2." Around Gelsenkirchen, he encountered a way to deal with sewage he had not seen anywhere in the world: "I have investigated the hygienic situation in Naples, Palermo, and Constantinople. I know very well the hygienic conditions in Lisbon, Oporto, and Funchal, furthermore in some French and Austrian as well as in many German cities, but nowhere have I come across a situation as grave with regards to sewage, waste water, and faeces, as well as with regards to the degree and range of soil pollution as in those areas of the Emscher valley affected by the typhoid epidemic" (Emmerich 1906, pp. 145 and 168).[1]

This epidemic was the final straw and brought about the foundation of the Emscher Water Authority (Emschergenossenschaft), to solve the problem of sewage and industrial waste water. Earlier attempts to find a solution had failed, mainly because of the resistance of the mining companies, who were the main culprits but refused to finance their share of the costs. After years of talks and under the pressure of the typhoid epidemic, they gave in but insisted upon having the majority of votes in the Emschergenossenschaft. The Genossenschaft combined the towns and industry of the Emscher valley, and it was the first water authority in Germany which established self-government by those responsible for pollution. Towns and industry had to finance the cost, but at the same time they decided how much money to spend and what to spend it on (Emschergenossenschaft 1912; Rommelspacher 1987).

The first and main aim was to get the water flowing, not so much by cleaning it but by turning the Emscher into a sewage canal: its bed was straightened and filled with concrete to create a smooth surface. The small rivers and streams leading into the Emscher were also turned into sewage canals, whose water, as the Genossenschaft put it in 1910, "was of course not suitable for any further use" (Bach 1910, p. 17). Almost 90% of the sewage originated from the mines and other industries, and while it was known that industrial waste water contained toxic substances, these were held not to be dangerous for human health, in contrast to the sewage from the towns and villages, largely consisting of faeces and other

[1] Emmerichs description may be exaggerated. He was a pupil of Pettenkofer, who had stressed the importance of changes in the water table and of fermentation of decomposing matter in causing epidemic diseases; after the discovery of bacteria by Koch, however, Pettenkofers theories were no longer accepted, and Emmerich tried to use the Gelsenkirchen epidemic to reestablish Pettenkofer; see Evans (1973); Howard-Jones (1973).

organic material. According to most experts, only the latter, the communal sewage, was "so dirty, that it had to be cleaned, mainly mechanically, occasionally biologically as well. The industrial sewage is in general not polluted to such a degree that its purification requires a comparable expenditure" (Bach 1910, p. 54; Heinrichsbauer 1936; Kluge and Schramm 1986).

In 1909, state officials came to inspect the work of the Emschergenossenschaft. They planned to establish a standard for the purification works and a minimal limit of time that the sewage had to be treated; they also wanted to restrict the amount of coal sludge that was allowed to remain, and they argued that it was necessary to clean the water and not just to guarantee its flow. As a consequence, they pleaded for biological purification works and wanted to establish an independent control (Oberpräsidium Münster 1909).

The association objected and proved powerful enough to have its way. The standards and limits were defined in such a way that they had hardly any effect; an independent control was avoided and, with one exception, biological methods were not introduced. Instead, mechanical clearing systems were declared sufficient as long as they kept the water flowing. As a consequence, the Emscher and its tributaries became open-air sewage canals and the polluted water reached the Rhine without proper clearance. As a report from 1934 put it: "It is mainly the purification capacity of the Rhine which determines the degree to which the waste water from the Emscher is cleared" (Helbing 1934, p. 486).

The case of the Emscher is exceptional, not just by German standards. To my knowledge, there is no comparable example in another industrialized country where a river of approximately 70 miles length, and almost all its tributaries, were turned into sewage canals. Even in the Ruhrgebiet, the Emscher was an exception; since it had to carry most of the industrial and communal waste water, the other rivers, after initially widespread pollution, could be spared. In the south, i.e., along the Ruhr, water treatment plants were established, even though it was not until after the Second World War that acceptable standards were reached. In the north, in the Lippe area, another water authority was founded in the 1920s, this time with more effective control by the authorities; stricter limits were set and waste water had to be cleared before entering the Lippe, so that pollution could largely be avoided.

In the Emscher region, however, the system established just after the turn of the century prevails until today. While nowadays the Emscher water is cleared before it enters the Rhine, no such clearance takes place in the Emscher region itself; it is a very recent phenomena that at least some of the minor streams have been "renaturalized", mainly by removing the concrete from their beds. On the whole, the foundation of the Emschergenossenschaft has had disastrous consequences; in avoiding flooding and removing the mud and dirt, the Genossenschaft proved effective, but the costs were very high: in its area almost all normal aquatic life was and still is destroyed.

The case of the Emscher is characterized by some extraordinary features, but at the same time it is symbolic of a way of dealing with the environment that was and still is fairly common: polluting and noxious substances are not so much held back or removed but rather widely distributed, assuming that they will either disappear or that nature itself will somehow be able to get rid of them (Spelsberg 1984; Kluge and Schramm 1986); if distribution does not prove effective enough, pollution is regionalized, i.e., a region, a town, or a suburb is more or less given over to industry so that other areas can be spared; finally, costs are avoided as much as possible, arguing that the contribution of industry to the national welfare far outweighs all damages.

3 Air Pollution

Similar mechanisms prevailed in the case of air pollution. As early as 1831, the Prussian government had decreed that steam engines had to have a chimney with a minimum height of just under 20 m to make sure that dust, smoke, and other substances were widely distributed (Mieck 1967). Soon, however, it became clear that this regulation did not suffice; with the growing population more and more buildings were erected that exceeded 20 m, so that higher chimneys had to be built, but these proved to be no solution; in bad weather conditions, even high chimneys could not prevent the smoke and ash from affecting the neighbourhood. More important, in industrialized regions like the Ruhr, factories and chimneys were so numerous, that the whole area was polluted. A chimney may have been high enough to spare the immediate neighbourhood, but smoke and ash still came down a couple of hundred yards or some miles away (Hahn 1911; Reich 1917; Spelsberg 1984).

In 1912, a government survey stated that in the Ruhr, some places still were hardly affected by air pollution; most of the region however, was polluted to such a degree, that smoke and soot amounted to a plague. In addition, the chimneys were distributing sulphuric acid in such quantities that the growing of fruit and vegetables was severely affected and the tiny remaining areas of woodland in the north were in danger of disappearing. In Essen, e.g., flats had to be dusted at least twice a day, and smoke and soot often produced a situation "which could hardly be distinguished from the notorious London fog".[2]

The degree to which the air was polluted became particularly obvious when industrial production came to a standstill during the French occupation of the Ruhr. In 1923, French troops moved in to enforce the payment of war compensations. The occupation started in spring and continued until autumn; industrial air pollution stopped completely and did not interfere with the vegetation cycle. The results were astonishing.

Suddenly, everything started to blossom and bloom. To quote a contemporary observer: "The production of coal, coke, and steel had hardly ceased, when suddenly the air improved to such a degree, that everybody could notice the difference. The air was of the same quality as in nonindustrial regions. . . The leaves, which normally start to wither in early summer, stayed fresh and green

[2] Hauptstaatsarchiv Düsseldorf (HSTAD) Reg. Düsseldorf 38572, Schreiben des Kreisarztes Essen vom 1.11. 1912.

until autumn" (Bergerhoff 1928, 72/3). Potatoes grew bigger and the harvest in general was up to 50% greater than before. Fruits and vegetables, which normally were covered by a layer of dust, soot, and tar, stayed clean and delicious. Even the rings of the trees were wider than in the years before and after.

Soon afterwards the Siedlungs-Verband-Ruhrkohlenbezirk (SVR) started an initiative to save the remaining trees and woods. The SVR was a regional planning authority founded after the First World War to bring some order into the until then largely chaotic and unplanned growth of the region. In 1924 a commission was set up to investigate the problem of air pollution; it published two reports.

The reports stated several times that the main damage was caused not by the visible smoke and ash but by the invisible acid gases containing sulphur dioxide. The representatives of industry and the experts called in all agreed, however, that nothing could be done "to remove the sulphuric acid, which was so damaging to plants, from the smoke"; consequently, all proposals concentrated on ways "to remove or reduce the soot, the ash, etc." (Siedlungsverband 1928, pp. 1ff, 15).

It was suggested that stokers should be better trained or that the households be asked to be more careful when burning coal, and that coal with a lower content of sulphur be used. The last proposal was especially difficult to follow, since low-sulphur coal could be exported more easily to other parts of Germany and abroad, leaving the people in the Ruhr with coal which had a particularly high content of waste and sulphur. The installation of distance heating was also recommended in order to use some of the gases and the heat generated in industrial production: "Day-in, day-out the big iron- and smelting works simply burn and thereby waste the surplus gases they can not use themselves only in order to get rid of them" (Siedlungsverband 1928, p. 17).

These proposals promised a better future, but they offered little comfort in view of the almost desperate conclusion, "that the fight against air pollution caused by the big industrial companies seems to offer little chance of success" (Siedlungsverband 1928, p. 17). To make the situation worse, some of the proposals recommended contributed to a reduction of the visible pollution, but increased the output of the invisible, more harmful noxious gases. It was, for example, suggested that higher temperatures be reached, since these reduced the amount of soot and ash; or ammonia was to be removed from the smoke. However, in doing so, the quantity of acid gases increased, since higher temperatures set these gases free more effectively, and ammonia no longer served as a neutralizing base.

The situation was therefore rather complex and a promising solution was hard to find. In the end, the commission agreed upon a plan which was no solution at all: instead of demanding that industry undertake the changes necessary to reduce pollution in order to save the environment, the commission proposed to change the environment so that it would be able to cope with pollution. The final report stated it had been proved beyond doubt "that sulphuric acid was the substance which destroyed the regional pine forests and rendered the growing of fruit almost impossible". However, since no method existed to remove this gas from the smoke, "the only way to save the remaining woods was to plant acid-resistant trees, while all attempts to keep the sensible conifers offered no chance of success and had therefore to be given up" (Siedlungsverband 1928, p. 16). Decidous trees had proved more resilient than conifers, and some sorts were particularly suitable. Consequently, nurseries were opened to grow these trees and to sell them at a discount.

The statement that no method existed to remove acid gases is remarkable. For some decades, it had been known that lime could effectively be used to reduce sulphur dioxide (Andersen et al. 1986). Admittedly, it caused problems to develop an effective and practical method of using lime, but the real issue was the additional cost, especially when coal was burned. The industrialists, supported by the state, objected to any method that was not profitable, i.e., which did not produce enough sulphuric acid to sell it at a profit. With coal, this was not the case, since coal rarely contained more than 2–4% sulphur; moreover, most steam engines were employed by small factories or craftsmen, who could not afford the extra investment, so that the general introduction of such methods was hardly contemplated, let alone enforced. The big industrial plants along the Ruhr and Emscher, however, often had very powerful steam engines, and for them the additional expenditure would not have been too high, but during most of the period under consideration they proved powerful enough to resist all initiatives (Meldau 1926).

Methods to reduce pollution became especially necessary during the 1920s, when new techniques were introduced that allowed much bigger steam engines and power plants to be built; at the same time, it became technically possible to use coal with a higher waste content than before. Until then, a waste content of up to 8% could be tolerated, while now it rose to approximately 30%. In addition, the engines took a far greater load of coal and the temperatures reached were much higher, so that almost no slag remained; instead, almost all the gases and other waste matter were set free to escape through the chimneys, which for the most part had no filters installed. The consequences were often dramatic, as was the case in Sodingen, a suburb of Herne, where a new plant went into operation in 1929.

Within hours after the engines had been switched on for the first time, the whole town looked as if it was in deepest winter, covered with white ash; the nearby school had to close and the plant was ordered to shut down. In day and night shifts a filter was installed, but the concentration of acid gases was so high that its metal sheets disintegrated within a couple of days. The situation only improved after a special lime filter absorbed a proportion of the gases and ashes; this, however, took a long time. The school remained closed for more than a year and a half (Reerink 1981, p. 6).

Similar cases abound; new power plants were erected in great numbers and they caused a great deal of damage in the neighbouring areas, but they also had positive effects. For one thing, they generated electricity for households and industry, so that most of the often old, small, and dirty steam engines became superfluous and stopped working; secondly, given the high investment necessary to build the new plants, the additional costs for higher chimneys and effective filters were comparatively low, though it normally required a long struggle by the people affected to get these installed. In many cases, the national-socialist takeover in 1933 thwarted their efforts. In view of the priority of rearmament and preparation for war, it proved more and more difficult to fight pollution, and after the outbreak of World War II, all attempts were in vain (Brüggemeier and Rommelspacher 1990).

After the war, production came almost to a complete standstill, and for a while the future of the Ruhr region was uncertain. According to the Morgenthau plan, Germany was to be deindustrialized, particularly the region along the Ruhr. Consequently, research was carried out to find out how much land was needed to feed a family, providing a carefully worked-out plan was followed; a plot of land was rented and a family selected to prove that about one-third of an acre would be enough to provide a sufficient and varied diet. The experiment was successful. On October 1, 1948, hardly 2 years after the experiment had begun, the family was self-sufficient and handed back its rationing-cards, 17 months before they were officially abolished. It had been demonstrated that North-Rhine-Westphalia, of which the Ruhr is a part, could feed its population and did not need its industry (Koch 1983, pp. 20–25).

Soon, however, these experiments were completely forgotten. In view of the cold war, the Morgenthau plan was abandoned, and Germany's industrial base was too important to be destroyed. Instead, industrial production reached new heights, and with it air pollution. By the end of the 1950s, more than 600 000 tons of dust alone descended upon North-Rhine-Westphalia every year, 300 000 of which on the Ruhr (Radermacher et al. 1985). For 1961, it has been estimated that the inhabitants of the Ruhr had to put up with 1.5 M tons of dust, ashes, and soot, and 4 M tons of sulphur dioxide. In some places, more than 5 kg of dust per month per 100 m^2 was registered. According to experts, up to 3 kg were tolerable, but the population thought differently; for the first time widespread complaints were voiced and during the 1959 election campaign, the oppositional Social-Democrats promised a blue sky over the Ruhr (Brüggemeier 1988).

At first, this slogan was widely ridiculed, but in the Ruhr it proved very popular, and the national as well as the regional governments felt obliged to take up this issue. Over the next years, electric filters were installed to reduce the visible pollution caused by dust, ashes, and soot; these filters were so effective that pollution went down by approximately one-third in the period 1968–1980, and even further thereafter (Buck et al. 1982). As before, the fight against the invisible acid gases proved more difficult, and again, ever-higher chimneys were built. As late as 1977 the (by then socialdemocratic) government of North-Rhine-Westfalia planned to build chimneys of up to 300 m, to dilute the noxious substances more efficiently (Luftreinhalteplan 1977, p. 257). To a certain degree, this strategy worked; between 1965 and 1980 the sulphur dioxide emission in the Ruhr decreased by 50%, only to increase elsewhere. Remote, agricultural areas like the Black Forest or the Eiffel, until then hardly affected, experienced an increase in SO_2 and other substances, contributing to widespread acid rain and dying forests.

The chimneys had distributed the noxious substances more widely, but they had not brought about a dilution to harmless concentrations as predicted. The argument that the endless ocean of the sky was big enough to absorb all pollution, was proven wrong. This argument had been put forward for more than 100 years, and for a long time it had seemed to hold. With the increased production and the new chimneys after the Second World War, however, it could no longer be denied, that, as sceptics had warned before, the noxious substances did not disappear, rather, they descended almost everywhere and, to make things worse, accumulated in the soil, rivers, and oceans.

4 An Uphill Struggle: Fighting the Pollution in the Ruhr

The pollution of the Ruhr area as described in the sections above has been exceptional by German and possibly by any other standards; it makes one wonder, how such a pollution could take place, what the inhabitants of the region thought about it and whether any efforts were undertaken to do something about it. These questions will be dealt with in the following section; again, the argument will concentrate on the Ruhr, but the wider Prussian and national context also has to be taken into account: the laws, administrative procedures, etc., dealing with pollution came from Berlin; the Prussian government had the last say whenever a conflict arose, and the system established in Prussia served as a model for the rest of Germany (Mieck 1967).

The discussion on damages caused by industrialization and the rapid growth of cities had begun, long before the Ruhr became industrialized (Bergmann 1970). From the early decades of the 19th century onwards, there were numerous complaints about the pollution of air, water, and soil, which often were backed up by the local and even regional authorities (Mieck 1967). The Prussian government in Berlin, however, adopted a different approach: it seems that in the vast majority of cases the central government overrode local objections. In Berlin, the main interest lay in supporting industry and commerce, which in the first part of the 19th century were still in their infancy. The laws and regulations therefore, which developed in this period – especially the trade law – protected industrial interests, while they made it very difficult to do anything about pollution.

This bias in the laws became obvious, when in the second half of the 19th century, production, and with it pollution, reached new dimensions, and complaints became more widespread. The original objective of the laws to support an infant industry, no longer held true, but it proved impossible to change the laws, not the least because the ambition to become an imperial power and to rival or even surpass the British empire had come to dominate German politics. Again, industry enjoyed a privileged position, since it provided the base necessary to realize these ambitions.

Since the laws could not be changed, other mechanisms were employed to try to reduce pollution, e.g., new heating techniques and steam engines, and a whole variety of different types of chimneys were developed; police regulations tried to enforce the use of low polluting fuels; waste water was to be cleaned more effectively. On the whole, however, these methods were rather costly, so that in the end, of all the different procedures, two prevailed: polluting substances continued to be distributed through rivers and chimneys, and, since this approach often proved of doubtful value, pollution was concentrated in specific areas, be it certain parts of a town or whole regions like the Ruhr.

This approach obviously did not reduce pollution in general, but in many areas, it became possible to prevent pollution from increasing, or even to reduce it locally, while in the Ruhr the opposite was the case. In fact, in the Ruhr the existing laws did not so much protect the people and the environment, as rather establish an effective protection of industrial interests: the Ruhr area became a preserve for industry. That this happened in the Ruhr was the result of several factors.

More than in other areas of Germany, the inhabitants of the Ruhr were fascinated by the advances of industry, the huge new factories and technological progress. Nature, so it seemed, had to give way, and (as an inhabitant of Gelsenkirchen recalls) "it never crossed our mind that it might eventually disappear altogether. We were completely fascinated by the magnitude of industry, and the ingenuity and gigantic force of human efforts and we enjoyed the building of factories, streets and houses as much as we enjoyed natures' activities in the surrounding woods and fields. With some justification we were proud of the American-style growth of our native town and the neighbouring area and we felt we belonged to a purposeful community, hardworking and thirsty for action."

If, he continued, "poisonous smoke from a coke plant ruined a nearby wood, we accepted this as inevitable; hardly anyone ever complained" (Klose 1919, pp. 3–11).

It is true: in the Ruhr area there were fewer complaints than for example in Dahlem, a Berlin suburb for the upper classes who had moved there to escape the noise, dirt and smoke of the city; and it is also true that fascination by, and pride in, the rapid growth of industry played an important part, but these factors can not convincingly explain why such a widespread pollution took place. The notion of progress or fascination by industry are rather abstract categories, which do not give much comfort if the air is permanently filled with smoke and dust, if the washing gets dirty all the time, if trees disappear, and rivers are turned into sewage canals. In all these cases, the inhabitants of the Ruhr also complained. It was not for lack of complaints that pollution became widespread, rather, it was the fact that there was little chance of succeeding in the fight against pollution which characterized the situation in the Ruhr.

As already mentioned, at the onset of industrialization, the Ruhr area was only sparsely populated. Within a few decades, the picture changed completely; tens of thousands of people moved in every year, and apart from a declining minority of farmers and traditional artisans, who had lived there before, almost everybody else — be it steelworkers, miners, shop owners, lawyers or industrialists — had come to this region to find work and to make money. Their chances were good and they were willing to take them, even if pollution resulted. Furthermore, up to 80% and occasionally even 90% of the population belonged to the working classes, while bourgeois groups were grossly underrepresented (Crew 1979; Niethammer 1979; Vonde 1989). That holds true especially for the professional and academic middle classes, from which in towns like Berlin the reformist elements were recruited. The social base for conservationist ideas and organizations, therefore, was very weak, since workers were more or less prohibited from taking part in local and regional politics.

In Prussia, elections were held on the base of the so-called three-class suffrage. According to their share of the tax revenue, taxpayers were divided into three groups and each group elected the same number of representatives. Thus that part of the population which contributed the top third of the tax revenue had as many members in the state and local parliaments as the poorest third; furthermore, in local elections the vote was restricted to males who had not moved for more than a year, so that, given the enormous mobility among the working classes at that time, a large number of workers were disqualified. In Westphalia, finally, companies, too, had the right to vote as long as they paid taxes. Consequently, in most local councils, industrial interests prevailed: a third of the councillors often were elected by the companies, and up to another third by their leading employees, who also paid high taxes. Furthermore, considering that in cases of conflict the state normally supported industrial interests, it becomes obvious that there was little room for doing much about pollution (Croon 1960).

The Prussian trade code, which applied after 1871 to all of Germany, did not offer much help, either. Section 16 of this code stated that factories which could cause considerable disadvantages, dangers or nuisances for the owners or occupiers of the nearby property, had to apply for a licence; in practice, the term "considerable" proved to be of decisive importance. In many cases, it was conceded by industrialists and representatives of the state alike that damages, dangers or nuisances could result, but only rarely were these judged to be "considerable". This way, a few factories could be induced to move to less-populated regions, and quite often, special precautions or installations were demanded to reduce the noise, dirt, smoke, bad smell, etc. These demands, however, could prove to be of limited value and were difficult to implement.

For one thing, the growth of the population resulted in a rapid growth of the cities, so that the area around factories which had been opened in a safe distance from settlements, became populated, especially if these factories attracted new workers and additional business. Furthermore, due to the low state of technology, it was difficult, if not impossible, to lay down precise regulations for the factories to prevent pollution. In the case of air pollution, for example, numerous proposals to reduce the amount of ash or dust emitted from the chimneys existed, but very few of them proved effective. Enormous attention was given to the construction of furnaces and ovens which would reach higher temperatures and which thus guaranteed that the coal would be used more efficiently, so that less ash resulted (Meldau 1926). As to the merits of individual constructions, however, there was always an enormous debate and hardly any agreement, so that it was an easy task for industrialists to point at the doubtful value of most regulations (Hering 1888). To avoid extra costs, this weakness was readily exploited; for example, coke or particular sorts of coal could be used, resulting in less pollution, but these fuels were more expensive and their use could only rarely be enforced. Normally, the cost argument was accepted by the licensing authority, especially in the Ruhr, where coal was the base of the rapid industrial development. In most cases therefore, all that was asked for was a minimum height of the chimneys, which could have a positive effect on the immediate neighbourhood, but resulted in the pollution of ever-wider areas. In industrial regions like the Ruhr, there were soon so many factories that the positive effect of individual high chimneys was more than outweighed by their sheer number.

Section 16 of the trade act therefore offered limited scope for action; to make things worse, in combination with Section 26 of the same law it protected factories, once the license was given. Even if damages or widespread pollution resulted, the factory could never be ordered to close, as long as it did not violate the licensing conditions; since these conditions, as argued before, were not very precise and convincing, their violation could almost never be proved. As a consequence, all that could be asked for were measures to reduce pollution; companies were encouraged to introduce such methods, but only in so far as their economic existence was not jeopardized and as long as effective methods were known. Yet, in practice, both these qualifications resulted in an effective protection of factories once they had been given a license. In the end, neighbours could not do much to reduce pollution; instead, all they could do was to ask for compensation for losses suffered.

The question of compensation was dealt with in Section 906 of the Civil Code. It stated that the owner of a property could not prohibit the emission of smoke, gases, bad smell, noise, etc., originating from another property, as long as this emission interfered not at all or only insignificantly with the use of his own property or as long as the smoke, gases, etc., did not exceed the level that was common in the area. For upper-class residential areas or summer resorts this paragraph offered help. Here, it could be argued that it was not normal to have factories and that their presence reduced the value of properties. Generally, however, and in industrial areas in particular, Section 906 had other effects.

Damages could only be claimed if economic interests were affected. People who neither owned property nor ran a business were in an almost hopeless position. Theoretically, anybody could ask the authorities to intervene, if health risks were claimed, but in practice, this alternative hardly had any significance. In the case of air pollution, for example, a very complex discussion was conducted on its effects on health, but according to the vast majority of doctors no definite proof could be found. They agreed that the cumulative effects of different noxious substances, the reduced amount of sunshine, and other factors could be a risk to the very young, the elderly or the sick, but neither the exact mechanisms nor the particular contribution of the separate factors involved were known (Liefmann 1908; Fruböse 1927). A definite proof, however, was required, if individual factories were to be prosecuted. With regards to water pollution, the situation was somewhat different after the discovery of bacteria in the 1870s; here, organic materials like faeces or decaying food were seen as grave risks, and for drinking water a limit was set as to the number of germs allowed. Industrial production and its waste, on the other hand, were largely judged not to be dangerous to human health (Liefmann 1908; Fruböse 1927).

On the whole, it was only in spas or near hospitals that health risks could be cited to prevent the opening of new factories or occasionally force existing ones to install additional equipment or even to move. Given the low significance of health risks, conflicts were, with the odd exception, restricted to those cases where economic damage was claimed. At the beginning, it was mainly farmers and fishermen who objected to new factories and complained about damages. However, the reduction of the argument to a conflict of economic interests put them into a difficult position. It was easy to show that the contribution of new factories to the local or national wealth by far surpassed that of farming or fishing.

In 1890 this argument was explicitly put forward by Konrad Jurisch, a lawyer, in an article which expressed a widely held view. If, he argued, conflicts between, say, fishermen and industrial plants arose, the greater economic interest was to be protected. He held "that in Germany the economic contribution of industries producing waste water outweighs the contribution of inland-fishing in lakes and rivers by approximately a thousand times. If along a small river like the Wupper, the Emscher, the Boden or others, factories were to become so numerous that fishing in these rivers was impeded, fishing had to be given up. In these cases rivers are to be seen as beneficiary (*wohltätig*), natural canals to divert industrial waste water into the open sea" (Jurisch 1890, p 103).

In the 1920s this argument was carried even further. In the conflict about the pollution of the Wupper, just south of the Ruhr area, the industrialists wanted the term "industrial river" to be part of the new legislation. This term meant that when a river had taken industrial sewage for a long time, it had become its "natural" duty to continue to do so. The example of the Emscher had proven to be so cheap that in the case of the Wupper they wanted to establish the same procedure: waste water was only to be cleaned to such a degree that the river did not stop flowing and that imminent risks for the population were avoided. By now, however, the population put up more resistance, and it was supported by a majority of the political parties, which did not want a second Emscher; in the Weimar Republic, public pressure and political parties could exert a far greater influence than during autocratic Imperial Germany, so that this time the industrialists did not have their ways (Wey 1982; Brüggemeier 1988).

As in this case, it could happen that industrialists lost a conflict, but this did not necessarily lead to a reduction in pollution. Mostly, compensation was paid, so that the damages were covered; normally, the payment of compensation was cheaper than the installation of additional equipment. Quite often, a yearly sum was agreed so that production could continue unhindered.

In many cases, in fact, farmers and other property owners welcomed the advance of industry, since it could lead to an enormous rise in the value of their property (Kleffner 1960). After industry had moved in, however, the early enthusiasm faded; the chances of fighting pollution became ever slimmer and compensation too, was more difficult to enforce. For this, the Ruhr is a particularly good example. At the turn of the century, the Ruhr region had become an industrial area: air pollution was widespread; trees were about to disappear; crops were reduced and rivers loaded with waste and dirt. The effects of industrialization were so obvious that nobody denied them. On the contrary, the very existence of widespread damages was cited to justify pollution. Pollution had become so common that it was declared to be the normal state of affairs for the Ruhr.

In 1908, the mayor of Stoppenberg near Essen complained about the yellow smoke which the local coke plant "turns out almost every hour and which covers the whole region with a stinking fog". The vegetation, above all the trees, had almost completely disappeared, but in spite of 11 letters he had written to the mining-office responsible for this plant over the last 10 years, nothing had happened. The mining-office argued that elsewhere coke plants were situated much closer to houses, whose inhabitants therefore had to put up with even greater pollution. Furthermore, people living next to steel works had to suffer similar smoke and fog almost continually and not just every hour, so that overall the pollution caused by the factory in Stoppenberg was comparatively low.[3]

Some years later, when the owner of some fruit trees demanded compensation from a nearby coke plant, these arguments were given the blessing of the supreme court, which dismissed the claims. In its ruling the court asked how the complainant could expect his trees to grow fruit, given that they stood in an industrial area. There were in fact very few trees left in the area and it was not common in the locality to cultivate fruit trees. The court conceeded that the coke factory emitted noxious gases, but argued that other factories in the vicinity did the same. In fact, the court declared it normal for the area to have these factories, while it was not common to grow fruit trees.[4]

By referring to other factories, the court used an argument that had become of decisive importance. In addition to the problem that it was difficult to identify the substance which had caused the damages, it became more and more difficult to identify the factory or chimney it came from. In the conflict just cited, and in the Ruhr area in general, there were several possible sources that could have been responsible (Riehl 1907; Haselhoff et al. 1932). In 1928, summarizing the outcome of several court cases, it was argued that any hopes of success had to be forfeited and that juridicial procedures had to be avoided "if it was not possible to prove that individual factories were particularly damaging" (Bergerhoff 1928, p. 61).

At the beginning of the Weimar Republic, important changes took place, which looked promising in contributing to a reduction of pollution. The three-class voting system was abolished; workers, who suffered most from pollution, and their respective parties gained a much greater influence at the local and state level and new institutions were founded, above all the Siedlungsverband Ruhrkohlenbezirk (SVR) mentioned above. The SVR tried to put the further development of the region on a planned base, and it was the driving force behind the initiative to save the remaining trees and woods. However, due to the economic crises during most of the Weimar Republic, industry again enjoyed a special position, so that these new initiatives against pollution had little chance of success. Politically, industrialists and workers fought bitter fights, but the preference given to further production and industrial growth was not questioned. Given the high number of the unemployed, it is understandable that the cost argument could be used as effectively as before.

---

[3] HSTAD Reg. Düsseldorf 34220, Schreiben des Oberbergamtes Dortmund v. 1.8.1908 sowie des Bürgermeisters von Stoppenberg vom 2.3.1908.

[4] Zentrales Staatsarchiv Potsdam, RG-Urteile in Zivilsachen 16–30, Urteil v. 16.10.1915, Nr. V 178/15.

By and large, therefore, the situation hardly changed. If anything, it was the lower level of industrial output which reduced pollution, shown most spectacularly during the occupation of the Ruhr by French troops in 1923. But for the people living in the Ruhr, the lesson learnt from this occupation with its standstill of industrial production was all too clear. While it was true that nature recovered, it was also true that the population suffered; no work meant no wages, and despite governmental support, a situation of near-famine resulted. The consequence, therefore, was obvious: smoke and dust did not so much signify pollution as work and wages. In the words of a mayor, who in 1926 commented the variety of purple, brown, red, and other colours to be found in the Wupper: "We can only gladly welcome these colours, which the factories discharge into the river; for only as long as the Wupper is dirty, is there still work to be found".[5]

After the war, the established patterns of dealing with pollution prevailed well into the 1960s. In 1951, e.g., a citizen of Duisburg complained about the acid smoke coming from a nearby chemical factory. The local magistrate conceded that the smoke could be a nuisance to the neighbours, but he told the complainant that there was nothing he could do about it. The reasons given had hardly changed since Imperial Germany: the company was said to comply with the regulations laid down in the licence. A higher chimney would jeopardize the existence of the factory; above all, inhabitants of an industrial town like Duisburg "had to accept such nuisances, which had to be reckoned with", especially because a health risk did not exist.[6]

In the late 1950s, the situation began to change. The people living in the Ruhr were less and less willing to put up with the existing levels of pollution; medical opinion attributed a much greater importance to air pollution, while new techniques such as electrical filters had proved their effectiveness. Furthermore, there was a growing awareness of the problem of pollution and political pressure was mounting. Laws were changed to make it easier to oblige companies to reduce pollution and conflicts became more common, since the chances of winning them were not as one-sided as before. Pollution had become a political issue to a degree hitherto unknown, and it was obvious that something had to be done about it.

The development over the last 15–20 years has been contradictory. Air pollution has declined markedly, while the Emscher still is an open-air sewage canal; traditional industries, especially mines, have been closed down so that they are a less significant source of air and water pollution. But when their land was used on which to build new factories or houses, it came as a shock to realize that the soil had been contaminated to such a dangerous degree that some housing estates might have to be again pulled down.

On the whole it can be argued that the situation in the Ruhr has definitely improved. However, it is doubtful whether this improvement results from an overall reduction of pollution. On the contrary, it seems that pollution in general has increased, it is merely no longer concentrated in the Ruhr area to such a degree

[5] HSTAD Reg. D'dorf 48597, Verhandlung im Rathaus Elberfeld am 2.7.26.
[6] HSTAD BR 1015 I.

as before. The pollution has been exported to the rest of Germany and beyond: what was once normal in the Ruhr, has become normal almost everywhere. From an ecological point of view, this development may seem absurd, but historically speaking it is logical and seems to have been necessary. After decades in which large parts of Germany and its population could assume that industrial pollution was a problem of the Ruhr and other similar areas, the export of pollution has driven home with some force that all attempts to regionalize pollution have been a treacherous solution.

Furthermore, the attempt to distribute noxious substances has also been shown to be no solution; it has become obvious that these substances do not disappear, nor is nature able to get rid of them. As a consequence, filters and other equipment have been installed to clean the air and water before they are discharged into the sky, the rivers or the ocean. By and large, these methods have reduced emissions, but at the same time they have caused new problems, since they neither destroy nor dissolve the noxious substances, rather concentrate them and produce very toxic waste. At the moment a fierce debate is being conducted over the question of what to do with these waste products, and it is planned to build special waste-burning factories, where temperatures of up to 2000–3000°C are to be reached, in the hope that under such high temperatures the noxious substances will be burnt. This assumption, however, is hotly contested and it is argued that new and possibly more dangerous substances such as dioxins will result.

Whatever the merits of these arguments, in our context, another possible consequence of the plan to build these factories is worth noting. It seems that they will amount to the concept of regionalization being resurrected and creeping in through the back door. Given the widespread resistance to such factories, they will probably be concentrated in the Ruhr region: this region still is more polluted than other areas in Germany, so that additional pollution will not be so noticeable as, for example, in Bavaria. The rate of unemployment is in the Ruhr higher than elsewhere and bound to rise, since more of the traditional industries are to be closed down; and in spite of all the resistance environmental and other groups in the Ruhr have been putting up recently, larger parts of the population seem to be willing to accept such factories than is the case in more traditional and middle-class urban areas. The Ruhr may again become a region where industrial pollution is particularly concentrated, and in a way, history may therefore repeat itself, although on a more sophisticated and possibly more dangerous level.

References

Altkemper J (1905) Die Landwirtschaft der Kreise Recklinghausen und Gelsenkirchen unter dem Einflusse der Industrie. Bonn

Andersen A, Ott R, Schramm E (1986) Der Freiberger Hüttenrauch 1849–1865. Technikgeschichte 53:169–200

Avereck W (1913) Die Landwirtschaft unter dem Einfluß von Bergbau und Industrie im Rheinischen Ruhrkohlengebiete. Leipzig

Bach H (1910) Das gewerbliche Abwasser im Emschergebiet. Essen

Bergerhoff H (1928) Untersuchungen über die Berg- und Rauchschädenfrage mit besonderer Berücksichtigung des Ruhrbezirks, Diss., Godesberg/Bonn, Bonner Universitätsbuchdruckerei Gebr. Scheur
Bergmann K (1970) Agrarromantik und Großstadtfeindschaft. Hain Verlag, Meisenheim am Glan
Bette (1928) Von der Emscher. Gladbecker Blätter Orts Heimatkd 9/10:65–67
Brüggemeier FJ (1984) Leben vor Ort. Ruhrbergleute und Ruhrbergbau 1889–1919, 2nd edn. Beck-Verlag, München
Brüggemeier FJ (1988) Blauer Himmel über der Ruhr. Zur Wahrnehmung der Umwelt durch die Arbeiterbewegung. In:Faulenbach B, Högl G (eds) Eine Partei in ihrer Region. Klartext Verlag, Essen, pp 149–155
Brüggemeier FJ, Rommelspacher Th (eds) (1987) Besiegte Natur Geschichte der Umwelt im 19 und 20. Jahrhundert. Beck-Verlag, München
Brüggemeier FJ, Rommelspacher Th (1990) Umweltgeschichte des Ruhrgebiets. In: Köllmann W, Korte H, Petzina D, Weber W (eds) Handbuch zur Geschichte des Ruhrgebiets im Industriezeitalter. Patmos Verlag, Düsseldorf
Buck M, Ixfeld H, Ellermann K (1982) Die Entwicklung der Immissionsbelastung in den letzten 15 Jahren in der Rhein-Ruhr-Region, LIS-Berichte 18
Crew D (1979) Town in the Ruhr. A social history of Bochum, 1860–1914. Columbia Univ Press, New York
Croon H (1960) Die gesellschaftlichen Auswirkungen des Gemeindewahlrechts in den Gemeinden und Kreisen des Rheinlandes und Westfalens im 19. Jahrhundert Westdeutscher Verlag Köln/Opladen
Dueppe W (1939) Beiträge zur geschichtlichen Entwicklung der Kokereiindustrie. Poeppinghans Verlag, Bochum
Emmerich K (1906) Die Ursachen der Gelsenkirchener Typhusepidemie des Jahres 1901. München
Emschergenossenchaft (ed) (1912) Die Emschergenossenschaft. Essen
Evans AS (1973) Pettenkofer revisited. Yale J Biol Med 46:161–76
Fischer W (1965) Herz des Reviers. 125 Jahre Wirtschaftsgeschichte des Industrie- und Handelskammerbezirks Essen-Mühlheim-Oberhausen. Bacht Verlag Essen
Fruböse A (1927) Die Bedeutung der verunreinigten Luft für die menschliche Gesundheit mit besonderer Berücksichtigung der Großstädte und Industriebezirke. Verlagsbuchhandg R. Schoetz Berlin
Hahn M (1911) über die Ruß- and Rauchplage in den Großstädten. Beil Hyg Rundsch 21:96–113
Haselhoff E, Haselhoff W, Bredemann G (1932) Entstehung, Erkennung und Beurteilung von Rauchschäden. Gebr Boonträger Berlin
Heilmann A (1914) Neuzeitliche Wasserversorgung in Gegenden starker Bevölkerungsanhäufung in Deutschland. München Berlin
Heinrichsbauer H (1936) Die Wasserwirtschaft im Rheinisch-Westfälischen Industriegebiet. Glückauf Verlag, Essen
Helbing H (1934) Die Emschergenossenschaft in Essen. In: Brix J, Imhoff J, Weldert R (eds) Die Stadtentwässerung in Deutschland, Jena, pp 478–493
Hering CA (1888) Die Verdichtung des Hüttenrauches. Stuttgart
Howard-Jones N (1973) Gelsenkirchen typhoid epidemic of 1901. Robert Koch and the dead hand of Max von Pettenkofer. Br Med J pp 373–95
Imhoff K (1910) Die Reinhaltung der Ruhr. Haarfeld Verlag Essen
Jurisch K (1890) Die Verunreinigung der Gewässer. Berlin
Kleffner F (1960) Der Einfluß des bergbaulichen Grundbesitzes auf die Entwicklung der Gemeinden im Gebiet zwischen Emscher und Lippe. Bottrop
Klose H (1919) Das westfälische Industriegebiet und die Erhaltung der Natur. Borntraeger Berlin
Kluge T, Schramm E (1986) Wassernöte. Umwelt- und Sozialgeschichte des Trinkwassers. Alano-Verlag Aachen
Koch E (1983) Der Weg zum blauen Himmel über der Ruhr. Geschichte der Vorläufer der Institute der Landesanstalt für Immissionsschutz, Essen
Liefmann H (1908) Über die Rauch- und Rußfrage insbesondere vom gesundheitlichen Standpunkte und eine Methode des Rußnachweises in der Luft. F Viewegu Sohn Braunschweig

McCreary EC (1963) Essen 1860–1914. A case study of the impact of industrialization on German community life. PhD Thesis, Yale Univ
Meldau R (1926) Der Industriestaub. Wesen und Bekämpfung. VDI Berlin
Mieck I (1967) Aerem corrumpere non licet. Luftverunreinigung und Immissionsschutz in Preußen bis zur Gewerbeordnung 1863. Technikgeschichte 34:36–78
Niethammer L (1979) Die Unfähigkeit zur Stadtentwicklung. Erklärung der seelischen Störungen eines Communalbaumeisters in Preußens größtem Industriedorf. Syndikat Verlag Frankfurt
Niggemeyer H (1915) Die Beschädigung der Vegetation durch Rauch mit besonderer Berücksichtigung des rheinisch-westfälischen Industriegebietes. Diss Münster
Oberpräsidium Münster (1909) Bereisung des Emschergebietes durch die Ministerialkomission in der Zeit vom 27–29 Oktober 1909. StA Münster, Oberpräsidium Münster 1708, vol 1
Radermacher L, Prinz B, Rudolph H (1985) Bericht über die Ergebnisse der im Land NRW in der Zeit vom Januar 1984 bis Dezember 1984 durchgeführten Staub- und Schwermetallniederschlags-messungen. LIS-Berichte 63:1–72
Reerink W (1981) Aus den Anfängen der Staubmeßtechnik. In: TüV (ed) 50 Jahre Staubmeßtechnik und 25 Jahre Immissionsmessungen. TüV Schriftenreihe. Essen, pp 6–7
Reich A (1917) Leitfaden für die Ruß- und Rauchfrage. München Oldenbourg Berlin
Riehl J (1907) Über die Klagen zur Abwehr der Belästigungen durch Rauch, Gerüche, Geräusche, Erschütterungen und andere Imponderabilien. Verlag Franz Vahlen Berlin
Rommelspacher Th (1987) Das natürliche Recht auf Wasserverschmutzung. In: Brüggemeier FJ, Rommelspacher Th (eds) Besiegte Natur. Geschichte der Umwelt im 19 und 20 Jahrhundert, München Beck-Verlag, pp 42–63
Siedlungsverband Ruhrkohlenbezirk (ed) (1927) Denkschr Walderhaltung im Ruhrgebiet. Selbstr Essen
Siedlungsverband Ruhrkohlenbezirk (ed) (1928) Bisherige Tätigkeit des Ausschusses für Rauchbekämpfung beim Siedlungsverband Ruhrkohlenbezirk. Selbstr Essen
Spelsberg G (1984) Rauchplage. Hundert Jahre Saurer Regen. Alano-Verlag Aachen
Vonde D (1989) Grenzen der Stadtentwicklung. Klartext Verlag Essen
Vossiek W (1937) Hundert Jahre Krupp'sche Betriebskrankenkasse, 1836–1936. Verlag Paul Schmidt Berlin
Wey KG (1982) Umweltpolitik in Deutschland. Kurze Geschichte des Umweltschutzes in Deutschland seit 1900. Westdeutscher Verlag Opladen
Wiel P (1970) Wirtschaftsgeschichte des Ruhrgebiets. Tatsachen und Zahlen. Siedlungsverband Ruhr-Kohlenbezirk Essen

Environmental Consciousness

The Evolution of Environmental Sensitivity 1750–1950

F. Walter[1]

1 Introduction

This chapter proposes to tackle the relationship between history and ecology (ecological history) from the angle of ideology and mentality. What interests me is the composition – the component parts – of sensitivity towards the environment. The current ecological movement, with us now for some decades, is the most recent manifestation. I am not, therefore, going to write a history of the protection of nature, I am going to seek to understand the evolution of ecological sensitivity. Respect for equilibrium in the natural environment is a concern which is the result of a long evolution. This is understandable, providing one restores nature to a place in a social context where it has both a symbolical and economical meaning.

The departure point is the hypothesis that environmental sensitivity changes. One is aware of various attitudes from what has been called "The Ecological Crisis" or even "The Ecological Age", and from the new environmentalist ideology which developed in the 1950s and 1960s. This recent phase brings together various heterogeneous and sometimes contradictory paths: fear of upsetting the social order (by overpopulation, economic growth, exhaustion of resources, etc.), challenging the consumer society, pessimistically casting doubt on the possibilities of technological development, progressing in the scientific understanding of ecological mechanisms. All that constitutes the new "ecological conscience" of the western world which shows itself by symbolical practices (a taste for natural authenticity, rusticity, nostalgia for folk-dress, etc.) and political movements (the "Greens" have attracted particularly strong support in certain countries).

2 The Relationship with the Natural Environment

The present approaches have nothing new in themselves, rather it is their combination which is original, forming a model for the relationship with nature which one describes as ecological. The historian has therefore to operate in two

[1]Université de Genève, Département d'Histoire Générale, Les Bastions, 20, rue de l'Ecole de Médecine, 1205 Genève, Switzerland

Brimblecombe, Pfister (Eds.)
The Silent Countdown

ways: (1) he begins with a classical geneological approach which consists of identifying innovating practices and promotional speeches. (2) A systemic approach follows, using historical studies at different levels in mutation. changing between different periods. Every epoch has its model for environmental relationships. It is built up from a variable complement of approaches: I shall limit my examination to five components whose contents I shall try to characterize before applying them to a case study: (1) emotional; (2) techno-economic; (3) scientific; (4) ideological; and (5) utopian.

2.1 The Emotional Approach

The emotional approach is the attitude closest to us. It consists of finding esthetic enjoyment in the spectacle of nature, and is essentially visual. The promotion of visual perception dates from the Renaissance. It assumes the education of the aesthetic senses, which results in an awareness of class. A taste for nature is in the first instance a sign of being part of an elite. However, the notion of a beautiful landscape or of the beauty of nature has made some progress. The idea of a planned landscape, prized in the classical era (symmetry and regularity), gave way in the 18th century to an attraction to untamed nature and for the picturesque, taking the form of a romantic perception of nature. This esthetic relationship, developed in England at the end of the 17th century and propagated on the continent particularly by Rousseau, still influences the neoromantic attitudes which exist even today. References to a visual culture suggested societies for the protection of the countryside, picturesque places, and historical sites and heritage, particularly at the end of the 19th century.

2.2 The Techno-Economic Approach

The techno-economic approach is at first sight equally banal. It consists in putting the exploitation strategies of natural capital to work. The idea that the environment is a rare good is a recent variation. Throughout the centuries, societies have done little else other than to codify their rights to use the natural environment according to their needs. Legislation, some of it very ancient, on hunting, fishing, forestry, and use of the pastures was interpreted in this way. The point which interests us in particular here is that the countryside itself became merchandise. Financial investors and the tourist industry gave themselves the task of transforming it and giving it an exchange value. For the first time, the esthetic and the utilitarian began to compete with each other in a period where technological development (industrialization) pushed the contradiction between the necessity of economic growth and the defence of esthetic values, as well as those morals inherent in civilization, to its paroxysm.

2.3 The Scientific Approach

The scientific approach is essential in order to develop a theory of the risks of anthropological action. One has to admit that Darwinism, dominant in the natural sciences during the 19th and 20th centuries, does not readily lend itself for use here. In the 18th century, rationalism had developed an evolutionist concept of human history. Perfected by progress, each people passes through various stages of development. With Charles Darwin (1809–1882), the finality of evolution lost its intentional character and became laicized. From then on, the various species transform themselves by accident. It is the environment which assures the success of the best adapted. The environment becomes the great regulator of evolution. Adapted to man, the simplified and simplistic vision of Darwinism declares that history should accept that the most adaptable "cultures" have become part of their environment, that is, those which have best mastered their environmental constraints. Science, in explaining the processes, has as its objective the subjection of the natural world and does not yet show itself to be very sensitive toward the negative aspects of man's impact on the environment. One retains the paradigm of progress from Darwin's teachings.

It is symptomatic that it was the naturalist Ernst Haeckel who contributed to the spread of the new theory in Germany. It is also he who in 1866 invented the concept of "ecology" in order to study the relationship between living beings and their environment, never exposing the breaches of harmony. Karl Moebius introduced the concept of biocenosis in 1877. In 1865, the German physicist Rudolf Emanuel Clausius defined the notion of entropy in thermodynamics. Essential to the understanding of the global functioning of a system, his impact was extremely limited and the other scientific fields hardly noticed its result. Ecology integrated these contributions only in the 1950s. In addition, those ecological concepts which opened the way to a more global conception of the environment were late in coming. The decisive notion of an ecosystem was introduced by the English naturalist A.G. Tansley, designating "not only the organism-complex, but also the whole complex of physical factors forming what we call the environment". The concept acquired all its potential with Lindeman (1942, cited in Kormondy 1965). He refers in particular to McIntosh (1985). See also Schramm (1988).

The sciences most concerned with the analysis of the relationship between man and his environment have been, since the 19th century, anthropological geography and human geography. They were also inspired by Darwinist presuppositions in the natural sciences. Both F. Ratzel in Germany and Vidal de la Blache in France insisted on the positive capabilities of adaption of life-forms to their surrounding conditions (Raumolin 1984). By contrast, those who analyzed the negative role of man in the transformation of nature were little heard. I have in mind the *Raubwirtschaft* economists such as the German Ernst Friedrich, who around 1900 stressed the increasing aggressiveness of human intervention. The *Kulturvoelker* (civilised peoples) appeared to be much more destructive than the *Naturvoelker* (primitive peoples). That which Friedrich calls a "predator-characterized economy" is essentially linked to the contemporary civilization and its world-wide extension through colonial imperialism (Friedrich 1904). The pioneer work of the American geographer George Perkins Marsh, who in 1864 also stressed the results of man's acts on the environment, did not much influence thought outside the Anglo-Saxon world. It was, however, in the United States that systematic observation of the effects of catastrophes which affected cultivated lands (spectacular phenomena of erosion) would contribute toward an active policy of land conservation. For this reason, ecological sensitivity is particularly strong in the United States. In other words, the concepts which are essential today for an ecology-minded conscience only became operational relatively late.

Nevertheless, scientific circles played an important role in promoting the nature conservation movement. Let's remember a few milestones. It was in the United States that Yellowstone, the immense nature reserve and first national park, was created in 1872. This served as the model for later institutions of that kind. Equally, in Great Britain, scientific interest was foreseen in the activities of the first preservation societies such as the Society for the Preservation of Birds (1889). The debate on the necessity for nature reserves also goes back to the 1880s. On the continent, Germany followed the Anglo-Saxon model beginning in the end of the last century. In 1906, Hugo Conwentz convinced the Prussian government to create a *Staatliche Stelle fuer Naturdenkmalpflege* (Central Office for the Protection of Natural Monuments), which had a consultative role. Similar ideas were observed elsewhere in Europe. The pioneering activities of scientific circles (notably Paul Sarasin of Switzerland) were the origins of the First International Conference on the Protection of nature (Berne 1913), and the First International Congress for the Protection of nature (Paris 1923).

2.4 The Ideological Approach

The fourth approach is the ideological one, implying an imaginary relationship between individuals and their real conditions of existence. I should like to mention that vast compensatory movement which values nature as a counter-balance to urban life in general and to the industrial civilization in particular, a problem well-studied in Germany through the works of Bergmann (1970) and Sieferle (1984). One of the manifestations of this current is what D. Worster calls "Arcadianism" (Worster 1985). The ideal of harmony between man and nature became a real ethic with the holders of the new attitudes of humility toward and coexistence with the natural world. This ideology was illustrated at first in France by the Societe pour la Protection des Paysages, founded in 1901. The same inspiration was behind the founding of the Bund Deutscher Heimatschutz in Germany (1904) and the National Trust for Places of Historic Interest or Natural Beauty in Great Britain (1904). The foundation of the national societies was evidently the result of numerous local and regional initiatives. It would be too time-consuming to describe their development in detail here. The various organisations met each other at the international congresses in Paris (1909) and Stuttgart (1912), and at the Institut international d'art public (founded in Bruxelles in 1898) which had similar objectives, but was more directly orientated toward the problems of architectural esthetics, according to the culturalist view inspired by John Ruskin. At the international meetings, the protection of sites was thoroughly discussed.The first international congress for the protection of the countryside took place in Paris in 1909. The first legislation in various countries which attacked advertising by regulating the sites of advertisements, the latter being a sort of symbol for the consumer society, happened within a short time (in Prussia, the 1907 law; in Great Britain, also a law of 1907; in France, 1909; in Belgium, 1909).

In a certain manner, these institutions prolonged the most ancient discovery in historical heritage. Sensitivity towards historical monuments is contemporary with the French Revolution, whereas the notion of natural heritage was invented at the end of the 19th century. France was therefore already innovatory in the 1830s, with her Inspectorate of Historical Monuments. In Great Britain, the first legal text dates from 1882 – the Ancient Monuments Protection Act (see Fawcett 1976). The end of the 19th century saw the notion of heritage joined to that of the rediscovery of nature by integrating the traditional architecture into the landscape.

Since heritage is the memory of a country, patriotic connotations are essential. They showed themselves particularly vigourously in the age of *Heimatschutz* (protection of picturesque countrysides). Let us remember that the concept of *Heimatschutz*, forged by Ernst Rudorff in the 1880s, and used systematically from the end of the century onwards, was particularly rich in symbolical implications. At the Congrès de Paris in 1909, the German delegate stated that the movement was both a "tribute of admiration for the monuments of one's country", and action to protect "the whole country, from which nature evidently could not be excluded". In France, similar opinions were voiced on the occasion of the discussions relating to the 1906 law for the *Protection des Sites et Monuments naturels*, the first of its kind in Europe. A small extract from the parliamentary debate on this law:

"Patriotism, Gentlemen, is not only a moral entity, an abstract conception, a geographic or historical expression. It is, in a certain way, the material and visible representation of the country itself, its particular physical character and its diverse elements, with its mountains, forests, plains, lakes, rivers and river banks, the varied and multiple aspects of its soil, such as have been transmitted by the slow succession of the centuries. Certainly, patriotism is an innate and instinctive sentiment, but that which contributes most toward fortifying it and engraining it indelibly in the peoples' souls is the attachment to their natural soil, to their preferred horizons, and the memories which this evokes. To alter little by little this filial sentiment by neglecting to conserve all its *raisons d'être* is to extinguish the purest source of the love of one's country of which the love of one's native soil is the granite foundation." (Declaration of Maurice Favre, Journal officiel, March 28th, 1906, Sénat p. 282).

The German 1935 law on the protection of nature was commented on in similar terms by its promoters: "In addition, the German soul and the ethnic identity of the German people is formed by the German landscape and the preservation of its natural character." (Weber and Schönichen 1936).

Thus, in different contexts, the countryside was invested with a massive ideological meaning. Such a discussion was revived by conservative elements who were nostalgic for the society of the Ancien Régime. It served as a point of reference for the conservative revolution of which F. Stern says: "The term conservative revolution . . . denotes the ideological attack on modernity, on the complex of ideas and institutions that characterize our liberal, secular and industrial civilization" (Stern 1961). From this point of view, the city seemed to incarnate all the evils of contemporary society. The back-to-nature ideology idealized the eternal peasant, the harmony of a people, strong in their physique and in their morals, with their soil. The idealization of rural life had political implications and served the corporate defense movements such as the great agricultural societies at the end of the century. In France, it was Méline and his *retour à la terre*, followed by the myth of the peasant farrier during the Vichy régime. In Germany the slogan *Blut und Boden*, forged in the 1920s, had a patriotic basis which had racist tendancies, duly diverted for its own purposes by National Socialism.

Another point I should like to mention, again with the purpose of showing the ideological importance of the return to nature, is the frequently mentioned relationship between Protestantism and an awareness of nature (Viard 1985). The sensitivity towards the negative consequences of human acts is much stronger in Great Britain, the United States, in the Scandinavian countries, in Germany and in Switzerland. From the wide range of available literature one should mention particularly Hays (1968), Olwig (1984) and Brüggemeier and Rommelspacher (1987). According to Andre Gide, there is a "Helvetico-Protestant" image of nature. The conception of a mankind more responsible and more committed with regard to the possession of land could be the result of being accustomed to conquer vast territories. If this can be shown to be true in North America, precociously ecological, the relationship is less pertinent in Europe, and one can ask if it is not simply a north-south division. A nature which is, *prodigious* and *luxuriant*, or at least understood to be so? That leads to a wide appreciation of the renewable productive capacities of the soil. In the process of the industrial revolution, the centre of gravity of the industrial and urban Europe was placed in the northern and western regions, and the less industrialized regions remained to the south and east of this kern. What, therefore, would be astonishing about the northern and western regions being more sensitive toward environmental degradation?

2.5 The Utopian Approach

The final approach is that of utopian thinking. The French term *uchronie* refers to an utopia situated in an imaginary time, but not necessarily situated in a fictitious place. Since the 18th century, the utopian discourse has changed its location to exotic places in an imaginary time, under the effect of the general belief in progress. The myth of progress had been one of the obligatory components of the movement for sensitivity toward the environment, until the 1960s when it was severely breached. For some, degradation of one's environment is a normal accident and a problem which technology will solve in the future. For others, utopia is to be reversed and situated in a hypothetical return to the values of the past in the realization of their social ideal. Such a progressive/regressive dialectic was incarnated in the vast reformist current to which the Germans gave the name *Lebensreformbewegung* Movement for the reform of man's daily life (Krabbe 1974; Frecot 1976). The movement began during the first quarter of the 19th century in Great Britain and Germany, reaching its peak at the beginning of the 20th century, whereas in the Latin countries experiments were only being very timidly made with certain "alternative" ideas. In effect, rupturing with certain accepted ways of life and technology of urban civilization, the various reformist currents encouraged a return to a more natural way of life: natural medicine, natural food (health foods), biological agriculture, freedom from the bodily restrictions of clothes, modern dance, nudism (naturism), sexual liberation, social reform through cooperative and community idealism, and housing reforms (the garden city as opposed to the city of blocks of flats and dense housing areas).

There are some very strong links between the five components which we propose to study in order to understand the evolution of sensitivity toward the environment. For example, the Heimatschutz of 1900 allied an esthetic concept to economic concern, that of ensuring the continued economic viability of agriculture and of ensuring the continuing tourist attraction of the countryside, without disdaining scientific arguments (for example in respect of the disappearance of fauna). They combined a patriotic ideology with projects for an ideal city (the garden city). Nevertheless, it is in the degree in which their components combine that the attitudes toward the environment differ from one another in time and space. Every society and every epoch models its relationship with nature according to its own conceptions and strategies. In fact, the environment is a social stake. The complexity of this relationships is well shown in Lübbe and Ströker (1986). One distinguishes, therefore, the periods where the contemplative attitudes of admiration, ignorance, fear and ethical respect are dominant, and others, on the contrary, where attitudes implying voluntary action transforming or protecting nature appear to be essential in order to define the social link with the environment.

3 Switzerland and Environmental Sensitivity

In the third section of this chapter, I have decided to illustrate the preceding sections by showing how the different components of the relationship with the environment make themselves heard in Swiss society. At first sight, this country is more a part of the group of nations engaged from the start in a compensatory process in the face of wounds which industrial and urban society has inflicted on the environment. To write a history of the environment implies a specification in time of the relationship with nature. We isolate several periods, each of which experiments with a different model, or an original enumeration of the components described previously. The developments which follow are succint presentation of a more important work (in preparation), where precise details concerning the sources and literature can be found.

3.1 The Contemplative Model (18th Century)

The rationalist offensive of the classical age has not saved the Swiss landscape. The anthropocentric and individualistic perspective supports the thesis of the superiority of man over creation (animals, trees, vegetation, and soil); human ingenuity must affirm its mastery of the natural environment. This logic, and not some "ecological" necessity, underlies the codification of fishing, hunting, and forestry. It dispossessed the peasant communities of their rights in the natural habitat to the exclusive profit of aristocratic urban elites, who controlled power in the principal cantons and whose priority was profitability in the use of resources. A heavy hand laid on nature is not without risk, as is proven by the hazardous rerouting of the Kander, a torrential river in the Bernese Oberland. Undertaken in 1711–1714, this "correction" transformed itself into an ecological disaster with, among other things, the forging by the water itself of a 21-m-deep trench and a delta measuring some 5 million m^3 in less than 2 years!

Nevertheless, Switzerland found itself at the beginning of the formation of a new regard for nature. The rediscovery of the mountain as an untamed place, as opposed to the too-well cultivated nature of the plains, transformed the Swiss territory into a laboratory for the state of individual souls. After the initiatory travels of Albrecht von Haller (1728 and 1732), the mass of European travellers, many of them readers of J.J. Rousseau, was going to transform a hitherto subjective viewpoint into an integral part of reality (Walter 1984). The phenomenon is, for example, flagrantly apparent when one associates the image of the landscape with an anthropological conception of a pure pastoral society, in the same way in which, at the same epoch, the North American Indians were presented as the survivors of primitive peoples. The "homo alpinus", sheltered from the corrupting effects of social life, lived in an environment which encouraged the stereotyped images of a rediscovered Arcadia. Apart from anthropological rarities, the Alps revealed mineral and vegetable novelties to curious amateurs of science. A scholar such as Horace Bénédict de Saussure pioneered the concept of a natural monument for erratic blocks, "these precious monuments", he said, "of the great revolution to which the surface of the globe owes its present state". Without exception, men of the 18th century had an esthetic kind of relationship with nature and with the mountain in particular. It was based on a new perceptive code which replaced the beauty of symmetry with disorder: from cultivated landscapes to savage nature, from formalism to the picturesque. Parallel to this, the esthetic sentiment was accompanied by a growing mysticism, even a deification of nature, making the landscape a "spiritual resource" (Thomas 1983). It was therefore a contemplative model of nature which prevailed.

3.2 The Promethean Model (Until the 1880s)

The secularization of ideas, the emancipation of science from moral values and the fashion of individualism opened the door to a new conception of nature, accentuating even more confidence in the unlimited potential of knowledge oriented toward the mastering of one's environment. A summary can be found in Lepenies (1983). In fact, that meant that the dominant concept was still fully functioning: it was a question of protecting oneself against nature, not defending nature itself. If one proposed to save the forest and replant the mountainside with trees, it was in the first instance in order to preserve oneself from the devastating consequences of natural forces (in particular from inundations). Fashionable scientific conceptions relied on the Darwinist paradigm.

As elsewhere in Europe, it was the forestry employees who were among the first to be alarmed at the consequences of man's actions. Their arguments can be classified as preecological. Without a doubt, the Swiss practitioners did not escape the mythology of deforestation, to which all natural calamities were attributed. Legislation introduced in the 1870s gave the Federal authorities considerable rights of supervision in high-altitude regions, where ecological anarchy was considered responsible for devastations which affected the plains, such as floods. The inhabitants of mountain regions were therefore seen as unknowing predators, before being promoted in the 20th century to the rank of "Gardners of the Alps", and protector of an ecosystem.

During the same period, this territory open to human exploitation was given new values. It should not be forgotten that Switzerland, as a result of many historical phases which cannot be described here, ceased to be an amalgam of territories of varying status and became a state. Since the end of the 18th century, patriotic images have been very largely mixed with images of the landscape. The territorialization of patriotic sentiment is essential in order to understand the initiatory role of the Alps in Swiss mental perceptions. It is important to distinguish the internal employment of alpine stereotypes from those, more conventional, intended for external use. One knows that the Alps became, in the period 1820–1830, the chosen country of romanticism. The tourists created their own individual territory, their vision of nature, where poetic lyricism was the neighbour of morality and sensibility. The grandeur of nature supported a moral beauty and the values of liberty. The cliche of the Helvetian landscape realized an amazing popularity, thanks to the graphic arts and paintings by less well-known masters.

The Swiss themselves, as well as allowing their collective memory to acquire new national stereotypes, became rapidly aware of the mythical roles of the latter. Already, certain literature describing travels, such as the works of R. Töpfer (1844) ridiculed the mystical tourist. Instead of being aware of man's aggressions in the countryside, still relatively moderate, the corrupting effects of industrial urban civilization on man were stressed. The avaricious mountain dweller who "exploited" the tourist was discovered. It is, however, evident that the era favoured the conqueror. The first ascent of Mount Cervinia in 1865 attracted extraordinary interest. The Promethian image of the mountain, fuelled by the growing popularity of mountaineering, is also found among the engineers responsible for constructing the means of communication across the Alps. The St. Gotthard was pierced between 1872 and 1880, the apogee of an optimistic century fascinated by progress. The press described the event in the following terms: the "pioneers of progress" and the "workers of science", wrote a journal in 1882, "have succeeded in lowering the alps and piercing mountains", "a gigantic undertaking which honours human genius even more than the alpine crossings, well commemorated by history, of Hannibal and Napoleon".

4 The Esthetic-Patriotic Model (End of the 19th to Beginning of the 20th Century)

The awareness of the dangers which human activities could cause to the natural environment were successful in a period which is described as preecological. That was the moment of progression from a purely contemplative attitude to a will for action, not as yet to manage resources rationally, but at least to protect specifically, while there was still time, those areas which were regarded as being particularly menaced. For this period, see Walter (1989).

The second breath of the Industrial Revolution, which governed the rhythm of social life at the end of the 19th century, resulted in particularly rapid modifications in the countryside. It was alpine nature which paid the heaviest tribute. The hydroelectric equipment in the valleys had hardly begun to operate when the electro-metallurgical industry was installed at Chippis, in the Rhône valley. The peaks which hitherto had been reserved for the slow approach of mountaineers were now covered with rack railways (more than 60 lines built between 1870 and 1910). The hotel industry turned the mythical image of the mountains into profit: about 100 hotels existed in 1880, in 1912 there were 3585!

In addition, Switzerland experienced an unprecedented urban explosion (+3.6% per year) during the two decades 1890–1910. The three agglomerations of Zurich, Basel, and Geneva exceeded 100 000 inhabitants (certainly modest by European standards). The question of urban hygiene, a better understanding of organic phenomena, in particular of the contamination process in drinking-water, attracted attention to the problem of water quality. In Basel, in particular, the authorities were alerted to chemical pollution in the 1870s. In 1888/9, the Confederation issued a regulation which indicated the precise quantities of liquid and solid industrial wastes which were allowed in watercourses. However, the dominant representation according to which society should protect itself against nature continued to function normally. Laws regulating the hunting of animals distinguished between animals which were regarded as useful and those which were regarded as pests so successfully that one could triumphally exterminate the last wolf, lynx, and bear. In 1902, Switzerland ratified an international convention whose objective was to protect such birds as were "useful to agriculture".

As regards scientific theories which might take into consideration the consequences of action by man, they still commanded but little attention. Even if numerous Swiss naturalists had been the precursors of ecology, rare were those who were active outside their individual fields. Such was the case of the limnologist. F.A. Forel who, before 1900, already described trophic chains. This was also true in the case of Jean Brunhes, Professor of Geography at the universities of Fribourg and Lausanne (1896–1912), one of the founders of human geography. Under the influence of Friedrich's theories of the *Raubwirtschaft*, he attempted, in a large work published in 1910, a globalizing approach to the relationships between man and the environment by applying the laws of thermodynamics to the degradation of energy.

Even if perception of the consequences of the industrial and urban society was alive at the end of the 19th century, it was as yet on an esthetic basis. It was the refusal to accept the ugliness created by modern utilitarianism which motivated action. In 1915, Marguerite Burnat-Provins launched an association against vandalism and prostitution of the countryside. It was an immediate success and the movement became the "League for the Conservation of Picturesque Switzerland" (Schweizerische Vereinigung für Heimatschutz). Becoming famous through its national campaign against the Cervinia railway, the Heimatschutz declared war on advertising posters, and found considerable support in the cantons. Several cantons passed laws prohibiting posters where they would prejudice the natural beauty of sites. The same care for esthetic matters on this occasion motivated the Naturschutz.

From 1906 onwards, the Schweizerische Naturschutz commission of the Swiss Society for Natural Sciences brought together experts in the natural sciences, engineers, and foresters. Beginning in 1909, the Commission worked together with the Swiss League for the Protection of nature (Schweizerischer Bund für Naturschutz). Their principal work was certainly the Swiss National Park in Lower Engadine, created with Federal support in 1914. The clearly affirmed objective was the detachment of a part of the national territory from "the fury of destruction which the modern world has unleashed". Although it was spectacular, the creation of the National Park should not make us forgetful of the revealing arsenal of legislation which resulted from this new preecological conscience. It would be too time-consuming to enumerate here all the measures taken by the cantons and communes to protect historic monuments, natural curiosities, the aspects of the various sites (*Landschafts- und Ortsbild*), and plants and beautiful trees.

If the esthetic component is closely linked to Heimatschutz and Naturschutz, the model for action in respect to nature during the Belle Epoch functioned, in addition, through a fundamentally patriotic concept. More particularly, since Swiss patriotic feeling had as much need for landscapes as for heroes. Geography and map-reading were more important than history as educational methods, in order to ensure that schoolchildren and future soldiers developed a love of their country. In addition, natural objects which were protected by law corresponded to an equation between the homeland (*patrie/Vaterland*) and the natural landscape. This is the case when fine trees are considered to be witnesses of the past; it is the case with bizarre blocks of stone, themselves witnesses of many ages; it is also the case of a flower such as the edelweiss. The edelweiss, by nature of its smallness and its resiliance in a hostile environment, is the incarnation of a Switzerland which created the identity of a small state confronted by the imperialism of great powers. The great national exhibition of 1896 in Geneva constituted in this respect a summit in the theatrical presentation of the alpine scene. In offering a reconstruction of a Swiss village and a mountain of plaster, populated with authentic inhabitants, the promoters intended to produce the "real symbol of our Swiss homeland". The Swiss village was surrounded by an artistic atmosphere which had for its motto "Art, Mountain and Homeland".

In default of scientific arguments, the preecological current defended in esthetic terms the nonproductive, nonutilitarian and nonpredatory use of natural spaces, sheltered from human action. For the Swiss Society of Natural Sciences, it was a question of (according to a report in 1911) "saving that Switzerland menaced by industry and invasion by foreigners from its imminent destruction and restoring it to its former state". The alps were a museum of landscape-relics, a conservatory of traditions and Swiss values. These ideological connotations appear extremely strong. It was, in effect, a period which attempted to invent alternatives to which one could escape, when faced with a disquieting economic and social revolution. Valuing nature and rurality went therefore hand in hand, for example, in end of the century literature, which illustrated a Helvetian authenticity, "Swiss-ness", the *Schweizerisch* (Linsmayer 1987). Opposed to this was the modern city which endangered collective values and menaced the national identity. In Switzerland, anti-urban sentiment concentrated on the foreigner, symbol par excellence of urban life, one-third of known foreigners lived in the three great cities. The massive presence of foreigners (14.7% of the resident population in 1910) was such that around 1910, one spoke of "Überfremdung", became a symbol of materialism and decadent moral values. In other words, it was the authenticity of the indigenous race which appeared to be compromised, and some people regretted that one could not "construct territorial reserves where things such as the splendid alpine flowers and the ancient qualities of our race could be cultivated and safeguarded" (de Montenach 1910).

Politically, sensitivity toward degradation of the environment and the myth of the return to nature have been mostly retained by various conservative and nationalist currents of opinion, strongly tainted by outdated nostalgia. The Union Suisse des Paysans, founded in 1897 on the model of the Bund deutscher Landwirte, propagated an anti-urban ideology much influenced by the analogous debate in Germany. On the contrary, it was not astonishing that the Social-Democrat party was the only one not to be a supporter of the National Park Project in 1914, arguing that there were more urgent social tasks in need of support.

But Switzerland participated equally in that great Western current called *Lebensreform*. Here, criticism of the materialist civilization and of liberal capitalism nourished the currents combining the regressive aspects of conservatism and an Utopian outlook. It would be impossible to review here all those organizations and publications where such orientations were debated as had as a common purpose the propagation of life-styles considered closer to nature.

Natural medicine, corporal hygiene, healthier food, and reformed clothing all had their periodicals such as *Die Gesundheit. Zeitschrift für gesundes Körper- und Geistesleben*, founded in 1900 by Pastor W. Stern. The *Wandervogel* movement, which originated in Germany, had a Swiss counterpart from 1907 onwards. This movement of young people sought to rediscover a taste for nature by journeying on foot and offered a culture which was a mixture of neoromantic traits in Heimatschutz style and the most modern reformist tendancies. Ticino and the Ascona region were, around 1900, a laboratory for such ideas. There was a community of "pre-hippies", long-haired vegetarians, living in huts open to the sun, adherents of nudism, close to theosophic thought and mystic cosmology (Green 1986). These few examples show how closely Switzerland had become linked to the common evolution of European countries. There were slight differences, of course. Environmental problems were quantatively less spectacular than was the case in the great industrial regions. On the other hand, the symbolic investment of nature and landscape was more decisive than elsewhere, because of a massive territorialization of the national sentiment from the end of the 18th century onwards.

5 Nature as a Social and Economic Stake (World War I–1950)

The esthetic-patriotic model for intervention in nature could not remain intact after the shock of the Great War. And yet, in several aspects, the period between the two World Wars prolonged the preecological wave dating from the end of the previous century. The continuity is remarkable, but the conditions under which the model now functioned brought the objectives face-to-fact with the hard realities of the economic, social, and cultural crises of Europe resulting from the war. The land and the landscape, as always symbolically invested, had now to reply to the different needs of social and economic viability. The vaguely sentimental prewar patriotism had now to anchor itself more deeply. Heimatschuz and Naturschutz sought to acquire a new autonomy with reference to a Germanic model, now compromised. Helvetianism had to be reinvented.

The anti-urban and xenophobic attitudes (as was the case with agrarian nostalgia) were going to serve as a new springboard. Hostility towards the city was at its peak between 1918 and 1923, a period when diverse measures were introduced limiting freedom of establishment and residence in the cities. The inconvenience of a large presence of foreigners seemed exerbated by the war, and in the 1920s, long debates culminated in a more restrictive attitude in the granting of naturalizations, and the first measures taken against the excessive foreign population. Nationalist attitudes took on the most unexpected accents. Thus the forestry employees strongly criticized the fashion for "exotic" perfumes and offered the viewpoint that for esthetic and climatic reasons, the indigenous perfumes were best adapted to the Swiss countryside.

The shortage of housing in the 1920s is a prime example of how nature served as an argument for economic and social reasoning. The ideal of a small house as opposed to barrack-like flats, access to property, contact with the soil (thanks to a small garden) appeared to be alternatives to roundly condemned Muscovite and Viennese solutions. This stimulated preference for the more-or-less watered-down and fragmentary forms of the garden-city. In 1916, one of the partisans of "disurbanization" wrote: "Green is the favourite colour of the people . . . . who adopt and prefer simple poetry – the green of nature". The examples of projects which take the Howardian model of *Lebensreform* dynamics to the limit are rare. Only at Hannes Meyer's Freidorf (Muttenz, Basel 1919–1921) was it possible to demonstrate the rejection of private property and capitalist viability as well as the promotion of the cooperative community ideal so dear to the social reformers. Investment in the countryside by the flux of urban desires remained one of the ways in which the ecological relationship to nature could progress.

Other alternative forms of a return to nature developed during the 1920s and 1930s. Their vitalist inspiration was close to that of *Lebensreform*. Some of them had already an institutional anchor, such as the youth movements and new "active" teaching methods where corporal expression and "lessons with things" (direct physical contact with nature) played an important role. The dietic principles of Dr. M.O. Bircher-Benner, a partisan of vegetarianism, begun to be known outside the circle of his Zurich clinic. Certain forms of returning to nature continued to be objectionable to the "common sense" of the day and found favour with marginal elements of society. A typical example was the moral revolution proposed by the psychiatrist August Forel, author of a constantly reedited work *Die sexuelle Frage* (1905). The development of the first Swiss nudist (naturist) organizations caused occasional scandals, in particular Der Schweizerische Lichtbund and its magazine, "*Die neue Zeit*". The work by Werner Zimmermann, *Lichtwarts*, appeared in 1922 and was reedited several times. It insisted on cosmic relationships between man and earth. [The symbolic importance of light (*licht*) and of the benefits of the sun in the names of these movements should be noted.]

In a general way, Swiss political culture centered on rural norms. The peasantry incarnated the values of *rein schweizerisch* (pure Swiss values). The maintainance of a strong rural class became the guarantee not only of Swissness but of the economic and social system. The world of the countryside was the incarnation of the social order. The language of discourse surpassed the usual arguments of group interests. More than a simple ideological support, it was the expression of a veritable doctrine. As late as 1940, the powerful secretary of the Union of Peasants Ernst Laur could write in his journal *Le Paysan suisse:* "The development of cities and industrialization leads irrevocably to the numeric weakening of the Swiss people, and also, unfortunately, to a qualitive degeneration (. . .) Urbanization equates with sterility and in consequence constitutes a danger for the existence of our people". It should be noted that such *Blut und Boden* accents never had the racist connotations found on the other side of the Rhine.

Nevertheless, the earthbound interpretations of Helvetian reality impregnated all representations which Switzerland produced about itself. An author such as C.F. Ramuz recalled in *Besoin de grandeur* (1937) that in Switzerland, in the *grande opposition entre l'homme et la nature*, that it was nature which won, in other words the victory of the peasant over the city-dweller. This debate clung narrowly to an economic reality. Agricultural development and resistance to the depopulation of the mountain regions was the order of the day during that period. Numerous initiatives contributed to the definition not only of a sectorial agricultural policy but also of a veritable regional policy in favour of the mountain regions. This context favourable to the peasant question was directly linked to those arguments which had hitherto served to join the esthétic-patriotic action directly to economic and social questions. This approach banalized to a certain extent the protection of heritage and nature.

In consequence, dynamics were displaced from the echelon of private initiatives to that of the political authorities. Without following the elaboration of the legal base in the cantons, the national debate is sufficiently revealing. In effect, apart from an article in the 1907 civil code, the sole base for intervention can be found in those articles called *Heimatschutz*, referring to laws pertaining to the use of hydraulic forces. That of 1916 in particular stipulated "The beauty of the sites should be retained. It should therefore be conserved intact if a major public interest demands it. The factories (works buildings) should not infringe on or infringe as little as possible on the landscape." In other words, it is precisely through the impetus of hydroelectricity that the contradiction between industrial development and protection of the landscape erupted into the process of sensitivity toward nature.

The phenomenon of the electricity-generating industry entering the camp of the nature-protectors came relatively late in Switzerland. The conflict of interests proceeded at the same pace as technical progress. High-tension power lines, "scars on the beloved face of the homeland" and, above all, the construction of high-level water reservoirs in the 1940s resulted in sympathetic opposition. The filling of valleys with artificial lakes was regarded with disfavour not only from an esthetic viewpoint, but also because it implied a displacement of the population. In its 1940 bulletin, the Heimatschutz spoke of the "deportation of a small people", the mountain dwellers. It is interesting that those who opposed the construction of a generating station on the Rhine (the Rheinau initiative of 1953/4) saw nuclear power as an acceptable alternative! "There are great Swiss scholars who are working to harness nuclear power for peaceful purposes. . . .and in doing so will save our Swiss landscapes." Such a statement was what one could read in a brochure of the period. Concerning the legal basis of intervention by public authorities, one had as yet made little progress. The legal systemization of protective attitudes had to wait for the 1960s (a constitutional article in 1962 and a law on the protection of nature and the landscape in 1966). This type of institutional delay is currently common in Switzerland and is explainable by the complexity of appropriating responsibilities in the federal system. Because of this a federal law had been discussed on several occasions in the 1930s without tangible results. In addition, a Federal Commission for the Protection of Nature and the Picturesque, formed in 1936, would henceforth examine all requests for construction permits which implied damage to the countryside and landscape.

I shall finish by mentioning the integration of nature in cultural politics developed on the eve of the Second World War under the name of "Spiritual Defence". It appears to me to be the logical development of the massive ideological investment in the coupling of the ideals countryside/homeland. The message of the Federal Cultural Council in 1938 defined Swiss-ness using a small German phrase, very rich in German thought but very difficult to translate into English: *Das Schweizerisches Wesen ist schollenverbunden und bodenverwurzelt* (The Swiss soul is rooted in the soil). During the war, the necessities of military defence led to inevitable compromises. Protective idealism was discreetly realist vis-a-vis military installations (anti-tank barriers, fortifications, etc.). The constraints of obtaining essential supplies exacted a heavy tribute from the over-exploited forests, while the extension of cultivation brought with it the destruction of numerous biotopes, the elimination of hedges which were the home of insect-eating birds, thus creating much business (affirmed by the most critical) for the manufacturers of insecticides. It was of course out of the question for the Naturschutz/Heimatschutz groups to attack the military and the war economy. Nevertheless, the reflections on the consequences of such degradations permitted the abandonment of a narrow ideological vision to which resistance was dangerous, to the profit of a new kind of argument of the ecological type. The more clairvoyant had already understood in the mid-1930's, the importance of a more global perspective. Instead of fighting for a particular detail of the landscape, instead of trying to turn portions of nature into a museum, would it not be better to envisage the dislocation of the countryside viewed as a whole? The link began with a reflection on town and country planning, which had just come into being, and permitted, after 1950, a digression from the narrow impasse of esthetic-patriotic protective discourse (Walter 1985). Equally, this way made possible the integration of ecological scientific discoveries and the systematic functioning of the environment.

The formula which has progressively integrated the environment into a wish for global and concerted town and country planning has taken a long time to mature. The socio-economic model for action on nature which characterized this period was an important step. This model had the particular characteristic of closely articulating the social function of nature and economic necessities. A constituent of the Helvetian identity, nature guaranteed the social order. Parallel to this, nature is made economically viable by a policy of promoting tourism, productivity improvements in forestry and agriculture, and by the management of hydraulic resources. In essence, it is a sort of precarious reconciliation between nature and the economic necessities which prevailed at the time. The national exhibition in 1939, the so-called *Ländi*, was the successful expression of this. According to the concept of the exhibition, the countryside was a part of the primary materials. Section 1 evoked the Helevetian *Lebensraum* in the form of an expression of "natural conditions" and "spiritual heritage of the nation". The *Dörfli* was not factitious as was the Swiss Village of 1896. It had become a place where traditions and a perfected agriculture cohabited. In other words, the society of the 1930s succeeded in presenting itself as a completely harmonious unit. This is an important fact. Far from attributing a hypothetical ecological equilibrium only to preindustrial societies, it was necessary to envisage several homeostatic levels. After the excerbations of the contradictions between industrialization and esthetic and moral values (19th century), the period between the two World Wars succeeded in providing a very provisional conciliatory formula.

In its own turn, our time is trying to find a new ecological equilibrium. We find ourselves, in effect, since the 1950s, in the centre of a fifth period which one calls the age of ecology. During this period, the relationship to the environment functions according to a new ecological type of social model. We experiment daily with its dominant traits. It is history which will no doubt say one day how precarious our solutions and illusions have been.

References

Acot P (1988) L'histoire de l'écologie. Presses universitaires de France, Paris

Bergmann K (1970) Agrarromantik und Großstadtfeindschaft. Meisenheim am Glan

Brüggemeier FJ, Rommelspacher Th (1987) Besiegte Natur. Geschichte der Umwelt im 19. und 20. Jahrhundert. Beck, München

Fawcett J (ed) (1976) The future of the past. Attitudes to conservation 1174–1974. Thames and Hudson, London

Frecot J (1976) Die Lebensreformbewegung. In: Vondung K (ed) Das wilhelminische Bildungsbürgertum. Zur Sozialgeschichte seiner Ideen. Vandenhoeck und Ruprecht Göttingen, pp 138–152

Friedrich E (1904) Wesen und geographische Verbreitung der Raubwirtschaft. Petermanns Mitt 50:68–79 and 92–95

Green M (1986) Mountain of truth. The counterculture begins – Ascona 1900–1920. Univ Press of New England, Hannover London

Hays SM (1968) Conservation and the gospel of efficiency: the Progressive Conservation Movement, 1890–1920. Univ Press, Cambridge

Kormondy EJ (ed) (1965) Readings in ecology. Prentice-Hall, Englewood Cliffs

Krabbe WR (1974) Gesellschaftsveränderung durch Lebensreform. Strukturmerkmale einer sozialreformerischen Bewegung im Deutschland der Industrialisierungsperiode. Vandenhoeck und Ruprecht, Göttingen

Lepenies W (1983) Historisierung der Natur und Entmoralisierung der Wissenschaften seit dem achtzehnten Jahrundert. In: Markl H (ed) Natur und Geschichte. R Oldenbourg, München Wien, pp 263–288

Linsmayer Ch (1987) Die Eigenschaft 'schweizerisch' und die Literatur der deutschen Schweiz zwischen 1890 und 1914. In: de Capitani F, Germann G (eds) Auf dem Weg zu einer schweizerischen Identität 1848–1914. Editions Universitaires, Fribourg, pp 403–426

Lübbe H, Ströker E (eds) (1986) Ökologische Probleme im kulturellen Wandel. F Schöningh, Paderborn

McIntosh RP (1985) The background of ecology. Univ Press, Cambridge

de Montenach G (1910) Un musée vivant. Le parc national suisse. La Liberté 23.03.1910

Olwig K (1984) Nature's ideological landscape. Allen and Unwin, London

Raumolin J (1984) L'homme et la destruction des ressources naturelles: la'Raubwirtschaft'au tournant du siècle. Ann ESC 39:798–819

Sieferle RP (1984) Fortschrittsfeinde? Opposition gegen Technik und Industrie von der Romantik bis zur Gegenwart. Beck, München

Stern F (1961) The politics of cultural despair. A study in the rise of Germanic ideology. Univ Press, Berkeley

Thomas K (1983) Man and the natural world. Changing attitudes in England 1500–1800. Penguin Books Ltd, Harmondsworth

Viard J (1985) Protestante, la nature? In: Cadoret A (ed) Protection de la nature. Histoire et idéologie. L'Harmattan, Paris, pp 161–173

Walter F (1984) Perception des paysages, action sur l'espace: la Suisse au 18. siècle. Ann ESC 39:3–29

Walter F (1985) Fédéralisme et propriété 1930–1950. Les attitudes face à l'aménagement du territoire en temps de crise et de pleins pouvoirs. DISP 82:21–27

Walter F (1989) Attitudes towards the environment in Switzerland, 1880–1914. J Hist Geogr 15:287–299

Weber W, Schönichen W (1936) Das Reichsnaturschutzgesetz vom 26. Juni 1935. Berlin

The Obscure Problems: Rationalization, Power, and the Discovery of Environmental Problems

J. Thelander[1]

1 What Do We Do with Environmental Problems?

In a welfare society it is a duty to understand something of the well-being of the people. In Sweden the standard of living has been measured twice, once at the end of the 1960s and again at the beginning of the 1980s. In neither of these studies was the environmental situation considered a matter of concern despite public interest in environmental problems during the 1960s and the political pressure triggered by the United Nations' conference about the global environmental situation in Stockholm in 1972. Thus environmental problems could still be missing in measuring the welfare of one of the most developed countries in the world. One must ask if it is a coincidence that there was also another sphere of concern missing in these studies: the important changes in the moral climate during the latest decades (Allardt 1984).

Shall we interpret the lack of interest in the environmental situation as an expression of the fact that environmental problems do not concern the central values in a welfare society? Or is this absence to be interpreted as an expression of institutional incapacity? What we can say with absolute certainty here is that the environmental problems have been completely obscured.

In the Swedish parliamentary election in 1988, the Green party took seat in the parliament for the first time. The headline of a newspaper article questioned the future of the new parliamentary members by not believing in the election return for their constituencies because: "nobody threatens the welfare unpunished" (Rytterborg 1988).

People seriously engaged in doing something about environmental problems are seen as a threat! A threat towards what and towards whom? Obviously a threat towards central or dominating values in a welfare society. Noteworthy here is not what is said but what makes it possible to talk in this manner. It is an obscuring game, i.e., one plays upon hidden and self-evident values to which others are not expected to pay respect. The manner in which one discusses the environmental situation and the ameliorative measures needed is an example of a very common method of argumentation. In a deeper sense it is a subtle power-play, where the conventional defense of the welfare society seems to have all the advantages on

[1] University of Lund, Department of History, Box 2074, 220 02 Lund, Sweden

Brimblecombe, Pfister (Eds.)
The Silent Countdown

its side. The burden is laid upon the opponent of being irresponsible for the consequences of his actions.

We have seen how environmental problems may be completely neglected and how they may be talked about. But how are they in fact treated? Also, how are they for the most part discovered?

If we take four types of environmental problems and make a study of them over time, e.g., from the beginning of our century to the 1970s, what are we going to find? Is it possible to distinguish any distinct pattern in how they were handled? Water pollution, mercury, PCB, and some examples of environmentally induced cancer were the objects of study. The geographical area of the first three mentioned was Sweden; examples of environmentally induced cancer from the U.S. and Great Britain were used. The pattern that emerged had the following characteristics:

1. The problems were rediscovered, sometimes several times with much loss of time before treatment. Fifty years and more were not unusual.
2. The problems were transferred from one area to another, e.g., from lakes to the sea.
3. The loss of time converted situations to more critical predicaments.
4. The threatening images produced were not supposed to be really alarming to the populace involved.
5. There was a need of accidents to initiate responses.

How is this pattern to be interpreted? Everything points to a model of repair working behind these cases. A troublesome environmental situation is recognized as a problem only when damage is seen both in human terms and in the environment. It is a retrospective process which results in treatment.

Who discovered these environmental problems? Were any special persons or groups responsible for the discoveries? If we talk about an environmental problem as a societal problem, the study underlines that discoveries must be made on several social levels in order to be a public problem. We are accustomed to talk about discoveries in the natural sciences, but if we mean something to be a societal problem discoveries are to be made not only among researchers, but among administrators, politicians, and not least, journalists as well. A problem discovered by only one of these groups will have a very reduced chance of becoming a societal problem. This arises as a result of an intricate interplay between researchers, administrators, politicians and mass media (Lundgren and Thelander 1989). But where is the layman? Doesn't he or she discover anything important in the environmental sphere? Is it an area for experts of different kinds? This raises the all-important question about whose problems are to be considered in the environmental field.

Is there anything in the pattern of handling environmental problems which has connections with "the democracy of discoveries"? There are good reasons to discuss this matter later on. The question we have to answer is: has the layman's role become obscured in the orthodox treatment of environmental problems?

2 What Obscures Environmental Problems?

It may seem a little surprising that studies of the standard of living in Sweden so completely neglected the environmental situation. Is it an individual characteristic of the Swedish society to have developed a self-understanding stamped by a rather homogeneous view of the future during the decades after the Second World War. In that view, with its dominating value structure, the environmental problems had no natural place. On the contrary, such problems seemed to break up the whole established image of the future. In these circumstances, it is so much simpler to stick to established expectations.

Does this explanation also fit the model of repair? Against such a background, it also seems self-evident to talk about all deviating opinions as threats towards the welfare society. Are the reactions to the environmental problems unique for the Swedish society? Probably not, but there is a great need of comparative studies as to how different countries have responded during different time periods. However, if the reaction pattern is not unique for Sweden, it will indicate that a part of the explanation may be sought beyond the national system.

Are environmental problems systematically treated in a different way than other societal problems, for example political, economic or social problems? Doesn't it belong to the nature of the matter that we are usually left with the alternative of formulating problems and treating them afterwards, in a retrospective view. Wasn't Hegel right in saying that Minerva's owl takes off in the dusk?

Shouldn't we, in other words, be surprised over the reaction pattern in form of a model of repair which we have found? Is it rather what should be expected?

Environmental issues as a problem category of its own is of recent origin. It is a latecomer in Western history, where problem-spheres, e.g., the political and economic ones, emerged several hundred years ago. In this respect the differentiation of a special sphere called "environmental problems" follows a pattern established early in Western history. It means that the new category of problems should fit into a certain Western way of organizing reality, i.e., in order to master reality, it is classified into problems. What does this highly rationalized organization of reality mean for our way of discovering and treating environmental problems? What are the consequences for other problem spheres?

Perhaps it is here we have to seek a fundamental obscuring process in relation to environmental problems.

3 Environmental Problems and the Crisis of Reason

Reason and rationality are highly honoured concepts in Western societies. But the more we are able to experience how reason and rationality diverge, the more we are left with a feeling of an emergent crisis. Something may very well be rational; this does not necessarily mean that it is also reasonable. Every day before our eyes we are able to experience how supposedly rational actions and problem solutions are evidently unreasonable. We may add that all reasonable actions also are rational by definition, using these concepts (von Wright 1986). But what is the kernel in the crisis? By the instrumental use of reason, man has made himself master of the world. But now he is faced with a new puzzle, reason's own existence. One of the leading philosophers of the Enlightenment, Denis Diderot went to the crux of the matter when he wrote that nature must be described in such a way that man's own existence becomes intelligible (Prigogine and Stengers 1984). More than two hundred years later, the meaning of Diderot's remark is repeated in the following words: "we have little conceptual tools available to include the phenomenon of man in a unified picture of the world" (Hägerstrand 1988).

The kernel of the crisis of reason is not that man has been left out, but that he has been obscured. Therefore we must ask: what are the consequences for our understanding of environmental problems? One very evident consequence is the existence of the two modes of thinking, the humanities and the natural sciences. An environmental problem is not only a problem for the natural sciences; as a societal problem, it is to the same degree a concern for the humanities. But the two systems of description do not coincide. On the one hand there is a description of Nature, on the other hand one of man. The environmental problem has a double nature but it must be treated as a unified problem.

Another consequence is the split conception of the research process, overly encouraged during this century by Logical positivism, separating the research process into a "context of discovery" and a "context of justification". The former is the domain of subjectivity and cannot lay any claim to rationality. The latter deals with the proofs of science and is consequently the place where the rationality of science is given its fullest expression. In other words: the opposition between subjectivity and objectivity is here disguised within two different contexts. Man in his subjectivity is classified in a special category, which has nothing to do with the rationality of science. Man in his subjectivity is obscured by a limited concept of rationality (Thelander 1982).

The development of specialized professions in both the humanities and the natural sciences have strengthened the split conception of the research process. Questions raised by the layman or the amateur have no status at all. Who is then expected to make discoveries in the environmental sphere? Diderot's remark in the 18th century has ramifications in our own time when environmental problems are producing an even greater pressure on democracy.

Our discussion up to now may thusly be summarized: What is obscured in our discoveries and our treatment of environmental problems and how does it take place?

4 The Differentiation of Problem Spheres and the Scientific Method of Problem Solving

Men have always used some form of calculating means in order to reach some goal in everyday life. It is not something that arises with the advent of Western capitalism. Even the differentiation of some autonomous problem spheres occurred before the rise of capitalism, e.g., law and politics. When the natural sciences became a source of power and a dominating paradigm for understanding the world, ongoing rationalization processes were penetrated by a tendency to reduce everything to a science. To the means end reasoning was added a more precise theoretical calculation.

When we look at rationalization, it is important not to see it as an all-embracing process in the form of a linear development. Max Weber has warned of that in his studies of the rationalization of the Western world. What we have to look for, is how specific areas have been rationalized in specific directions (Weber 1920).

What we have to analyze are two different processes. On the one hand, rationalization occurs through which a societal sphere becomes differentiated from other spheres and reaches a status of a autonomous problem sphere. Autonomy means that a problem cannot be reduced to another type of problem. An environmental problem cannot be seen only as an economic problem, making it possible to do calculations within that sphere. Autonomy demands us only to consider known circumstances and to disregard conditions outside the sphere.

On the other hand, we have to determine what science means for the problem process. Throughout history, when confronting a task, humans have asked: what to do, how to do it, and when to do it, a simple form of strategic thinking and behavior. What happens when this mode of acting is overrun by a scientific way of viewing things?

I will try to give a short outline of the differentiation of some large problem spheres since the Renaissance. (How did the political, the economic, and the social spheres emerge before the growth of an environmental sphere. Which are the main lines in this development?) Then I would like to show in what respects the problem process has changed. The main point being the interplay between the two processes.

5 Was Machiavelli a Bad Historian?

The emancipation of the political sphere as an independent problem is embodied in the person of Machiavelli. His works are a very good illustration of some main lines in the development before scientific methodology began. Machiavelli was a strategic thinker, not a scientist, making him especially interesting for our purpose. Merely asking the question whether Machiavelli was a bad historian means that we have decided in favor of a scientific view, rather than an effort to understand Machiavelli on his own terms. We are in no position to realize the fundamental break that occurred between the time of Machiavelli and the emergence of the economic sphere as an independent problem area.

What is astonishing is really not the fact that politics became an independent sphere of action in the Italian city-states during the Renaissance, but rather why it did not exist in the ancient Greek city-states? For the Greeks, politics were intimately connected with morals; the latter was seen as the fundamental basis for the former. When Machiavelli viewed politics as a sphere of its own, he had to break completely with Christian ethics. This was the main reason for his bad reputation in Western history. But he did not put forth any claim that politics should be without moral considerations. On the contrary, he suggested that politics is concerned with morals, but not necessarily the Christian version. Instead, he reestablished the connections with more ancient conceptions of morals.

It is noteworthy that the rationalization of a sphere is compatible with one type of morals but in total opposition to another. As Weber has shown, Protestant ethics are not only compatible to, but even a presupposition for, capitalistic rationality within the economic sphere. The conclusion is not that rationality needs no morals; that is a misunderstanding of its role in history.

5.1 Machiavelli as a Strategic Thinker

The distinguishing mark of a strategic thinker is the intimate relationship between thinking and doing. The dividing line between Machiavelli as a strategist and as a scientist may be seen in his attitude to experience, action, chance, and expectations as shown in Fig. 1.

Experience is the basis for our expectations of what will happen in the future; however, certain things may seem more likely to happen than others. Our need of action is therefore intimately connected with our estimation of how probable an event may be. It is therefore unlikely that changes in one of the components will not influence our conception of the other ones. We are able to see it in the case of Machiavelli and in the changes that took place after him.

The Greeks and Romans had a conception of change in history which did not permit them to imagine totally new things. They could talk about renewing, but not about intrinsic newness (Romilly 1977).

In the time of Machiavelli, history was seen as a storehouse of human experience. Through examples in history, something could be learned about future courses of action. But Machiavelli was no historical scholar when interpreting examples from the past. His fundamental experiential basis arose from his own experiences as a princely legate and as an acute observer of the intricacies in the ongoing struggle for power among the different city-states. This experience was the starting point for his subsequent reflections about events in earlier history.

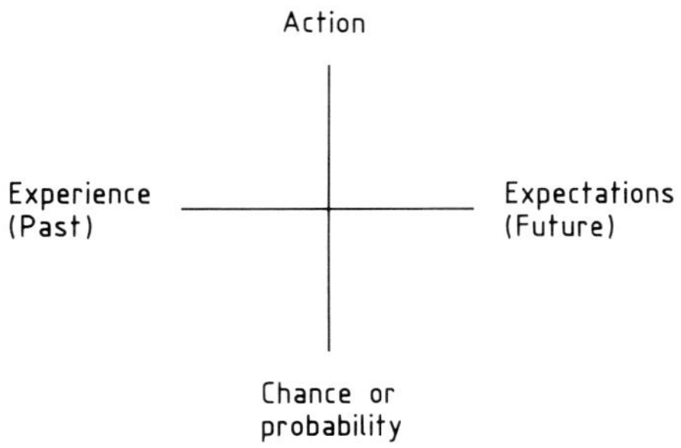

Fig. 1. Dimensions in the change from strategie to scientific behaviour

His personal experiences were the fundament upon which other constructions could be built (Whitfield 1947).

Machiavelli set himself the task of exploring the possibilities and consequences of introducing innovations to the political field. That meant a new horizon of expectations.

In another respect, Machiavelli conserved an ancient heritage. Action was for him closely connected with chance in the figure of the ancient goddess Fortuna. She was unpredictable but could be conquered under certain circumstances. A high moral standard among the population and a decisive will to seize the opportunity of action were the only options for man as an agent (Pocock 1975).

Machiavelli introduced a completely new conception of uncertainty into human affairs. The important point is that he did not weaken the component of action. On the contrary, he laid rather more emphasis on action even when facing a genuine uncertainty.

5.2 After Machiavelli

The emancipation of the political sphere in which Machiavelli took part was a deep-going rationalization of a part of the society, but did not involve scientific methodology. For the latter to occur, some fundamental changes in the conceptions of experience, expectations, actions, and chance had to happen. That is part of the history of how the strategic attitude to reality was substituted by a scientific one.

The biggest milestone in this development was, of course, the success of Newtonian science. It had a decisive influence on the conception of experience: experience was overridden by theoretical knowledge. The point of reference for action was no longer examples from the past. In an investing society, like England around 1700, it was opinion that counted. The displacement is clearly seen in the difference between Hobbes and Locke. The former was still interested in historical examples, manifested in his translation of Thucydides, while Locke was absorbed in his interest for the law of opinion. The main emphasis of interest had moved from the past to the present and future.

Another milestone was the introduction of probability calculations during the 1660s by the school of Port Royale, with Pascal as one of the leading figures. It is still something of a puzzle how probability as a concept could emerge so suddenly (Hacking 1975). The new conception of chance as something which could produce stable relative frequencies meant a total break with Fortuna in its ancient and Machiavellian forms. The important link between action and chance was cut off.

All the key concepts shown in Fig. 1 have thus changed its contents and the relationships between each other. We may also say that we have witnessed a transition from political to economic rationality. Action is seen in a new context. It is no longer dependent on an "unpredictable woman", i.e., Fortuna, but rest on a safer ground of economic predictability.

6 The Emergence of the Economic and Social Spheres

While the princes were the protagonists in the rationalization of the political sphere, it was the middle class which laid pressure upon the rationalization of the economic sphere. They were not alone. The recently established national states in northwestern Europe had also great interests in the power to manipulate a central societal sphere. The principles of economic rationality collided with old political morals, where civic virtues had been valued as the most important. In England, Adam Ferguson wrote about the potential danger of decaying political morals at a time when everyone seemed to be only interested in making money. Once again we can establish how the rationalization of one sphere generates tensions to other spheres: a new rationalization demands new morals.

Within the economic sphere, the tendency towards scientific methodology has been very strong; economics may even look like a completely scientific domain. As in the case with politics, there is a tension between rational thinking and science. Rationalization does not exclude a close relationship between experience and action, but science has no implications for action. There is a hidden struggle between a strategic and a scientific attitude in every autonomous problem sphere.

The economic problem emerged during the 18th century as a sphere of its own. If political rationality was connected with the formation of the modern states, economic rationality is interwoven with Western capitalism and the market.

The social problem sphere became autonomous during the 19th century. In Germany it was carried forward by two-opposite interests, on the one hand, the authorities, on the other hand, a rising working class. The authorities took measures on the so-called social question, anxious to prevent the working class from taking initiatives. In public debate, the social question was often called the socialist question. The emerging social rationality was squeezed between two problem spheres, embodied in the state and the market, which were already independent. The growing importance of the social sphere during this century has also meant that it has been more and more penetrated by scientific methods. In none of the other spheres has the conflicts between rationality and morals and values been so clear as in the social field. Value nihilism seems to be the only moral ground compatible with a social rationality.

Summing up the remarks about the political, economic and social problem spheres we may say that all of them have gone through a development where the emphasis is more on theoretical knowledge than on personal experience. Likewise there is a displacement from the past to the present and future. And most importantly, the pressure on individual or collective action has diminished. In all three spheres, the scientific attitude has come to dominate over the strategic one; however, in this respect there are differences between the spheres as to the degree of domination.

It is noteworthy that the greater emphasis on the future is coupled with a diminishing pressure on action. How is that to be explained? A part of the explanation may be found if we look more closely at the problem process as one expression of scientific learning.

6.1 What is a Problem?

The word problem comes from the Latin *problema* and from the Greek *problēma*, which at its root has *proballein*, meaning "to throw forward". It is interesting to note that it is the noun we have inherited in the first place, not the verb. In Webster's dictionary, a problem is a question proposed for solution or consideration. It is also a matter, situation or person that is perplexing or difficult. It may, finally, be something required to be done or requiring the doing of something (Webster's Dictionary 1978). If we combine these different meanings with each other we will get a problem square as shown in Fig. 2.

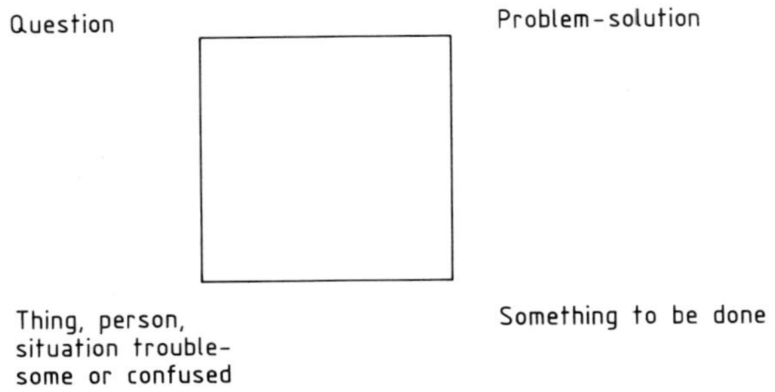

Fig. 2. The problem square

A look in some other dictionaries discloses an interesting fact. "Problem" is defined as an answer or solution to a question (Meyers Konversations lexikon 1909). It means that from the beginning the different corners of the square have not been given the same weight. More emphasis is laid on "problem solution". What does that indicate? Are the dictionary definitions an indicator of an already accomplished scientific process? In that case, we will have to try to get behind that process and analyze its meaning.

6.2 What Happens to the Problem Process?

What lies behind the definition of problem as a "problem solution"? In order to understand this, we will get help from the problem square on which the different steps in the scientific process can be illustrated.

First, we have to convert the static square to a dynamic form by adding arrows which show an idealized picture of the problem process. The starting point is the experience of a troublesome situation about which we begin to raise questions. Then the search for solutions begins. With some sort of solution we then move down to the point of action (Fig. 3).

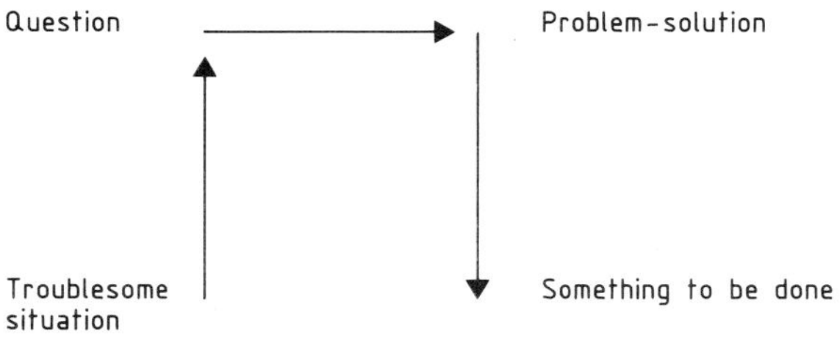

Fig. 3. From problem square to a problem process

In the strategic attitude we may as well draw a line directly between "troublesome situation" and the "point of action" in order to underline that experience, which is not articulated in questions and solutions, plays an equally important role.

The scientific method of the problem process means that in the upper half of the square, questions and solutions were seen in a particular way: (1) they became exposed to a theoretical articulation according to the model of Newtonian science. (2) There a demarcation line was drawn between the upper and lower halves of the square. The upper half was defined as producing certain, secure knowledge, and for that purpose it had to be produced according to certain rules, which required that the proof of the obtained knowledge was the most important part of the research process. Questions were connected with subjectivity and were not subjected to strict rules. "I feign no hypotheses", Newton is supposed to have said (McMullin 1967).

The success of Newtonian science is accompanied by a crusade against the dangers of subjectivity. In other words, science seems to be threatened by man himself. A control of subjectivity could only be obtained by splitting the research process into two parts, where at least the part dealing with proofs could be controlled (Thelander 1982).

With the matter of subjectivity obscured, Newtonian science concentrated its efforts on shedding light on the explanation of causal factors. Causal factors could be defined and controlled in a way which was totally lacking in relation to subjective factors. The success of Newtonian science as a model in the Western World can be traced in the following two steps of the scientific problem process; see the Figs. 4 and 5.

The second step is clearly mirrored in the dominating view today of what makes a situation problematic. It is only when something is known about the causal factors that the given state can be modified (Daele et al. 1977). Questions and actions are not the focus of attention. Instead, we can see how the causality functions as a lever below situations and determines which ones should be seen as problematic and in what way. The way to action goes through an identification of causal factors at work in a situation. The rationalization of problem spheres resulted among other things in an emphasis on the present and future and diminishing importance for action. Now we are able to see why such a tendency

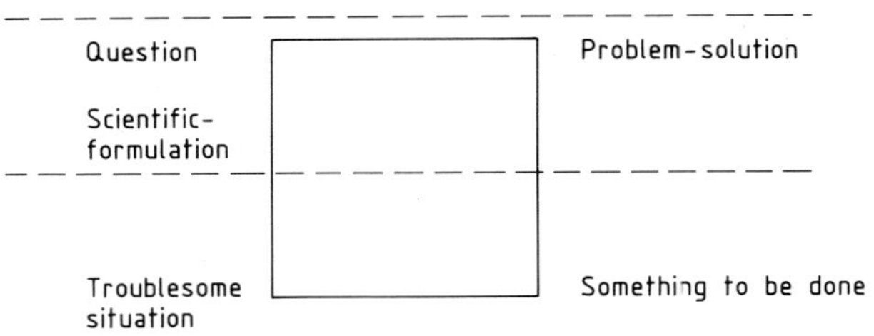

Fig. 4. The first step in the scientific method of the problem-solving process

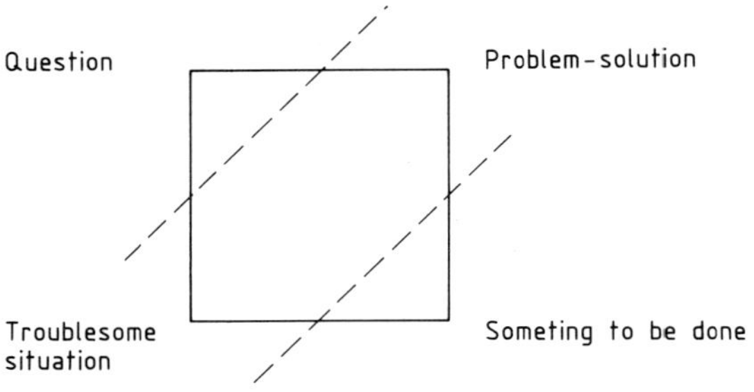

Fig. 5. The second step in the scientific method of problem-solving

has been firmly established and what the consequences may be. An imperative which says that no situation can be defined as really problematic without a secure definition of the causal factors at work has been interposed between questions and actions. It means that the time perspective of the future is drastically reduced; it is reduced to the identification of causal factors. However, the decisive drawback in this perspective is that the imagination of the future and even the present is cut off from questions which are based on values. That is the reason behind the fact that we so often are given seemingly neutral reports on environmental problems and their future. The so-called ticking problems are clear cases of this drawback. We know, for instance, that much waste has been buried in the soil, which sooner or later will give trouble. But often we do nothing until it has been proved that certain damages already have occurred.

Without a basis of values, it is impossible to take any moral responsibility for the distant future (Odén 1988). The rationalization and scientific formulation of societal problem spheres that have been underway in the Western World since the Renaissance show a similar pattern. The differences are not in kind but in degrees. The environmental sphere must be seen not as a new starter but as a sphere determined to a high degree by what has occurred in the past. What is needed is a recognition of that fact. Man has been obscured in all of the problem spheres. We have now seen some of the consequences. The determination of a problem, as a result also becomes obscured. But most important of all: man as a value-producing being has been obscured.

"The problem of life is to utilize resources economically" wrote the distinguished Chicago economist Frank H. Knight in the 1930s. How far is life rational, how far are problems reducible, using given means to achieve given ends, Knight asks himself. The scientific view of life is a limited view. He continues:

"... life is at bottom an exploration in the field of values and attempt to discover values rather than on the basis of knowledge of them to produce and enjoy them to the greatest possible extent. We strive to "know ourselves", *to find out our real wants,* more than to get what we want." (Knight 1935).

This is the kernel of the environmental problematic as a societal problem.

Knight underlines the deepest meaning of the obscuring of environmental problems. As a result we have to count on three different ways of determining the problematics of a situation (Fig. 6).

In Fig. 6A we have the known tendencies described above, showing the case when we have strongly established problem solutions which control the definition

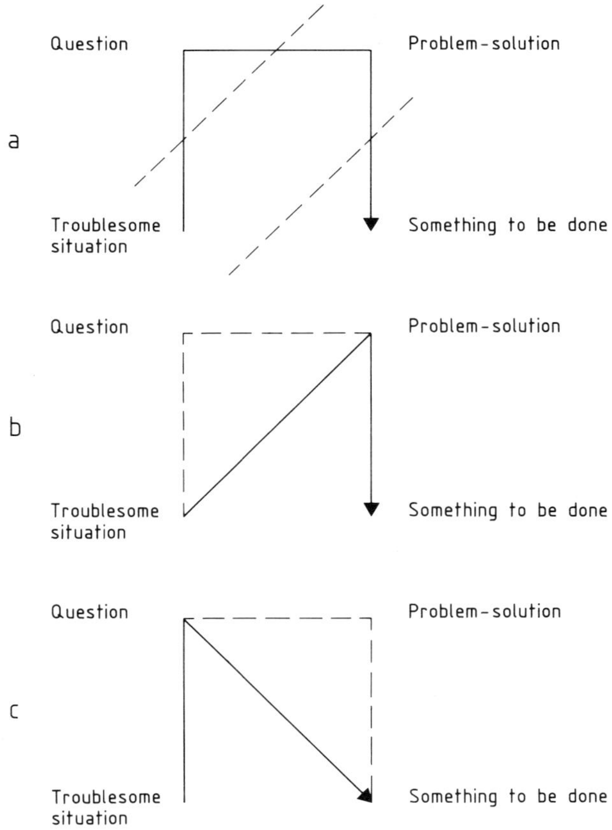

Fig. 6a-c. Three profiles of the problem-solving process. **a** The profile of the natural sciences; **b** the technological profile; **c** the humanistic profile

of the situation. Questions are not directing the solutions; here it is the other way round. The technological profile is a simplification of the first one.

What has been obscured in determining problems by scientific methods is the third profile: the humanistic one. It takes its starting point from values and questions in defining a situation as problematic. Problem solutions are not neglected, but they are not permitted to determine the definition of the situation and measures necessary to take. In that respect it is more of a strategic model than a scientific one.

The three profiles are applicable to all of the described problem spheres but they have most relevance in the environmental sphere. What are the reasons for that? What distinguishes the environmental sphere from the other ones? The difference lies in the circumstance that political, economic, and social problems are definable within society, while the environmental problem is not. The latter is a double problem. It must be understood both in relation to nature and society; that produces a fundamental value conflict, which determines the nature of the environmental problem. For which element in the relationship between nature and society is to be subordinate. Shall nature dominate society or shall society dominate nature? Society may circumscribe nature and nature may circumscribe society, but the two positions are opposites and irreconcilable; a third alternative does not exist.

The choice of one or the other position here is fundamental for our way of defining the problems in the environmental situation. The importance of values are easily forgotten or neglected in the problem profiles of natural sciences or technology.

7 Conclusions

Have the original questions in this chapter been answered? They dealt about why and how a model of repair seems to dominate the problematics and treatment of environmental issues.

In this respect this chapter has shown how the determination of obscure problems lies in the process of rationalization and scientific formulation of societal problem spheres, i.e., the retrospective way of defining problems is built into our understanding of rationality. The crisis of reason therefore opens up a necessary questioning of the different uses of rationality.

Up to now many interests in the environmental game have been able to play upon hidden presuppositions understood as self-evident cultural resources. That advantage in the power game of the environment must be revealed: that is the least which can be demanded in a democracy.

Acknowledgement. This chapter was part of the project "Traditions of problematization in Swedish environmental policy 1958–87" supported by the Swedish Environmental Protection Board.

References

Allardt E (1984) Missar de nya problemen. Dagens Nyheter 1.8
Hacking I (1975) The emergence of probability. Univ Press, Cambridge
Hägerstrand T (1988) Human responsibility in the face of global change. In: Human response to global change. Swed Counc Plann Coord Res: Report 88:3
Knight FH (1935) The ethics of competition. Allen and Unwin London
Lundgren L, Thelander J (1989) Nedräkning pagar. Hur upptäcks miljöproblem? Vad händer sedan? Naturvardsverket förlag. Stockholm (At present only available in Swedish. In translation: Counting down. How are environmental problems discovered and what happens then?)
McMullin E (1967) Empiricism and the scientific revolution. In: Singleton Ch S (ed) Art, science and history in the renaissance. John Hopkin Press, Baltimore, p 358
Meyers grosses Konversationslexikon (1909) 6th edn. Bibliographisches Institut, Leipzig und Wien, vol 16, p 363
Odén B (1988) Natur och politik. Sydsvenska Dagbladet 27.4
Pocock JGA (1975) The machiavellian moment. Univ Press, Princeton
Prigogine I, Stengers J (1985) Ordning ur kaos. Bokshogan, Göteborg
Romilly J (1977) The rise and fall of states according to greek authors. The Univ of Michigan Press Ann Arbor
Rytterborg U (1988) Håller sambadansen mer än mandatperioden ut? Välståndet hotar man inte ostraffat. Arbetet 8:10
Thelander J (1982) Historia, teori och kunskapsutveckling. Om frågandets konst i forskningsprocessen. Scandia 48:2
Weber M (1920) Vorbemerkung. In: Gesammelte Aufsätze zur Religionssoziologie. JCB Mohr/P Siebech Tübingen, p 12
Webster's New 20th Century Dictionary of the English Language W Collins & World Publ Co (1978) 2nd unabr edn.
Whitfield JH (1947) Machiavelli. Blackwell Oxford
van den Daele W, Krohn W, Weingart P (1977) The political direction of scientific development. In: Mendelsohn E, Weingart P, Whitley R (eds) The social production of scientific knowledge, Dordrecht-Holland. D Reidel Publ Comp.
von Wright G (1986) Vetenskapen och förnuftet. Månpocket: Bonnier fakta förlag, Stockholm

Subject Index